Stormwater Management for Smart Growth

Stormwater Management for Smart Growth

Allen P. Davis

and

Richard H. McCuen

Department of Civil and Environmental Engineering
University of Maryland
College Park, Maryland

 Springer

Library of Congress Cataloging-in-Publication Data

A C.I.P. Catalogue record for this book is available
from the Library of Congress.

ISBN-10: 0-387-26048-X   ISBN-10: 0-387-27593-2 (e-book)
ISBN-13: 9780387260488   ISBN-13: 9780387275932

Printed on acid-free paper.

Printed in the United States of America.

9 8 7 6 5 4 3 2 1   SPIN 11424451

springeronline.com

PREFACE

Land development to support population increases and shifts requires changes to the hydrologic cycle. Increased impervious area results in greater volumes of runoff, higher flow velocities, and increased pollutant fluxes to local waterways. As we learn more about the negative impacts of these outcomes, it becomes more important to develop and manage land in a smart manner that reduces these impacts.

This text provides the reader with background information on hydrology and water quality issues that are necessary to understand many of the environmental problems associated with land development and growth. The variability of runoff flows and pollutant concentrations, however, makes the performance of simple technologies erratic and predicting and modeling their performance difficult. Chapters on statistics and modeling are included to provide the proper background and tools. The latter chapters of the text cover many of the different technologies that can be employed to address runoff flows and improve water quality. These chapters take a design approach with specific examples provided for many of the management practices. A number of methods are currently available for addressing the problems associated with stormwater runoff quality from urban areas; more continue to be developed as research is advanced and interest in this subject continues to surge. Traditionally, techniques for the improvement of runoff quality were borrowed applications from water and wastewater treatment, such as large sedimentation ponds

Recently, increased interest has been placed on using natural systems to improve water quality. This includes grassy filters, forested buffers, wetlands, and bioretention areas. These natural areas can slow flow, store water, filter sediments, and promote physical, chemical, and biological processes that can attenuate pollutants.

Finally, novel approaches are being developed to improve stormwater quality by minimizing surface flow volumes and rates, and overall pollutant production in the first place. These ideas are receiving considerable interest throughout the U.S. and Europe where severe and haphazard development has lead to negative environmental impacts on local water bodies. On-site roof gardens, rain gardens and bioretention, permeable pavements, and specialty landscaping can minimize runoff, hold pollutants, and promote evapotranspiration and infiltration. Integrating these techniques into new land development and retrofitting existing developed areas provides new challenges to professionals who are concerned with urban water quality.

This book is intended both as a textbook for classroom use and as a guide for professionals who are responsible for mitigating the detrimental effects of land development. This includes engineers, hydrologists, land use planners, natural resources managers, and environmental scientists. Those responsible for the development of stormwater policies may also find value in the approaches to design discussed herein.

On the cover: Bioretention is a stormwater management technology that can be implemented in variety of land development situations. (clockwise from top left): 1) In a parking lot median. 2) Accepting runoff from a roadway. 3) A schematic of bioretention. 4) Bioretention in a condominium development. Bioretention is discussed in Section 9.4.

ACKNOWLEDGMENTS

Allen P. Davis would like to express his appreciation to the various agencies that have supported his work on stormwater and stormwater management technologies. These include The Prince George's County Government, The Maryland State Highway Administration, The Cooperative Institute for Coastal and Estuarine Environmental Technology (CICEET), The Maryland Water Resources Research Center, and The District of Columbia Department of Health.

A heartfelt thanks is also offered to his wife Dolores for her support during the preparation of this text.

CONTENTS

1 INTRODUCTION **1**
  1.1 URBAN SPRAWL: THE PROBLEM 1
      1.1.1 A Historical Perspective 1
      1.1.2 Characteristics of Urban Sprawl 3
      1.1.3 Pollution of Waterways 4
      1.1.4 The Effects of Urban Sprawl 5
      1.1.5 Difficulties Faced in Improving Stormwater Quality 6
  1.2 SMART GROWTH: THE SOLUTION................................. 8
      1.2.1 Urban Sprawl or Smart Growth 8
      1.2.2 Alternative Perspective on Smart Growth 8
  1.3 PROBLEMS 9
  1.4 REFERENCES 10

2 WATER QUALITY PARAMETERS **11**
  2.1 INTRODUCTION 11
  2.2 MASS, CONCENTRATION, AND LOADING 14
      2.2.1 Mass Balances 14
      2.2.2 Concentration-Flow Relationships 16
  2.3 FACTORS NECESSARY FOR LIFE 18
      2.3.1 Dissolved Oxygen 19
      2.3.2 pH 19
      2.3.3 Temperature 20
  2.4 WATER POLLUTANTS 21
      2.4.1 Suspended Solids 21
      2.4.2 Oxygen Demanding Substances 23
      2.4.3 Nitrogen Compounds 24
          2.4.3.1 Nitrogen Chemistry 24
          2.4.3.2 Nitrogen in the Environment 25
      2.4.4 Phosphorus 27
      2.4.5 Microbial Pathogens 27

2.4.6   Heavy Metals   28
2.4.7   Oils and Grease   29
2.4.8   Toxic Organic Compounds   29
        2.4.8.1 Pesticides   29
        2.4.8.2 Polycyclic Aromatic Hydrocarbons   30
        2.4.8.3 Solvents   30
2.4.9   Trash   30
2.5   WATER QUALITY INDICES   31
2.6   TOTAL MAXIMUM DAILY LOADS–TMDLS   32
2.7   PROBLEMS   34
2.8   REFERENCES   36

3 STATISTICAL METHODS FOR DATA ANALYSIS   37
3.1   INTRODUCTION..   38
3.2   POPULATION AND SAMPLE MOMENTS   38
      3.2.1 Mean   38
      3.2.2 Variance   40
      3.2.3 Standard Deviation   40
      3.2.4 Coefficient of Variation   41
3.3   PROBABILITY DISTRIBUTIONS   42
      3.3.1 Probability   42
      3.3.2 Types of Random Variables   42
      3.3.3 Uniform Distribution   43
      3.3.4 Normal Distribution   44
             3.3.4.1 Standard Normal Distribution   45
             3.3.4.2 Log-Normal Distribution   47
      3.3.5 t Distribution   48
3.4   A PROCEDURE FOR TESTING HYPOTHESES   48
      3.4.1 Step 1: Formulation of Hypotheses   49
      3.4.2 Step 2: The Test Statistic and its Sampling Distribution   50
      3.4.3 Step 3: The Level of Significance   50
      3.4.4 Step 4: Data Analysis   51
      3.4.5 Step 5: The Region of Rejection   51
      3.4.6 Step 6: Select the Appropriate Hypothesis   51
      3.4.7 Summary of Common Hypothesis Tests   53
3.5   OUTLIER DETECTION   56
3.6   PROBLEMS   59

4 STORMWATER HYDROLOGY   63
4.1   INTRODUCTION   64
4.2   THE HYDROLOGIC CYCLE   64
      4.2.1 Water Quantity Perspective   64
      4.2.2 Water Quality Perspective   66
4.3   PRECIPITATION   67
      4.3.1 Depth-Duration-Frequency   67
      4.3.2 Rainfall Maps   68
      4.3.3 Intensity-Duration-Frequency   68

　　　　4.3.4 Development of a Design Storm 69
　　4.4 WATERSHED CHARACTERISTICS ....… 76
　　　　4.4.1 Watershed: Definition and Delineation 77
　　　　4.4.2 Drainage Area ....…......... 78
　　　　4.4.3 Watershed Length ....… 78
　　　　4.4.4 Watershed Slope 78
　　　　4.4.5 Land Cover and Use ....…..... 79
　　　　4.4.6 Surface Roughness 80
　　　　4.4.7 Channel Cross Sections ....…. 80
　　　　4.4.8 Channel Roughness 81
　　　　4.4.9 Runoff Curve Numbers 82
　　　　　　　4.4.9.1 Soil Group Classification 83
　　　　　　　4.4.9.2 Hydrologic Condition 84
　　　　　　　4.4.9.3 Curve Number Tables ....…...... 84
　　　　　　　4.4.9.4 Estimation of CN Values for Urban Land Uses 84
　　　　　　　4.4.9.5 Effect of Unconnected Impervious Area on Curve
　　　　　　　　　　　Numbers 88
　　　　4.4.10 Time of Concentration 88
　　　　　　　4.4.10.1 Velocity Method 89
　　　　　　　4.4.10.2 Sheet-Flow Travel Time 90
　　4.5 RATIONAL FORMULA 95
　　4.6 TR-55 GRAPHICAL PEAK-DISCHARGE METHOD 97
　　4.7 PROBLEMS 98
　　4.7 REFERENCES 104

5 INTRODUCTION TO MODELING 105
　　5.1 INTRODUCTION ....….... 106
　　5.2 UNIVARIATE FREQUENCY ANALYSIS 106
　　　　5.2.1 Population versus Sample 107
　　　　5.2.2 Regionalization ....…... 107
　　　　5.2.3 Probability Paper 107
　　　　5.2.4 Mathematical Model 108
　　　　5.2.5 Procedure 109
　　　　5.2.6 Sample Moments 109
　　　　5.2.7 Plotting Position Formulas 110
　　　　5.2.8 Return Period 110
　　　　5.2.9 The Normal Distribution 110
　　　　5.2.10 The Log-Normal Distribution 111
　　5.3 BIVARIATE MODELING 111
　　　　5.3.1 Correlation Analysis 112
　　　　　　　5.3.1.1 Graphical Analysis 112
　　　　　　　5.3.1.2 Bivariate Correlation 113
　　　　5.3.2 Regression Analysis 113
　　　　　　　5.3.2.1 Principle of Least Squares 113
　　　　　　　5.3.2.2 Zero-Intercept Model 114
　　　　　　　5.3.2.3 Reliability of the Regression Equation 115
　　5.4 MULTIPLE REGRESSION ANALYSIS 117
　　　　5.4.1 Correlation Matrix 118

5.4.2 Calibration of the Multiple Linear Model 119
5.4.3 Evaluating a Multiple Regression Model 119
5.5 NONLINEAR MODELS 121
5.5.1 The Power Model 121
5.5.2 Transformation and Calibration 122
5.6 PROBLEMS 125
5.7 REFERENCES 129

6 STORMWATER QUALITY ....… 131
6.1 INTRODUCTION 132
6.2 POLLUTANT LEVEL DETERMINATIONS 132
6.2.1 Grab Sample Measurements 133
6.2.2 Composite Sample Measurements 134
6.3 STORMWATER RUNOFF QUALITY DATA 136
6.3.1 pH 137
6.3.2 Suspended Solids and Oil and Grease 137
6.3.3 Organic Carbon/Oxygen Demand 137
6.3.4 Nutrients 137
6.3.5 Metals 141
6.3.6 Toxic Organics 141
6.3.7 Pathogens 141
6.4 POLLUTANT MASS LOADS 142
6.5 THE FIRST FLUSH 143
6.5.1 Defining the First Flush 144
6.5.2 First Flush Measurements 148
6.5.3 The Antecedent Dry Weather Period 149
6.6 PARTICULATES IN STORMATER RUNOFF 151
6.6.1 Physical Characteristics 151
6.6.2 Chemical Characteristics 152
6.7 POLLUTANT SOURCES 153
6.7.1 Contributions from Different Land Uses 153
6.7.2 Specific Pollutant Sources 155
6.8 EMPIRICAL HIGHWAY RUNOFF MODELS 159
6.9 STOKES LAW 159
6.10 UNIVERSAL SOIL LOSS EQUATION 163
6.11 PROBLEMS 168
6.12 REFERENCES 172

7 IMPROVEMENT OF STORMWATER QUALITY ....…... 175
7.1 INTRODUCTION 175
7.2 BEST MANAGEMENT PRACTICES 176
7.3 PROBLEMS 182
7.4 REFERENCES 184

8 STORAGE AND FLOW CONTROL **185**
  8.1 INTRODUCTION 186
      8.1.1 Effects of Urban Development 188
      8.1.2 SWM Policy Considerations 188
      8.1.3 Elements of SWM Structures 189
      8.1.4 Analysis versus Synthesis 190
      8.1.5 Planning versus Design 191
  8.2 WEIR AND ORIFICE EQUATIONS 191
      8.2.1 Orifice Equation 192
      8.2.2 Weir Equation 194
  8.3 DETENTION PONDS 196
      8.3.1 Storage Volume Estimation 196
      8.3.2 Sizing Riser Structures 197
          8.3.2.1 Sizing of Single-Stage Risers 199
          8.3.2.2 Sizing of Two-Stage Risers 206
  8.4 EXTENDED DETENTION 213
  8.5 WATER QUALITY IMPROVEMENT 215
      8.5.1 Conventional Sedimentation Theory 215
      8.5.2 Other Water Quality Issues in SWM Basins 219
      8.5.3 Case Study Information 219
  8.6 PROBLEMS 221
  8.7 REFERENCES 224

9 VEGETATIVE CONTROL METHODS **225**
  9.1 INTRODUCTION 226
  9.2 VEGETATED BUFFER STRIPS 228
      9.2.1 Performance Characteristics 229
      9.2.2 Buffer Strip Design 230
      9.2.3 Design Sensitivity 233
      9.2.4 Water Quality Performance 236
  9.3 VEGETATED SWALES 236
      9.3.1 Design Constraints 237
      9.3.2 Design Procedure 238
      9.3.3 Simple Water Quality Models 240
      9.3.4 Water Quality Performance 241
  9.4 BIORETENTION 241
      9.4.1 Design Characteristics 242
      9.4.2 Design Procedure 243
          9.4.2.1 Effect of Vegetation 243
          9.4.2.2 Storage Volume 246
          9.4.2.3 Infiltration 247
          9.4.2.4 Underdrain Design 248
      9.4.3 Flow Routing 250
      9.4.4 Performance Characteristics 254
          9.4.4.1 Pollutant Removals 254
          9.4.4.2 Suspended Solids and Filtration Theory 256
      9.4.5 Maintenance and Sustainability 257
  9.5 LEVEL SPREADERS 260

9.6   CHECK DAMS 261
9.7   GREEN ROOFS 263
9.8   PROBLEMS 264
9.9   REFERENCES 268

10 TRAPS, BASINS, AND FILTERS 271
    10.1 INTRODUCTION 272
    10.2 SEDIMENT TRAPS 273
        10.2.1 Trap Efficiency 273
        10.2.2 Design Procedure 274
        10.2.3 Design Considerations 276
    10.3 SEDIMENT BASINS 278
        10.3.1 Design Procedure 279
        10.3.2 Integrated Trap Efficiency 281
    10.4 INFILTRATION TRENCHES 283
        10.4.1 Considerations in Design and Construction 283
        10.4.2 Sizing of Storage-Trench Dimensions 284
        10.4.3 Sizing of Rate-Trench Dimensions 286
        10.4.4 Siting of an Infiltration Trench 288
        10.4.5 Considerations in Selecting the Filter Cloth 289
        10.4.6 Strategies for Increasing the Infiltration Potential ... 289
        10.4.7 Water Quality Considerations 290
        10.4.8 Regulatory Considerations 290
    10.5 CISTERNS 290
    10.6 SAND FILTERS 292
        10.6.1 Sand Filter Configurations 292
        10.6.2 Design Procedure 294
        10.6.3 Maintenance Considerations 295
        10.6.4 Filter Performance 297
    10.7 PROBLEMS 298
    10.8 REFERENCES 302

11 WETLANDS 303
    11.1   INTRODUCTION 304
    11.2   WATER BUDGET 305
    11.3   STORAGE ACCUMULATION METHOD 306
    11.4   RESIDENCE TIME IN WETLANDS 308
    11.5   PERMANENT POND DEPTH ESTIMATION 310
    11.6   ACTIVE STORAGE DESIGN 312
    11.7   FOREBAY DESIGN 313
    11.8   IN-STREAM WETLANDS 316
    11.9   OFF-STREAM WETLANDS 319
    11.10 OUTLET CONTROL STRUCTURES 321
        11.10.1 Uncontrolled Outlets 321
        11.10.2 Controlled Outlets 323
    11.11 WATER QUALITY IN WETLANDS 325
        11.11.1 Modeling Water Quality Improvement 326
        11.11.2 Wetland Design for Water Quality 328

11.11.3 Water Quality Performance 329
11.12 PROBLEMS 331
11.13 REFERENCES 336

12 LOW IMPACT DEVELOPMENT 337
12.1 INTRODUCTION 337
12.2 REDUCING IMPERVIOUS SURFACE 340
12.3 INTEGRATING LID PRACTICES 343
12.3.1 Porous Parking Swales 344
12.3.2 Design Parameters 346
12.3.3 Design Procedure 346
12.4 LANDOWNER ISSUES 349
12.5 POLLUTION PREVENTION 350
12.5.1 Material Substitution 350
12.5.2 Public Awareness 351
12.6 MAINTENANCE AND LONG-TERM STABILITY 353
12.7 RETROFITS 353
12.8 CASE STUDIES AND MODELING 354
12.9 PROBLEMS 354
12.10 REFERENCES 356

APPENDIX 359
A.1 NORMAL STANDARD DISTRIBUTION 360
A.2 STUDENTS t STATISTICS 362
B.1 PHYSICAL PROPERTIES OF WATER 363

INDEX 365

1

INTRODUCTION

1.1 URBAN SPRAWL: THE PROBLEM

1.1.1. A Historical Perspective

As society has developed, the living conditions of people have evolved to meet their needs. The living units have changed from individuals to small gatherings to the present cities, towns, and suburbs. As these cities and towns have emerged, the inherent characteristics of the original land and the surrounding areas have been altered. In many cases, what was once forest and open space, is now houses and manicured lawns, driveways and roadways, and commercial and industrial areas. This transition in land use has caused a number of changes directly to the local environment and has had significant implications to the local ecosystems. Many of these changes have had adverse environmental impacts.

Some of the worst effects have resulted from sprawl development. The word sprawl connotes something that is spread out carelessly in an unnatural manner and distributed or stretched out in an irregular distribution. In the context of urban sprawl, the term implies that urban facilities are distributed spatially in a way that appears to be unplanned and nonoptimal from the perspective of the inhabitants of the community, the surrounding infrastructure, and the environment.

Sprawl is not the result of urban development; it is the result of poor methods of development. The first step in solving the problems associated with urban sprawl is to understand the problems from the viewpoint of the underlying physical, chemical, and biological processes. In terms of hydrologic engineering, this means the important physical processes that are part of the hydrologic cycle. Over 2 million acres of open space are claimed by sprawl each year in the U.S. (Mitchell, 2001). This translates into a significant change in local and regional hydrologic processes. Processes that relate to watershed storage such as interception, infiltration, and depression storage are greatly modified by urban development. Runoff timing is also adversely affected by grading and cut-and-fill practices.

Modifications to the hydraulic characteristics of small watersheds, especially the elimination of first-order streams, are also a major problem. The associated problem of eliminating natural stream buffer areas has important ecological implications. Other elements of the hydrologic cycle are also adversely influenced by urban development. Specifically, methods of land development that are insensitive to the effects on the physical hydrologic processes contribute to sprawl.

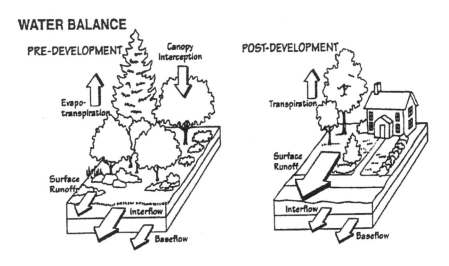

Figure 1.1. Change in Water Balance from Land Development

As pristine areas are developed, the various hydrological pathways change in importance, resulting in significant modification of both the runoff flow and quality characteristics (Figure 1.1). Development clears trees and vegetation; the local topography is graded such that water-holding depressions are eliminated. Development will increase the imperviousness of an area in the form of roadways, parking lots, and rooftops. While forested and grass areas allow 70 to 90% of the rainfall to infiltrate, roofs of houses and other buildings do not allow infiltration. This is also true with sidewalks, driveways, and roads, which allow only a small amount of infiltration due to cracks and small openings. As a result, developed areas produce more direct runoff from a site at the expense of infiltration. The fraction of precipitation that runs off is a composite of all of the different types of land surface: roofs, sidewalks, driveways, lawns, patios, and roads. The amount of precipitation that runs off ranges up to 80% in land use areas with a very high density of housing to 10% with single homes on very large lots; ultra urban areas may approach 95%.

Because of the increase in surface runoff volumes, the change from undeveloped to developed land increases the opportunity for rainwater and runoff to detach, entrain, and transport increased loads of many different types of pollutants.

These pollutants are transported in the runoff to nearby streams or storm drains. The storm drains ultimately discharge the polluted water into local streams and waterways where the in-stream ecosystem is often unable to cope with the increased loads of pollutants. Stormwater runoff into waterways is classified as a non-point source of water pollution as pollutants are generated at almost every point within a drainage area. For example, washoff from roadways includes motor oils and grease, while lawns are the source of suspended solids and fertilizers. These nonpoint sources are in contrast to point sources, which are the discharge of wastewater (usually treated) from wastewater treatment plants or industries.

To better appreciate this focus on the changes to the physical hydrologic and water quality processes, it is important to recognize that over the last couple of decades many urban areas have doubled in spatial extent even though the population has increased very little, maybe even decreased. As much as 60% of this increase in land devoted to new development is associated with the automobile, i.e., impervious parking spaces, driveways, and roadways that are used only for a small part of the average day. Furthermore, most of the residential development has been in the form of single-family homes on large lots, with wide streets and cul-de-sacs. From the standpoint of the community ecosystem, society could not have selected a type of development that was more environmentally adverse. Unfortunately, the advances in stormwater and water quality mitigation practices have not kept pace with the advances of the outer boundaries of urban and suburban areas.

1.1.2. Characteristics of Urban Sprawl

It is unlikely that a comprehensive, yet concise, definition can be provided for urban sprawl. It can be viewed from many perspectives, and each perspective adds something new to the definition. However, the perspectives are not independent of each other. Three perspectives are especially relevant to this book: land use planning, transportation, and hydrologic management. By identifying some problems or characteristics of each of these perspectives, a broad definition of urban sprawl should emerge.

A major contributor to sprawl is land use policies that allow development to take place such that the area developed per capita is high. Very often, low density development is allowed to take place in communities that are located far from the urban center. This sets a precedent, and often creates a ring around the urban center such that the distribution of the values of homes is low in the urban center and increases with distance away from center city. This pattern of development also lessens the chance of adequate open space per capita. The negative effects of urban sprawl would be less if clustered land development patterns were the rule, with expanses of open space separating clusters of high density residential and commercial development. Forward thinking land use planning also incorporates corridors for future mass transit, as the under utilization of mass transit is a characteristic of urban sprawl. Low density land development patterns, as opposed to cluster patterns, contributes to poor accessibility to mass transit depots, even where it exists. Lack of accessibility also increases automobile usage.

Land use planning that favors commercial and industrial development over residential contributes to sprawl. Commercial/industrial lands provide higher tax bases than residential. Residences require services, especially school systems, which are more costly for the municipalities, that induce these decisions. Lack of adequate affordable housing drives residential areas further from commercial centers.

Since transportation related imperviousness represents more than 50% of all imperviousness, transportation facilities and attitudes about transportation have been central to most of the prime examples of urban sprawl. Whether talking about urban sprawl or smart growth, important transportation related metrics include the roadway length per capita, the average distance to mass transit connections, the average street width, the width of parking spaces, the number of parking spaces per vehicle, the number of parking spaces per square foot of commercial square footage, the number of vehicle miles traveled per capita, and the average trip length. Relatively high values of each of these metrics are associated with urban sprawl. To initiate smart growth in a community requires the adoption of an attitude towards reducing the value of each of these metrics.

Since pollution levels are often positively correlated with imperviousness and the amount of impervious cover is largely associated with transportation related activities, the environmental perspective is closely related to the transportation perspective. Urban sprawl is associated with a lack of control of surface stormwater runoff and a simultaneous reduction in the amount of natural storage within the watershed and streams. Instead of controlling the water at its source through natural and designed storage, urban sprawl results in runoff being quickly conveyed away from the source. The removal of natural storage allows the dispersal of pollutants, which subsequently makes pollution reduction more difficult to address. Urban sprawl is associated with the removal of buffers from the perimeter of streams to allow for development closer to the water's edge, and very often urban sprawl efficiency is increased by replacing the headwater streams with a pipe. Urban sprawl creates a major reliance on automobile use, discharging pollutants from a variety of sources. Adoption of a smart growth philosophy on these environmental metrics is necessary to counter attitudes towards increasing urban sprawl (McCuen 2003).

1.1.3. Pollution of Waterways

Many streams, rivers, and lakes around the world have been polluted by various discharges over many years. A variety of sources have contributed to this pollution. For the longest time, and even now in a number of areas, point source discharges represented the largest input of pollutants. Point source discharges are direct pipelines of wastewater from factories, industries, and municipal wastewaters. For many years, these wastewaters did not receive treatment, and the impact on the receiving streams, in many cases, was severe. One of the major pollutants was organic waste that microorganisms degraded when the waste was discharged into water bodies. This microbial waste degradation consumed all of the oxygen in the streams, killing the fish, creating severe odors, and making the streams unsuitable for just about any other use. Toxic compounds, such as organic solvents and heavy metals, were discharged in wastewater from industries in their manufacturing of consumer products such as automobiles, electronic devices, and fuels. A third set of pollutants, nutrients, which are primarily compounds of nitrogen and phosphorus, resulted from both municipal and industrial wastewaters.

In the early 1970's, U.S. Federal regulations were established to address water pollution in rivers and lakes. The Clean Water Act (CWA) was enacted in 1977 as an amendment to the Federal Water Pollution Control Act of 1972 with a goal to restore the "swimability, fishability, and drinkability of our nations waters." The result of the CWA was the creation of a permit program for point source discharges, the National Pollution Discharge Elimination System (NPDES). NPDES permits require treatment and removal of major pollutants before wastes can be discharged. Since that time, implementation and enforcement of the NPDES has significantly reduced water pollution from point sources.

Nonpoint sources of pollutants, however, have not been addressed to the extent that point sources have. Nonpoint sources are primarily stormwater runoff, from agricultural and developed areas. As pollutant discharges from point sources have decreased, a large fraction of the pollutant load becomes contributed by nonpoint sources, particularly from runoff. The nature of the runoff depends on the characteristics of the drainage land. Runoff from agricultural areas can include solids, nutrients, and pesticides from fields, but also animal wastes and feed from pasture areas. Generally, runoff from agricultural lands has not been regulated, but programs have been established to encourage practices that minimize runoff and the pollutants that it carries.

As more and more land is developed for residential, commercial, and industrial use, stormwater runoff from these urban land covers becomes a more significant pollutant source. Research has been completed over the past two decades to try to understand the characteristics of urban runoff and its affects on stream morphology and water quality. In the last few years, a significant amount of information has become available about these issues. Also, as the detrimental effects of urban stormwater runoff on the environment have been recognized, different devices and practices have been developed, designed, and installed for stream water protection to deal with higher runoff flows, but with a more recent emphasis on improving water quality.

As one example, the small detention pond has been a commonly used solution to the increased flooding associated with land development. While it helped to mitigate the effects of urban development on the magnitude and frequency of flooding, it is not able to adequately address the water quality issues that surface with land development. Thus, receiving stream ecosystems continue to be degraded as development increases. An array of water quality control methods is needed to address complex land development environmental concerns.

1.1.4. The Effects of Urban Sprawl

Urban sprawl is associated with the development and utilization of land at a rate faster than the increase in the associated population. Such a scenario indicates a low-density use of land. With low density development, the physical infrastructure, i.e., roads, water and sewage pipelines, and power and communication systems must efficiently transverse wide land areas. Land development removes forested land and replaces it with developed, impervious areas. When less water infiltrates, the volume of surface runoff increases. Additionally, the frequency of flood flows increases (Dunne and Leopold 1978). Studies have shown that the number of problem runoff magnitudes, such as those that cause undesired ponded water, may increase three fold in a developed basin compared to conditions prior to development. Higher runoff velocities can increase the erosion of streams and destroy stream ecosystems. The eroded sediments are deposited at areas downstream where the flows are slowed, filling in streambed areas and burying active ecosystems on the stream bottoms.

The introduction of buildings, automobiles, people, and pets also increases the loading rates and diversity of pollutants that are discharged from the developed areas. Oils and grease, metals, pesticides, fertilizers, and pathogenic microorganisms are common products of urbanized land uses. The increases in both the magnitude and frequency of runoff flow rates over impervious surfaces increases the supply of these pollutants that are rapidly transported to receiving streams. Increased stream degradation is highly correlated to increase in impervious area and population in a watershed.

It is not necessarily the use of the land that makes urban sprawl undesirable; it is the unintended hydrologic and environmental consequences that are the culprits. If reliable management practices were available to mitigate these negative consequences, then many of the undesirable environmental conditions associated with development and sprawl could be minimized.

1.1.5 Difficulties Faced in Improving Stormwater Quality

Looking back over the last half century, the future effects of decisions about land development were not evident at the time when the decisions were made. Addressing urban stormwater quality issues was and is an inherently difficult problem. Our understanding of the dominant physical, chemical, and biological processes is not complete. Additionally, many are inherently influenced by uncontrollable factors. That is, both the quality and flow quantity vary temporally and spatially and are not readily predictable. Focusing on just the hydrology and runoff flow characteristics of an urban area, various rainfall events will have different durations, intensities, and frequency characteristics. The spatial layout of the watershed will control the time of concentration and the amount of infiltration; antecedent moisture will be important. Water quality parameters will depend on the density, type, and usage of automobiles, rainfall intensity, soil and vegetation characteristics, and, in many cases, choices made by individual residents. Development itself is often much like a random process as many regions of the country have lacked institutional and regulatory control of land development.

Because of the lack of understanding of the consequences of development, most of the traditional stormwater control methods have been inadequate for mitigating the effects of sprawl development. Most stormwater control systems have been designed to remove and convey runoff from the site as quickly as possible, notably with curbs, gutters, and storm drains. Designs were based on extreme, high flow events to minimize the risk of flood damage in local areas. The storm drains were piped together and discharged to nearby streams. In retrospect, these control methods passed the problem on to other parts of the watershed, possibly creating additional problems and delaying the development of reliable methods of control.

Later, the effects of urbanization on streams were more fully recognized. The increase in the magnitude and frequency of flows caused major problems in streams where stormwater was discharged. The larger flows increased shear stresses on the river banks. Thus, stream bank erosion increased, which changed their shapes, gouging and eating away land, felling trees along the river banks, and carrying eroded soils into the receiving waters down stream. Cumulatively, significant long-term damage to river ecosystems resulted.

In addition to the wide variability of storm and flow characteristics, the difficulty in dealing with the problem of urban sprawl has been compounded by the myriad issues associated with problematic water quality and its measurement and estimation. Each pollutant can originate from several sources, be transported to a nearby drain, and travel to a receiving stream. Thus, it is difficult to associate a stream quality measurement with its origin. Sources, and associated pollutants, depend on the land use and the specific makeup of the area. When a pollutant moves with the flowing water in a stream, it may be unevenly distributed in the water. Therefore, simultaneous measurements at one point in a stream at one point in time will differ. This makes it difficult to accurately characterize the concentration of the pollutant and compounds the problem of associating the level of pollution with the degree of urban sprawl. By itself, the year-to-year variation in rainfall amounts and distribution can introduce a significant level of uncertainty in water quality measurements.

Pollutants can be trapped at any point between the source of the pollutant and the point of measurement. This will reduce the measured concentration at one point in time but if the trapped material is later released, a down gradient measurement may actually inflate the measured value above the representative value for that point in time. Given these measurement complexities, it is difficult to both make forecasts of future pollution levels or evaluate the effects of specific urban sprawl activities on pollution levels.

Increased knowledge about the connections between urban sprawl and the environment will result from both theoretical research and empirical analysis of data collected in planned experiments. As we learn more about pollutant sources in urban areas, society and the profession will be better equipped to address the pollution problems. Pollution prevention and waste minimization, as applied to urban stormwater, means reducing the pollutants produced, keeping them on-site and treating them there, and reducing runoff velocities and volumes that detach and transport pollutants. More community designs that take these concepts into consideration are being developed. However, costs, current regulations, and public acceptance can discourage these "green" designs from being implemented.

It is easy to say that urban sprawl has created a myriad of environmental problems. It is difficult to develop comprehensive, reliable solutions. A single method is unlikely to be adequate alone. Developing a strategy for the improvement of runoff water quality is a formidable task. This strategy must be aimed at reducing the negative effects of as many different pollutants as possible. It will most likely need to be targeted to the characteristics of a specific land use area. Different pollutants have different physicochemical properties and require different provisions for their removal. With so many different target pollutants in runoff, such as suspended solids, nutrients, metals, pesticides, oil and grease, temperature, and others, it is unlikely that a simple treatment device would have high efficiencies for all pollutants. Additionally, any treatment device must be able to provide efficient treatment over a very wide range of flows and pollutant concentrations. Flows should not be impeded under extreme conditions such that flood safety is compromised. A treatment device should be inexpensive, not require a large amount of land, not present a public safety hazard, and not be aesthetically displeasing. Maintenance requirements must be infrequent, inexpensive, and easy to perform. These are the challenges of stormwater quality improvement.

1.2 SMART GROWTH: THE SOLUTION

1.2.1 Urban Sprawl or Smart Growth

Is smart growth the answer to urban sprawl? The public is experiencing the effects of urban sprawl. The Sierra Club reports that American drivers spend 443 hours per year behind the wheel, mostly commuting. While traffic jams are the most obvious effect, environmental effects are just as significant. For example, if smart growth is to be intelligent growth, changes to the hydrologic and environmental processes related to urban sprawl must be identified and considered along with the effects of other aspects of sprawl such as the transportation-related issues. The loss of small streams to environmental degradation and the lack of public open space in traditionally-developed communities have negative societal consequences. If smart growth is to be the answer to urban sprawl, problems related to environmental and water quality issues must be part of the smart growth solution. A multi-perspective view of smart growth is necessary for it to be an effective solution to sprawl.

1.2.2 Alternative Perspective on Smart Growth

Urban sprawl is complex to define, as any one definition cannot address all of the perspectives from which sprawl problems can be viewed. It is also similarly difficult to provide a definition of smart growth, as a definition that satisfies the land use planner may be inadequate for the environmental engineer, and certainly the definition for environmental engineers will not satisfy the transportation planner or engineer. Just as the problem of sprawl was viewed from the perspectives of planning, transportation, and the environment, a multi-perspective view of smart growth as a possible solution to sprawl will be required.

Comprehensive land use planning is a necessary force in smart growth development. It seems reasonable that the density of development should receive special attention. While it seems that cluster development would be favored, single-family detached units are not precluded from a smart growth development plan. Clustered developments are more conducive to mass transit layouts and to environmental control through best management practices.   With greater concentration of open space, small headwater streams will be placed under less of the pressures associated with large-lot-size development patterns.

From a transportation infrastructure perspective, smart growth development should require smaller areas of paved roadways, as the collector streets will not need to be as wide and the number of feeder streets can be reduced. Fewer sidewalks will be necessary, and pedestrian walkways to mass transit depots will be shorter. With higher concentrations of development, parking areas will be used more frequently, thus reducing the need for parking spaces that are unused a large part of each day. With shopping areas located nearby within each cluster, the number of vehicle trips and the average length of each trip will be lower than with low-density, single-family development patterns. The transportation metrics for smart growth are largely low values of the same transportation related metrics used in evaluating sprawl.

In terms of the environment, with specific focus on water quality, it seems that smart growth will be defined by the extent that best management practices are used to limit the environmentally detrimental aspects of development. The loss of natural storage must be minimized, and any storage that is lost must be replaced, where possible, with storage that more closely simulates the physical processes of the natural system. In-stream pond storage, while in some cases necessary, can be reduced or eliminated if more natural storage such as bioretention and even on-site extended detention is used. The storage should be located on-site, rather than off site, to minimize the dispersal of pollutants. Vegetated buffer strips should be emphasized early in the planning process, as these reduce the likelihood of pollutants being dispersed through the stream system. Pipes are a necessary part of any development, but smart growth development should not be based on a narrow pipe-and-pond perspective. These systems would provide poor values of any on-site-treatment metric. Grass swales provide alternate conveyance pathways. Understanding the role of best management practices is complicated because of the general lack of knowledge as to how effective these practices are in actual settings.

1.3 PROBLEMS

1.1 Look around at the development in your neighborhood. Describe the land use and how it relates to stormwater management.

1.2 If it rains on the roof of your building, where does the water go? Does it stay on the property of your residence, or is it directed elsewhere?

1.3 Locate a small stream in you locality. Follow the stream for a half of a mile and note the land uses that drain into the stream. Specifically examine parts of the stream where the cross-sectional configuration changes, such as a wider bottom or shallower sides. Notes changes in the contributing watershed at those points. .

1.4 Look at the storm drains on your street or in a nearby neighborhood. What is the source of the water entering the inlets (e.g., local driveways)? Is a stormwater management practice such as a pond located in the neighborhood?

1.5 In your immediate locality, identify a cluster of houses, an apartment complex, a small shopping or strip mall, a public school, and a large shopping district. For each of these, estimate the fractions of impervious cover and parking space. Discuss the results as they relate to the amount of runoff expected.

1.6 Obtain and article from a local newspaper or newsmagazine that discusses some aspect of urban/suburban sprawl. Analyze the article on the basis of factors such as the scale of the assessment (e.g., individual lot, neighborhood, watershed), the problem (e.g., water quality or water quantity), recommended solutions, or the role of modeling.

1.7 Perform the analysis of Problem 1.6 using an article from a professional journal.

1.8 Obtain definitions of bias, precision, and accuracy, as they relate to modeling, from a dictionary and from books on statistics. Interpret the definitions as they relate to models for predicting the concentration of a water quality parameter.

1.9 Obtain an article from a professional journal that describe the development of a water quality model, either a simple regression model or a more detailed conceptual model. Assess the way that the model was evaluated for accuracy, including the criteria, the methods, (i.e., graphical, statistical), and the results.

1.4 REFERENCES

Dunne, T. and Leopold, L.B. (1978) *Water in Environmental Planning*, W.H. Freeman & Co., San Francisco.
McCuen, R.H. (2003) "Smart Growth: Hydrologic Perspective," *J. Prof. Issues in Engrg. Educ. And Pract.*, **129**(3), 151-154.
Mitchell, J.G. (2001) "Urban Sprawl," *Natl. Geogr.*, **200**(10), 48-73.

2

WATER QUALITY PARAMETERS

NOTATION

C = concentration
M = mass
Q = flow rate
t = time
V = volume

2.1. INTRODUCTION

It is difficult to provide a general definition for water quality, at least one that applies to all situations. The relative quality of a water depends on its purpose or use. To further complicate the matter, any water or water body will usually have many different uses. Communities may use a water body as a source for drinking water and for discharging treated wastewater. Individuals may use the same water body for fishing, for boating, and maybe even swimming. Certainly the desired quality is different for all of these uses. Also, water quality may need to be defined for species other than humans. Wildlife, such as various types of fish, require high quality water to thrive. Aquatic plants and smaller animal life have their own quality requirements and are ecologically important for the proliferation of higher-level trophic states in the food chain. Consequently, one of the fundamental problems in stormwater quality is to appropriately define water quality in a way that reflects all of the uses and users of the water.

The corresponding second water quality issue is then to determine the specific characteristics of the water that make the quality acceptable for each use. For example, a drinking water supply should have very strict requirements for receiving discharges, be of very high quality, and contain few or no toxic pollutants and organisms that can cause diseases. Healthy fishing areas may require high levels of dissolved oxygen, cool temperatures, clear water, and be devoid of high levels of toxic compounds.

Generally, a few benchmark physical and chemical parameters are used to define the water quality for a specific use. Using these parameters, a set of guidelines is developed. These guidelines may have a range in which the water must fall, a maximum value of a pollutant, or a minimum value of a necessary substance (such as dissolved oxygen). The quality of a water is then determined based on whether it meets all of the requirements. Table 2.1 provides an example of chemical water quality criteria.

Table 2.1. Typical fresh surface water quality criteria for general use

Parameter	Value	Parameter	Value
PH	5.0 - 9.0	Dissolved Oxygen	>5.0 mg/L
Total Ammonia Nitrogen	< 9.0 mg/L	Orthophosphate	< 0.01 mg/L
Fecal Coliform	< 200/100 mL		

Example 2.1 Assume that the quality of a river water can be discussed with respect to the criteria of Table 2.1. Also, assume that the characteristics of a river are as follows: pH (7.8); total ammonia nitrogen (3.2 mg/L); fecal coliform (8/100 mL); dissolved oxygen (4.8 mg/L); and orthophosphate (0.008 mg/L). The criteria of Table 2.1 have a range (pH), a minimum (dissolved oxygen), and three maximum values. The measured river water has ammonia-N, orthophosphate, and fecal coliform, all of which are below the required values. Also, the pH is within the required range. The dissolved oxygen level, however, is below the required value of 5.0 mg/L. This low value may stress the fish in the water and suggest that a program be implemented to improve the water quality if a healthy fishing ecosystem is desired.

It is not practical to list zero concentrations for any substances in setting water quality standards. First, it is probably impossible to not have any amount of a material present. Very, very low concentrations (µg/L, ng/L, pg/L) are likely present for one reason or another, maybe even due to natural events. Extremely low concentrations of most pollutants are generally not harmful to aquatic and human life. It is also necessary to realize that reliable detection and quantification of extremely low concentrations of substances in water are difficult, even with the most sophisticated instruments and techniques. Regulatory agencies realize that a limit of zero is not possible, so they try to set low, but realistic values as standards.

Second, a few substances must be present in the water or at specific values to allow desired aquatic animal and plant species to thrive. Thus, the availability of these substances is required for the water quality to be considered good. Dissolved oxygen (DO) is the most important of these substances. Nearly all aquatic animals need an adequate level of DO in the water to "breathe" and survive. The DO level in a water body is one of the basic traditional measurements used to describe water quality. Other water characteristics must be within a range amenable for aquatic life. These parameters include water temperature and pH. Aquatic life need moderate levels of both temperature and pH for survival, growth, and reproduction. As either high or low extremes of both temperature and pH are approached, the quality of the water decreases.

Through natural hydrologic processes such as rainfall, surface runoff, and ground water inflow to a water body, substances are dissolved and mobilized in these waters. A number of these materials and substances can be harmful to aquatic life. These include particulates and substances that cause the reduction of DO, such as organic waste and nutrients. Oil and many different chemicals such as pesticides, solvents, and heavy metals are toxic. Therefore, in defining water quality, it is necessary for the water to contain appropriate levels of some necessary substances (what we will call "factors necessary for life"), but have low levels of other substances that are potentially harmful. Both of these categories, necessary substances and harmful pollutants, can be discussed in detail as they relate to the problems of defining water quality.

Many different parameters must be considered when evaluating water quality. First, the water must be appropriately habitable for the various species that are to exist. Thus, adequate oxygen levels must be dissolved in the water. The pH and temperature must be acceptable. These are the factors necessary for life. Some pollutants can affect these factors, including organic waste material and nutrients. For example, as organic wastes are microbially degraded, the level of dissolved oxygen is reduced. Elevated concentrations of toxic compounds can be lethal to some aquatic organisms.

With both toxics and factors necessary for life, water quality is defined by setting threshold values for various important parameters. Measurements are made *in-situ*, or samples are collected and taken to a laboratory for analysis. Values are compared to guidelines or regulatory requirements, such as those in Table 2.1. This provides a snapshot of water quality based on the parameters used to define it.

A second method to classify the quality of a water is to evaluate the various plant and animal species that are present, which are referred to as *biological indicators*. The type and number of specific organisms in a water body depend on the quality characteristics of the water. Based on the number of organisms (usually insects or some other macroinvertebrate) or plants found, the habitatability of the water can be assessed. Finding abundant levels of organisms that cannot survive in water of poor quality indicate that the water quality is very good. For example, high amounts of bivalves, such as clams or mussels, snails, flies, and worms per unit length or area of stream can suggest good quality water. On the other hand, the absence of these species is a direct indication of poor quality water. The presence or absence of various species can give a comprehensive picture of the water quality. Using these biological indicators also gives a more pervasive view of the water, rather than the snapshot provided by chemical and physical measurements, which only provide information about the water at the exact time that the sample was taken. Proliferation of living species provides a historical view to the water quality and therefore a better idea of what can be expected in the near future. These surveys, however, do not reveal the cause of the water quality problem. If target organisms are not present, the water quality is likely poor, but the specific reasons may not be evident. The effect could be the result of low DO, a high level of salt, or something toxic. A subsequent detailed investigation is needed to identify the specific cause.

2.2. MASS, CONCENTRATION, AND LOADING

Before discussing specific water quality parameters and pollutants, it is first necessary to address issues of concentration and mass of materials. Mass and concentration measurements are necessary to quantitatively describe the materials present. Mass is a measure of the amount of a material present. The SI unit for mass is the gram (g) with derived units of kilogram (kg = 1000 g) and milligram (mg = 0.001 g). In traditional English units, rather than mass, weight is generally used, that is the common pound (lb).

The amount of mass per unit volume of fluid, i.e., water, is the concentration:

$$C = M/V \qquad (2.1)$$

where C represents concentration, M the mass, and V the volume. The typical units of concentration are mg/L. Measurements of water quality parameters are represented as concentrations. Generally, the toxicity of a pollutant towards various fauna is best represented by the concentration of that pollutant in the water.

Example 2.2 Mercury is a highly toxic substance. The Federal limit for mercury concentration in drinking water is 0.002 mg/L. If the mercury concentration in a large swimming pool is equal to this value, the mass of mercury dissolved in the pool can be determined. Assume that the pool is 16 m wide, 50 m long, and 2 m deep. The pool water volume is:

$$V = L\ W\ H = (50\ m)(16\ m)(2\ m) = 1600\ m^3 = 1.6 \times 10^6\ L \qquad (2.2)$$

Rearranging Eq (2.1):

$$M = CV = (0.002\ mg/L)(1.6 \times 10^6\ L) = 3200\ mg = 3.2\ g \qquad (2.3)$$

At a density of 13.5 g/cm^3 (mercury is very dense), this corresponds to a spherical drop of mercury with a diameter of only 0.77 cm.

2.2.1. Mass Balances

Concentration, however, is not a conservative parameter. In the mixing of waters, concentrations cannot be added or otherwise combined to produce the concentration in the mixed water. Mass is conserved and concentrations can be determined using a mass balance approach. In the simplest case, the conservation of mass says that total mass into a system from all sources is equal to the total mass leaving the system:

$$\Sigma M_{in} = \Sigma M_{out} \qquad (2.4)$$

In a flowing system, the mass into and out of a system per unit time, t, is:

$$\sum \frac{M_{in}}{t} = \sum \frac{M_{out}}{t} \qquad (2.5)$$

The flow rate of water, Q, is defined as the volume per unit time:

$$Q = V/t \qquad (2.6)$$

Therefore, combining Eq. (2.1), (2.3), and (2.4), the mass balance can be represented as:

$$\sum (QC)_{in} = \sum (QC)_{out} \qquad (2.7)$$

The input and output mass flows are summed for all contributions.

Example 2.3 A small stream that runs through a highly developed area has a phosphorus concentration of 8.5 mg/L at a flow rate of 40 L/s. This stream discharges into a river with a flow of 400 L/s and a phosphorus concentration of 2.0 mg/L. After the stream and river mix, the phosphorus concentration can be found employing a mass balance as in Eq. (2.7):

$$(CQ)_{stream} + (CQ)_{river} = (CQ)_{s+r} \qquad (2.8)$$

The sum of the stream and river flow rate is found from the continuity equation (which is based on a mass balance for the water itself):

$$Q_{s+r} = 40 + 400 = 440 \text{ L/s} \qquad (2.9)$$

Therefore:

$$C_{s+r} = \frac{(CQ)_{stream} + (CQ)_{river}}{Q_{s+r}} \qquad (2.10)$$

$$= \frac{(8.5 \text{ mg/L})(40 \text{ L/s}) + (2.0 \text{ mg/L})(400 \text{ L/s})}{440 \text{ L/s}} = 2.6 \text{ mg/L}$$

In more complex systems, material can accumulate in a system (e.g., sediment can settle), and chemical and/or biological reactions can degrade (or produce) pollutants. Considering these processes, a full mass balance expression is, in words:

$$\frac{\text{change in mass}}{\text{time}} = \text{total mass flow in} - \text{total mass flow out}$$

$$+ \frac{\text{mass produced}}{\text{time}} - \frac{\text{mass consumed}}{\text{time}} \qquad (2.11)$$

In most systems, we will need to include only a few of these terms, but a complex system will include all.

Example 2.4   A stream that contains 120 mg/L of sediment flows into a treatment pond at 20 L/s. The effluent stream leaves at the same flow rate, but with a sediment concentration of 20 mg/L. The sediment accumulation rate can be determined using a mass balance analysis:

$$\frac{\text{change in mass}}{\text{time}} = \text{total mass flow in - total mass flow out} \tag{2.12}$$

or, mathematically:

$$\begin{aligned}
\frac{\Delta M}{\Delta t} &= (CQ)_{in} + (CQ)_{out} \\
&= (120\,\text{mg/L})(20\,\text{L/s}) - (20\,\text{mg/L})(20\,\text{L/s}) \\
&= 2000\,\text{mg/s} = 1728\,\text{kg/day}
\end{aligned} \tag{2.13}$$

In many cases, the total mass, or load, of a pollutant that has entered or exited a system over a given time is of interest. This loading is found from Eqs. (2.1) and (2.6) by integrating over the time period of interest, T:

$$M = \int_0^T CQ\,dt \tag{2.14}$$

Total mass loads, and concentrations are used to characterize the total pollutant inputs into streams and rivers and the mixing of waters with different concentrations, as well as to evaluate the removal of pollutants using various treatment practices. Care must be taken when investigating loadings and removals in runoff situations because both the flow rates and the concentrations can vary significantly with time. The same mass of a pollutant can be moved into a water body either with a high concentration in a low flow rate or with a low concentration at a high flow rate.

Example 2.5. Data from a 12.8-acre urban area were obtained for part of a storm event (October 18, 1981) in Baltimore, MD. Table 2.2 gives the measured instantaneous flow rates and total recoverable lead concentrations to the storm sewer. The total load of lead can be computed using Eq. (2.14) and the trapezoidal rule. For the measured data, the total load of lead to the sewer is 105 lbs.

2.2.2. Concentration-Flow Relationships

Studies have shown that the concentrations of many constituents in a stream are proportional to the flow rate Q, with the power-model form commonly used:

$$C = aQ^b \tag{2.15}$$

This occurs because high flow rates can mobilize many substances. In this equation, a and b are fitting coefficients, which are commonly fitted using least squares analysis following a logarithmic transform of C and Q. Equation (2.15) is valid for continuous flows. For intermittent flows, such as with washloads from storm events, an alternative approach is used. Once the concentration is determined from Eq. (2.15), the load can then be estimated using Eq. (2.14). Some empirical formulas are designed to predict the load directly, while others provide predictions of concentrations.

Table 2.2. Estimation of Lead Loadings for Hampden Storm Sewer, Baltimore, MD (Gage No. 01589460) for Part of the October 18, 1981, Event

Time	Flow, Q (ft^3/s)	Lead conc., C (mg/L)	Load, CQ (lb/min)	Time increment, Δt (min)	Load, CQt (lb)
1807	2.30	780	6.71		
				7	32.2
1814	2.30	290	2.50		
				20	29.2
1834	0.95	140	0.498		
				20	8.9
1854	0.87	120	0.391		
				68	22.4
2002	0.80	90	0.269		
				30	6.7
2032	0.40	120	0.180		
				60	5.5
2132	0.01	80	0.003		
				60	0.1
2232	0.01	40	0.0015		
				Total =	105

Example 2.6 Table 2.3 gives measurements of the instantaneous streamflow and chloride concentration at the Conowingo, MD, station (01578310) on the Susquehanna River. The measurements were made over a period of four water-years. Streamflow rates varied from about 3,000 ft^3/s to almost 450,000 ft^3/s. Using a non-linear program, Eq. (2.15) was fitted to the data, with the following result:

$$\hat{C} = 113.7Q^{-.2071} \qquad (2.16)$$

in which $\hat{C}$ is the predicted value of the concentration C. The model had a bias (mean error) of -0.81 mg/L, a standard error of estimate of 5.29 mg/L $(S_e/S_y =$

0.918), and an adjusted correlation coefficient of 0.43 $(R^2 = 0.187)$. Thus, the model does not provide a good fit to the data. However, such scatter is typical of concentration-flow relationships.

Table 2.3. Instantaneous Streamflow Q and Dissolved Chloride C: Susquehanna River at Conowingo, MD (Gage No. 01578310)

Q (cfs)	C (mg/L)	Date
444000	7.5	2-17-84
265000	6.4	4-09-84
185000	18.0	2-13-81
173000	9.7	3-17-82
163000	6.1	2-27-81
81800	12.0	4-15-81
78600	9.5	1-12-82
76700	9.2	4-29-81
72800	9.5	3-12-81
68900	9.0	6-22-81
67800	7.7	11-9-81
67500	8.0	5-21-84
60200	27.0	11-04-80
60000	9.8	5-12-82
51000	16.0	12-10-80
41600	14.0	1-04-84
39900	15.0	9-09-81
34900	12.0	3-25-81
34300	7.5	5-27-81
32700	17.0	9-29-81
32400	24.0	10-23-80
31700	10.0	7-28-81
31400	22.0	11-24-80
31000	13.0	12-30-80
30800	13.0	9-19-84
22200	18.0	1-29-81
17000	16.0	9-22-82
5500	10.0	7-15-82
3370	25.0	11-07-83

2.3. FACTORS NECESSARY FOR LIFE

As briefly discussed above, having good water quality is important to healthy ecosystems. Several basic conditions must be met for aquatic life to thrive in the water. When these conditions are not optimal, species populations become stressed. When conditions are poor, organisms die. Dissolved oxygen, pH, and temperature are three parameters that are especially important to the growth, vitality, and healthy reproduction of aquatic life.

2.3.1. Dissolved Oxygen

Water, whether flowing in rivers and streams or stored in lakes, mixes with air above the water surface. Oxygen from the air dissolves in the water. Oxygen dissolved at the surface is typically dispersed throughout the water column due to turbulence in the water flow. However, water has a limit as to how much oxygen can be dissolved in it, with the capacity decreasing with increasing temperature. Water can hold a maximum of about 8 to 14 mg/L DO when exposed to the air, depending on the temperature. The maximum, or equilibrium, level is referred to as the *saturation* level. If the water is undersaturated, oxygen will be absorbed into the water through contact with the atmosphere. Water in rapidly moving, shallow streams with rocks and riffles will absorb oxygen quickly. Oxygenation is slower in slow moving, deep waters.

More than any other parameter, dissolved oxygen is used as an indicator of natural water quality. Standards are set for most water bodies such that minimum requirements for dissolved oxygen (DO) levels are maintained. Most popular freshwater fish need dissolved oxygen levels in the range of 6 to 8 mg/L. Fish need oxygen for respiration; it is transferred to the fish from the water through the gills. During sensitive development stages or for spawning, DO levels even higher than 6 to 8 mg/L may be necessary to achieve optimum survival rates. At DO levels below 5 mg/L, many fish become stressed. This is generally the standard that is set by regulatory agencies. When DO levels stay below 2 mg/L for an extended period of time, most fish life will die. Without oxygen, the water ecosystem is drastically different, containing different organisms, those that do not need oxygen will thrive. The overall chemical characteristics of the water will be different as sulfur and nitrogen compound are converted to their reduced chemical forms, which results in the production of malodorous sulfide and ammonia compounds.

Several pollutants will ultimately result in DO reduction in waters. The microbial degradation of organic materials will consume oxygen. Also, nutrients that stimulate algal grow ultimately cause DO reduction as the algae die and are decomposed, resulting in oxygen consumption. These processes are described in detail in forthcoming sections.

2.3.2 pH

The pH value of a water is a measure of the acid strength in the water. The pH directly measures the activity (approximately the concentration) of the hydrogen ion, H^+:

$$pH = -\log \{H^+\} \qquad (2.17)$$

It is a log scale, so a difference of one pH unit represents a factor of 10 in hydrogen ion activity. The lower the pH, the higher the H^+ activity and the more acidic is the water.

Most aquatic organisms require a pH near neutral, between 6 and 8. Acidic discharges into rivers can cause the pH to drop below 5 and directly kill fish populations. Additionally, water pH below 5 causes aluminum in soils to leach into the water at levels that can be toxic to plants and animals. Acid rainfall is an environmental problem that has produced widespread concern with respect to lowered natural water pH. The pH of acid rain can be lower than 4 because of the presence of sulfuric and nitric acids. These acids are formed from emissions of sulfur and nitrogen compounds during combustion processes in power plants and automobiles. When acidic rainfall runs off into receiving waters, the pH of the receiving waters can be lowered. Fortunately, the ground surface tends to have a high buffering capacity, and acid rainfall is usually buffered to pH values somewhat closer to neutral before entering the receiving streams. Discharge of acidic water from mines has destroyed the viability of many streams, due to lowered pH, but also due to toxic metals that are mobilized by the mine drainage.

2.3.3 Temperature

The water temperature is a measure of the heat content of the water mass and influences the growth rate and survivability of aquatic life. Different species of fish have different needs for an optimum temperature and tolerances of extreme temperatures. Additionally, the optimum temperature for a specific species of fish often varies with the stage of life. For example, the optimum temperature for brook trout to hatch is 8°C, while the optimum is 15°C for the growth of an adult trout. Generally, the most desired fish, such as trout, tend to do better in cooler or even cold waters. In some cases, increases in temperature have a direct affect on the organism, and their ability to thrive decreases in sub-optimum conditions.

Table 2.4. Saturation levels of dissolved oxygen in fresh water as a function of temperature.

Temp. (°C)	DO (mg/L)	Temp. (°C)	DO (mg/L)
1	14.2	16	10.0
2	13.8	17	9.7
3	13.5	18	9.5
4	13.1	19	9.4
5	12.8	20	9.2
6	12.5	21	9.0
7	12.2	22	8.8
8	11.9	23	8.7
9	11.6	24	8.5
10	11.3	25	8.4
11	11.1	26	8.2
12	10.8	27	8.1
13	10.6	28	7.9
14	10.4	29	7.8
15	10.2	30	7.6

High water temperatures have an indirect affect on water quality. The maximum level of oxygen that can be dissolved in a water is inversely related to temperature. Table 2.4 shows saturated DO levels for a fresh water. At the highest temperatures, the maximum DO levels are less than 8. Any process that consumes DO, such as an organic waste discharge, at these high temperatures can easily produce low DO levels, which will place corresponding stresses on fish populations.

For a water body exposed to summer solar radiation, the temperature of the water can increase significantly. Urbanization of land, however, plays a major role in increasing the temperatures of receiving streams and rivers to unacceptable values. Industries, such as power plants, use water for cooling. This water is discharged back to the river source, but at a higher temperature than when it was taken from the river.

The clearing of land for development often includes the removal of shading trees from the banks of small streams and rivers. As a result, these waters receive more direct sunlight. When a stream is small and has a low velocity, direct summer sun can cause measurable rises in temperature. Importantly, the transition from pervious infiltrating grass and woodland into roofs and asphalt can produce significant temperature impacts. Most roof materials and pavements are black, and adsorb and store a significant amount of heat energy from sunlight. As rainfall and runoff come into contact these heated surfaces, the heat is transferred to the excess runoff and then to the receiving waters, which causes measurable temperature increases. Lower natural DO levels in summer months, along with greater demands for cooling waters and washoff from hot, black impervious surfaces, can result in poor water quality and even fish kills from lack of adequate DO.

2.4 WATER POLLUTANTS

2.4.1 Suspended Solids

Probably the most basic of pollutants in a water environment are suspended solids. Suspended solids are primarily small particles of clay, silt, other dirt materials, small particles of vegetation, and even bacteria. The Total Suspended Solids (TSS) test involves filtering a fixed volume of water, typically 1 L, followed by drying and weighing the filter to determine the mass (mg) of particles. The ratio of particle mass per volume is the TSS measurement in mg/L.

Although most of the suspended solids themselves are not toxic directly, the sediment can cause many problems as it enters a water body. Large loadings of solids can be picked up by a fast moving overland flow of water during a storm event. However, when the flow eventually slows down, these solids settle throughout the water column. This material covers the natural bottom of the stream, making food unavailable for all of the organisms in the stream. Very large volumes of material can settle to the stream bottom and block the flow. Such material is loosely packed and can be resuspended during subsequent storm events.

The blockage of light is a second water quality problem associated with suspended solids. The penetration of light into natural waters is important to aquatic life for a number of reasons. Aquatic plants and other organisms that grow in these water bodies require light for photosynthesis. These plants are the food and shelter for higher level organisms. When high levels of solids prevent light penetration, the plant population decreases and fish and other species cannot find adequate food supply and habitat. By limiting the load of suspended material that enters water bodies to levels that are safe for aquatic life, the quality of the water body is not degraded because of the blockage of light.

Finally, the suspended solids can transport other materials that can cause water quality problems. Toxic compounds such as heavy metals and pesticides are often affiliated with particulate matter and are transported with the solids. Also, biodegradable organic matter can be affiliated with the solids, or organic solids themselves may be biodegradable. Degradation of these substances cause water quality problems, as will be discussed in Section 2.4.2. Finally, suspended solids can also include bacteria and other pathogens. These disease-causing organisms pose health risks to anyone or anything in contact with the waters.

Suspended solids in water originate from either the streambed or the surface runoff. The particles that enter with storm water runoff are most likely from areas of open ground, although areas with high fractions of impervious area contribute particles from automobiles, crumbling asphalt and concrete, and other sources. Grass, trees, and other plant coverings protect the ground from direct action of the rain and runoff and thus help prevent wash off. The kinetic energy of raindrops on bare soil causes a considerable amount of soil and accumulated materials to be dislodged. Without the vegetal covering, the energy of the moving water can dislodge additional soil particles and suspend them in the water runoff. Areas of new construction are especially susceptible to erosion. The equipment used to clear the land subjects the underlying soil to considerable energy, thus loosening considerable masses of soil. This produces high levels of suspended solids in the runoff from construction sites. As the vegetal coverings are removed during land development, the exposure of fresh soil allows for easy erosion and the transport of suspended solids via runoff. Muddy, turbid waters are indicative of high suspended solids levels.

Most jurisdictions require silt fences, ponds, straw bales, and other devices during new construction to minimize erosion and the transport of suspended solids from the site to local waterways. For land areas that will be exposed for more than a few days, regulators often require either seeding with fast-growing grass or a covering of straw.

In developed areas, suspended solids accumulate on impervious surfaces from everyday activities. Wind blows dust and dirt from open ground areas. When this dirt lands on impervious surfaces like sidewalks, roads, and driveways, it is easily washed off by the next rain event. Materials from anthropogenic activities also accumulate on these impervious surfaces. Minute particles from the wearing of automobile tires and brakes represent a substantial source of material. Paints, wood, asphalt, concrete, bricks, and roofing materials on buildings all gradually release particles as these materials weather and wear. In areas where snow and ice are common, the use of salt, sand, and abrasives on roadways create major water quality problems when these substances are later washed into the stormdrains and transported to receiving waters.

Because of the importance of suspended solids in determining water quality, even the most basic stormwater treatment device will usually address suspended solids. Usually these particulates are more dense than water and will settle in an appropriately designed treatment facility.

2.4.2 Oxygen Demanding Substances

Most organic substances in waters can be metabolized by aerobic microorganisms, primarily bacteria. This decomposition process also requires the consumption of oxygen. This process can be symbolized as:

$$\text{Organic Matter} + O_2 \xrightarrow{\text{microbes}} CO_2 + H_2O + \text{cells} \qquad (2.18)$$

The organic material can be of natural origin or result from some kind of biodegradable material or waste discharge. This process causes a decrease in the dissolved oxygen content of the water. Thus, biodegradable organic materials exert an "oxygen demand" on a receiving water.

The oxygen demand of an organic compound or mixture in a water can be measured in several ways, each with inherent advantages and disadvantages. The most traditional measurement is that of the Biochemical Oxygen Demand (BOD). BOD is measured by placing a sample of water in a specific type of glass bottle (BOD bottle) and, if necessary, inoculating it with a bacteria mixture. The initial amount of dissolved oxygen in the water is measured. The bottle is sealed and left in the dark at 20°C for five days, after which the amount of dissolved oxygen is determined again. The difference is the oxygen utilized by the microorganisms in degrading the organic material, corresponding to the oxygen demand of the waste:

$$BOD = DO_{initial} - DO_{final} \qquad (2.19)$$

Usually several tests are done at different dilutions in case the organic matter concentration is high enough to use all of the DO in the BOD bottle. If a dilution is used, the dilution factor (DF) is multiplied by the difference in DO measurements to obtain the BOD value:

$$BOD = DF\,(\,DO_{initial} - DO_{final}) \qquad (2.20)$$

The use of five days for the BOD test is traditional, and the BOD is more specifically known as the 5-day BOD, or BOD_5. The major problem with BOD measurement is that you must wait 5 days for a result.

Chemical oxygen demand (COD) is an alternative to the BOD test. COD uses chromate as an oxidant, high temperature, and a catalyst to chemically oxidize the organic materials in a water to CO_2. The amount of chromate required to oxidize the waste is determined, and this number is converted to an equivalent amount of oxygen. Both COD and BOD have units of mg/L of O_2. Since COD is a chemical process, it is a faster method than BOD. Results are available in a few hours instead of five days. It also tends to provide more reproducible results since it does not rely on the somewhat erratic characteristics of a microbial seed. Usually, values for COD are greater than those of BOD for the same water sample. While COD will oxidize all organic material, the microorganisms in the BOD test will only metabolize the most biodegradable portion of the organic. COD values that are two or more times higher than BOD are common.

A third analytical technique for organic material in water measures the total organic carbon (TOC) in the water. Most TOC instrumentation utilize a catalytic combustion of all of the organic material in the water. The CO_2 that results is monitored by the instrument. Measurements of TOC are given in mg/L of carbon and are not directly related to the biodegradability of the material or to the amount of oxygen required to the aerobic breakdown of the material.

2.4.3 Nitrogen Compounds

Nitrogen in water can be present in a number of forms and can cause several water quality concerns. The forms are dependent on the pH of the water and the water redox characteristics. Generally, the redox characteristics are controlled by the amount of dissolved oxygen present.

2.4.3.1 Nitrogen Chemistry

In addition to mass/volume (e.g., mg/L), concentrations of substances can be expressed as the number of molecule, or moles per liter. This is the molar concentration, designated as M. The molar concentrations multiplied by the molecular mass produces the mass concentrations.

Acid/base and oxidation/reduction (redox) chemistry are important in understanding pollutant fate and transport in stormwater systems. An acid dissociation reaction is written as:

$$HA \leftrightarrow H^+ + A^-$$ (2.21)

where HA is used to represent the acid and H^+ is the hydrogen ion, as discussed in Section 2.3.2. After dissociation, the remaining anion, A^-, is known at the conjugate base.

The equilibrium constant (acidity constant, K_a) for the acid dissociation reaction is given by:

$$K_a = \frac{[H^+][A^-]}{[HA]}$$ (2.22)

where the brackets represent molar concentrations of the respective species.

The oxidation state of a compound is important in determining a number of environmental properties, including toxicity. Under reduced conditions, in the absence of oxygen, chemical species become transformed to their reduced forms, which contain a high number of electrons and low oxidation state. Conversely, under oxidized conditions, chemical and biological processes convert redox-active species to their most oxidized form and highest oxidation state. The oxidation state of a redox-active species is determined by employing a standard set of rules. These include (1) setting the oxidation state of each atom in an uncharged diatomic molecule, such as N_2, equal to 1, (2) assigning an oxidation state of 1 to H, and (3) assigning an oxidation state of –2 to oxygen. The total of all oxidation states must equal the charge of the molecule.

Example 2.7. Using the rules above, the oxidation states of nitrogen species can be determined. For nitrogen, N_2, as described in rule 1, the oxidation state of both nitrogen atoms is 0. For ammonia, NH_3. which employs rule 2, each of the three hydrogens is assigned an oxidation state of +1, for a total of +3. In order to make the ammonia molecule have a total charge of zero, the oxidation state of the nitrogen must be –3. Nitrate, NO_3^-, has an overall charge of –1. Using rule 3, each oxygen is assigned an oxidation state of –2, for a total of –6. Therefore, the nitrogen has an oxidation state of +5.

Because of the many nitrogen forms, usually nitrogen levels in waters are presented as N. That is, the molar concentration is converted to mg/L as N by using the molecular weight of nitrogen (14) rather than the molecular weight of the different possible species.

Example 2.8 A water sample contains 3.8×10^{-4} M ammonium and 8.6×10^{-5} M nitrate. These concentrations are converted to mg/L as N by multiplying by 14, the molecular weight of nitrogen. Therefore, the ammonium concentration is 5.32×10^{-3} g/L, or 5.32 mg/L as N. The nitrate concentration is 1.2×10^{-3} g/L, or 1.2 mg/L as N.

2.4.3.2 Nitrogen in the Environment

The common reduced form of nitrogen in the environment is ammonium, NH_4^+, containing N in the -3 oxidation state. Ammonium is a weak acid, with an acidity constant of $10^{-9.9}$.

$$NH_4^+ \leftrightarrow NH_3 + H^+ \tag{2.23}$$

The conjugate base, ammonia, is a gas. Ammonia is toxic to many fish species.

In water, ammonia is microbially transformed to nitrite (NO_2^-), and eventually to nitrate (NO_3^-), in the presence of oxygen (aerobic conditions) via a two step process.

$$NH_4^+ + 3/2\ O_2 \leftrightarrow NO_2^- + 2H^+ + H_2O \tag{2.24}$$

$$NO_2^- + \tfrac{1}{2}\ O_2 \leftrightarrow NO_3^- \tag{2.25}$$

The overall process described by Eqs. (2.24) and (2.25) is known as nitrification. Both of these reactions result in the consumption of dissolved oxygen, thus categorizing ammonium as an oxygen demand. Also, the production of nitrate may be undesirable. Nitrite may be found in waters, but usually at concentrations less than nitrate. It is fairly quickly converted to nitrate.

Nitrate is a pollutant of major concern in water bodies. Nitrate is a nutrient that is easily utilized by plants and algae. Excess nitrate in a water body is utilized by algae, causing excess growth. As other nutrients become limiting, the algae die and release metabolic products that can cause odor problems near the algal outbreak and create problems if the water is used for drinking water supply. Importantly, the decaying algae also support excessive bacteria growth via the decomposition reactions of Eq. (2.18). The growing bacteria consume large amounts of oxygen, thus leaving the water with low DO levels and possibly initiating fish kills. This process of nutrient input, algal growth and decay, followed by bacterial predation and DO decrease is known as *eutrophication*.

Nitrate is also a regulated contaminant in drinking water. It has an MCL (maximum contaminant level) of 10 mg/L as N and is linked to *methemoglobinemia*, or blue-baby syndrome, in infants. The chemical properties of nitrate are such that its removal from drinking waters is a significant challenge. Chemically, it does not readily adsorb and does not form precipitates. Biological processes or advanced physicochemical operations are generally required.

In the absence of oxygen, microorganisms can utilize nitrate as an electron acceptor. This process, known as *denitrification*, also requires an electron donor, such as organic compounds. The nitrate is converted to nitrogen gas, N_2, shown here with methanol as the electron donor/carbon source (which is used in advanced waste water treatment):

$$6NO_3^- + 5CH_3OH \leftrightarrow 3N_2 + 5CO_2 + 7H_2O + 6\ OH^- \qquad (2.26)$$

Nitrogen is also a basic biochemical component. It is a critical element in amino acids, and correspondingly, is a major component of proteins. Proteins and other forms of organic nitrogen are found in waters due to waste discharges and natural decomposition processes. These organic nitrogen compounds decompose to ammonium. Organic nitrogen is measured using a high temperature digestion process that coverts all organic N to ammonium. The ammonium concentration is subsequently measured. This value is known as Total Kjeldahl Nitrogen (TKN). Since the original ammonium is measured in the analytical process, the TKN is the sum of organic nitrogen and ammonium:

$$TKN = organic\ N + NH_4^+ \qquad (2.27)$$

To find organic N, NH_4^+ is measured separately and subtracted from TKN.

Total nitrogen (TN) is the sum of all common nitrogen forms in water:

$$TN = organic\ N + NH_4^+ + NO_2^- + NO_3^- \qquad (2.28a)$$

$$= TKN + NO_2^- + NO_3^- \qquad (2.28b)$$

where, again, nitrogen species are quantified as mg/L as N.

These various compounds of nitrogen are found in stormwater runoff from sources that can include fertilizers, animal wastes, plant decay, and atmospheric deposition. Organic N from animal wastes is washed from both impervious and vegetated areas. Decaying grasses and leaves will release organic N and ammonium. Nitrate can enter stormwater from fertilizer applications, either from agricultural areas, or from fertilizers applied to homeowner lawns and to golf courses. Since nitrate is an anion, it will not attach to soils and soil particles. As a result, it is very mobile in water systems. Nitrate can be washed from fertilized areas and transported into receiving waters

Recent studies have shown that nitrogen compounds released into the air from automobile and industrial combustion processes can lead to elevated nitrate levels in rainfall and dry atmospheric deposition, representing another source of nitrate that enters into water bodies. During high temperature combustion processes using air, such as in vehicle engines and fossil fuel power plants, some of the nitrogen in the air is converted to various nitrogen oxide gases, NO_x (x= 1, 1.5, or 2). These NO_x compounds are emitted into the air, become oxidized to nitrate, and eventually are washed out of the atmosphere with rain events.

2.4.4 Phosphorus

Phosphorus is required for the growth of living things; it is a primary nutrient like nitrogen. Phosphorus, like nitrogen, is found in several forms, and therefore its concentration is usually given in mg/L as P. The molar concentration of the phosphorus compound is multiplied by the molecular weight of phosphorus, 31. High levels of phosphorus in waters can contibute to eutrophication.

Three forms of phosphorus are common in waters. Ortho phosphorus is the basic phosphate form, $H_2PO_4^-$ or HPO_4^{2-}, depending on pH. Phosphorus is also an important biochemical element and is found in various organic forms, or in the form of a polymer phosphate. Phosphates also have a strong affinity for soils, sediments, and other particulates and are transported with these particles. Total phosphorus is the sum of these three forms.

Phosphorus enters the runoff from the washoff of excess fertilizers, where it is added as a nutrient. Phosphorus is also released during the decay of vegetation, such as lawn clippings and leaves, and is present in animal wastes.

2.4.5 Microbial Pathogens

Pathogens are disease-causing organisms and include various bacteria, viruses, and protozoa. Common waterborne diseases caused by pathogens include cholera, typhoid, and cryptosporidiosis. These diseases generally produce vomiting and diarrhea and may be fatal it the illness becomes too severe, or if the person is very old, very young, or is already sick.

Determining the many different types of pathogenic organisms in a water sample is time consuming, expensive, and may be prone to errors. As a result, *indicator organisms* are commonly used to determine if a water has been exposed to warm-blooded animal wastes (including human). An indicator organism is one that is usually present when an array of other organisms is present. Its presence generally indicates that contamination has occurred and that other pathogens may be present. *Fecal coliform*, a bacteria species found in warm-blooded animals, is a commonly used indicator organism. It is relatively easy to measure and monitor. The presence of high levels of fecal coliform suggests significant exposure of the water to waste and a high likelihood that high levels of many different pathogens are also present. Conversely, low levels of fecal coliforms suggest a lack of contamination and that pathogens are absent from the waters. Coliform measurements are usually expressed as colony forming units per volume of water (cfu/100 mL).

Animal wastes and pathogens deposited on land can be washed off by rainfall and runoff to receiving streams. Certainly, this can be a major problem for water bodies near agricultural areas. However, even in urban areas, significant pathogen problems are found. Although many of these problems can be attributed to leaks and overflows from waste water collection lines, data have shown some high pathogen loads in urban stormwater runoff.

2.4.6 Heavy Metals

Many heavy metals are toxic at high concentrations to humans and other flora and fauna. These metals include cadmium, chromium, copper, lead, mercury, nickel, and zinc. Some metals, such as cadmium, lead, and mercury, are highly toxic and are regulated at very low levels--concentrations in the low μg/L (parts-per-billion) range. Different organisms have different tolerances for metals. For example, copper is a micronutrient for humans, and humans have a high tolerance for copper, so moderate concentrations are not toxic. Even low levels of copper in waters, however, are highly toxic to algae. Some fish are also adversely sensitive to copper, even at low concentrations.

Most of the metals are divalent cations, with an oxidation state of +II. These include Cd^{2+}, Cu^{2+}, Hg^{2+}, Pb^{2+}, Ni^{2+}, and Zn^{2+}. In natural waters, these heavy metals are usually complexed or bound to inorganic species, such as OH^- or CO_3^-, or organic compounds, such as natural organic matter. This complexation can affect the behavior of a metal in a treatment practice or can alter its movement through the environment.

Chromium, as an example, has a more complex chemistry than most other heavy metals; it can be found in the +III form or in the highly oxidized (and much more toxic) +VI form. In the latter case, the chromium is usually present in waters as an oxyanion, such as $HCrO_4^-$. Arsenic is a metalloid (it has chemical characteristics between those of metals and non-metals), which has a complex chemical behavior. Arsenic generally exists in the +III oxidation state as arsenite, H_3AsO_3, or as As(V), arsenate, $HAsO_4^{2-}$. Because these species are anions, their fate, transport, and treatment efficiencies are typically much different from the other cationic metals.

In the presence of suspended solids, metals become commonly adsorbed to, or somehow attached or affiliated with the particles. This occurs as the metals complex or bind with the inorganic or organic components of the suspended solids and become transported along with the solids. As a result, dissolved metal concentrations can be low, but suspended solids can contain high metal levels. When solids settle or are removed via treatment, metals can be removed with them.

Heavy metals do not degrade in the environment. Over time, changes in the chemistry of the surrounding water bodies may cause them to be released into the solution. Another possibility is that, as the captured suspended solids age and weather, adsorbed metals and other pollutants can gradually become incorporated into the crystalline structure of the mineral. This incorporation will render the metal strongly bound to the suspended solid so that it is not easily removed, thus greatly reducing the possibility that the metal can exert a toxic effect on some organism. For details on the environmental chemistry of metals, the reader is directed several books on water and soil chemistry (e.g., Stumm and Morgan 1996; Evangelu 1998).

2.4.7 Oils and Grease

Oils can produce several different problems in a water environment. They float, producing a strongly negative visual effect. They can coat parts of aquatic animals, such as the gills of fish, which reduces their ability to efficiently transfer oxygen. As oils degrade, they can exert an oxygen demand, as discussed above. Finally, oils contain many substances, both organic compounds, such as naphthalene and toluene, and inorganic species, such as heavy metals, that are toxic in the aquatic environment.

The typical measurement for oil and grease (O/G) is known as Hexane Extractable Material (HEM). In this procedure, hexane (an organic solvent) is added to a water sample. The two liquids are mixed and hydrophobic compounds in the water, like oils and greases, will dissolve and partition into the hexane. The hexane is then separated from the water and evaporated, leaving behind the extracted O/G, which is measured gravimetrically.

Oils, grease, and other fluids leak from automobiles onto roadways and parking lots. Motor oils are primarily heavy hydrocarbons, although they contain a number of additives for lubrication, corrosion protection, and as antioxidants. Zinc is a common motor oil additive used as an antioxidant. During the time the that oil is in the engine, the high temperatures break down and oxidize the hydrocarbons into smaller organic fractions. Other fluids that can adversely affect water quality can leak from automobiles, including transmission, power steering, and differential oils. These fluids contain mixtures of hydrocarbons and other organic compounds, many of which are toxic.

2.4.8 Toxic Organic Compounds

A variety of different organic compounds are toxic, with a wide range of properties that control their transport in runoff and their ultimate fate in the environment. Solubility, volatility, and hydrophobicity determine how these organic compounds interact with surrounding particulate matter and the air. The compounds also may be degraded by photolytic, hydrolysis, biologically-mediated, or other reactions. Thus, these compounds will partition to soils or volatilize once released to the environment. They may also degrade to other compounds that may be more or less toxic. Organic compounds that are not susceptible to reaction can be stable in the environment for many years.

Toxic organic compounds can end up in storm water runoff from oils, fuels, pesticide use, atmospheric deposition, spills and leaks from storage tanks, ash from burned vegetation, and other sources. The fate of these compounds and how they are mobilized and transported in runoff can be very complex.

2.4.8.1 Pesticides

Pesticides are chemicals specifically designed to kill or somehow alter the growth or reproductive characteristics of a plant or animal species. However, pesticides can be washed into a water body during the runoff process. When this happens, the lethal actions of the pesticide are placed into an ecosystem where they are not needed. Severe toxicity to fish, plants, or other species can result.

Pesticides can be a problem in urban areas. Pressures for lush, green lawns lead homeowners to apply pesticides to kill unwanted weeds and grasses and to keep away pests. Key issues in the management of pesticides is the amount applied and the time between application and the next rain. When rain occurs soon after pesticide administration, the pesticide can be easily washed away. Also, the amount applied must be carefully controlled. Evidence has suggested that commercial lawn services tend to be more careful in this regard than the individual homeowner.

2.4.8.2 Polycyclic Aromatic Hydrocarbons

Polycyclic aromatic hydrocarbons (PAH) are compounds with two or more common aromatic rings. The simplest is naphthalene, with two rings. Several 5-ring PAH compounds, such as the benzopyrenes, are very carcinogenic. The greater the number of rings in a PAH, the less soluble the compound. Multiple-ring PAH have very low water solubilities and tend to adsorb to sediments and soils rather than to be dissolved in water. Larger PAH also tend to be quite stable in the environment.

PAH are formed several different ways. They are minor components of hydrocarbon fuels. However, the heavier the fuel, the more PAH typically are present. Very heavy, thick petroleum products, such as asphalts contain high levels of PAH. Therefore, PAH will result in runoff from parking areas and roadways where fuels and oils are leaked. Trace levels may leach from asphalt paving, especially when it is new. Runoff from asphalt roofs can also be a source.

PAH are also produced by incomplete combustion of fuels. During the combustion process, if inadequate oxygen is present, or the temperature is too low, PAH can be formed from the carbon-based fuel. PAH discharged to the atmosphere from incomplete combustion processes can be deposited on the ground and eventually washed off into waterways with runoff.

2.4.8.3 Solvents

Organic solvents are used by industry and homeowners for cleaning, degreasing, and lubricating. Common industrial solvents are trichloroethylene and tetrachloroethylene. Other than illicit discharges, solvents are not usually a major problem in urban runoff water quality

2.4.9 Trash

One of the most visible pollution problems is trash or floatables. Paper and plastic materials such as cups, bottles, and wrappers, unfortunately, are discarded along roadsides. These light materials are easily mobilized by sheet and gutter flow during storm events. Without treatment, this trash is carried into receiving streams, where it is a major eyesore. Plastic and coated papers are very slow to degrade in the environment. Other trash materials such as leaves, pieces of wood, clothing, and metal containers can also end up in stormwater flow, especially during a large storm event.

2.5 WATER QUALITY INDICES

The designated use of a water body, whether it is for fishing, swimming, or drinking, will be dependent on the quality of the water. Unfortunately, water quality is a very difficult parameter to define. As discussed in the previous section, many different parameters define the quality of a water and can be responsible for water degradation.

In order to classify the quality of a water body, water quality indices (WQI) have been developed to consider many of the different pollutants and factors required for life. Of interest is an overall index that combines several important water quality parameters. The simplest WQI consider several simple water quality parameters such as dissolved oxygen, TSS, pH, and possibly some nutrients. Measurements of each of these parameters are taken and compared to a classification table, where the water is classified as excellent, good, fair, or poor. Each of these classifications carries a corresponding numerical value (e.g., Excellent = 1, Good = 0.5, etc.). Thus each water quality measurement has a quantitative value.

Water quality indicators employed in the Potomac River basin are given in Table 2.5. Eight parameters are included as indicators and the quality is divided into four different descriptions, ranging from excellent to poor.

Table 2.5. Water Quality Parameters used as Basis for Water Quality Indices in the Potomac River Basin.

Parameter	Excellent	Good	Fair	Poor
pH	6.9 - 8.0	8.1 - 8.4	8.5 - 9.5	> 9.5
		6.5 - 6.8	5.5 - 6.4	< 5.5
Dissolved Oxygen (mg/L)	8.0 - 9.5	6.0 - 7.9	4.0- 5.9	<4.0
Suspended Solids (mg/L)	< 25	25 - 80	81 - 400	> 400
Total Organic Carbon (mg/L)	< 5	5 - 20	21 - 35	> 35
Nitrate Nitrogen (mg/L as N)	< 0.20	0.21 - 0.60	0.61 - 2.0	> 2.0
Total Phosphorus (mg/L as P)	< 0.05	0.05 - 0.25	0.26 - 0.99	> 1.0
Chlorophyl-a (µg/L)				
Freshwater	0 - 24	25 - 49	50 - 99	> 100
Brackish Water	0 - 49	50 -99	100 - 149	> 150
Fecal Coliform Bacteria (MPN/100 mL)	< 200	201 - 500	501 - 1000	> 1000

Table 2.6. Water Quality Index Classifications.

Parameter Score	Classification
6.5-8.0	Excellent
5.25-6.49	Good-Excellent
4.00-5.24	Good
3.25-3.99	Fair-Good
2.25-3.24	Fair
1.25-2.24	Poor-Fair
0-1.24	Poor

The WQI is a composite number that considers the selected water quality parameters. Typically, the values for each parameter are added or averaged to give a single numerical value to represent the overall quality of the water body. A water quality designation (Excellent, Good, etc.) is then assigned to the numerical value to quantitatively represent the water body quality. Table 2.6 demonstrates a simple overall classification based on the water quality parameters from Table 2.5. Each of the eight parameters is measured and given a value (Excellent=1, Good=0.5, Fair=0.25, Poor=0). These values are summed to give an overall score and the corresponding classification is determined from Table 2.6.

While simple classifications can be useful, they do not usually tell the enitre story on water quality in a complex water body. By adding all of the parameters, the significance of a single parameter is minimized. Using the summation scheme, one parameter can be classified as poor, but if all other parameters are excellent or very good, the water can still be overall classified as excellent. Several different procedures have been suggested to address this issue, including basing the quality on the worst parameter, using geometric means for the overall score, and others.

2.6 TOTAL MAXIMUM DAILY LOADS–TMDLs

As the behavior and fate of pollutants in water bodies is better understood, the amounts of these pollutants that can be safely discharged to them will be better evaluated. Natural water systems are able to tolerate and assimilate certain pollutant loads without causing deterioration of the water body use functions. These may be ecological functions of the water body or desired human uses, such as drinking or swimming. Provisions of the Federal Clean Water Act, legislation that protects the quality of the waters of the U.S., support the estimation of these load determinations for use in calculating allowable pollutant discharges to the water body. Large rivers that are not subjected to human activities may be able to tolerate large pollutant loading. Small streams that are used for swimming, however, will be much more sensitive to pollutant loads. This calculated pollutant load is known as a *Total Maximum Daily Load* or *TMDL*, which is the pollutant mass (or weight) per year permitted to be discharged to a water body by all sources in the watershed.

Water bodies that do not meet certain water quality criteria require the establishment of TMDLs for the pollutants responsible for the impairment. These TMDLs estimate the pollutant capacity for a water body. This TMDL is then divided into contributions from both point and nonpoint sources, with inclusion of a factor of safety:

TMDL = Point Source Contributions + Nonpoint Source Contributions
        + Factor of Safety   (2.29)

Point sources are direct discharges from wastewater treatment plants and from industries. These discharges require permits and monitoring. Fairly reliable information on pollutant discharges can be obtained on these discharges. Nonpoint sources consider contributions from stormwater runoff from all land uses. Obviously, these pollutant loads are much more difficult to quantify. Runoff models, databases, and information on storm water Best Management Practice efficiencies must be included in this effort. The factor of safety is generally fixed at 5% of the TMDL.

If various pollutants exceed the allowable TMDL, point and nonpoint sources of the target pollutant(s) will have to be reduced. This can result in forcing more stringent discharge requirements to point sources, as well as reductions in nonpoint pollutant loads that will require the implementation of stormwater management practices that can remove pollutants. Under the most severe scenario, further development in a watershed, which may contribute both point and nonpoint source pollution, can be halted until pollutant loadings are reduced.

Example 2.9. A TMDL of 3000 lb/yr has been written for phosphorus discharges into a river. A small wastewater treatment plant discharges treated wastewater at an average flow of 0.2 million gallons per day (mgd), containing 3 mg/L phosphorus. A small meat packing plant discharges 1.5 mg/L phosphorus at 0.11 mgd.

The yearly phosphorus loads from the point source wastewater discharges can be calculated as:

$$M/t = C\,Q = (3\ mg/L)(0.2\ mgd)(8.34) = 5.0\ lb/day = 1830\ lb/yr \qquad (2.30)$$

The factor 8.34 converts $\dfrac{(mg)(million\ gallons)}{(L)(day)}$ to lb/day. A similar calculation for the plant provides a yearly loading of 500 lb/yr.

The phosphorus loading allocated to nonpoint sources is calculated by rearranging Eq. (2.29):

Nonpoint Source Contributions = TMDL - Point Source Contributions -
                        Factor of Safety   (2.31)
                = 3000 - (1830 + 500) - 3000(0.05)
                = 520 lb/yr.

Approximately 520 lb/yr total phosphorus discharge can be allocated to nonpoint sources in the watershed.

2.7 PROBLEMS

2.1 Three measurements of total phosphorus in a stormwater retention pond give values of 3.2, 2.8, and 3.3 mg/L as P. The pond is approximately circular, with a diameter of 160 ft and an average depth of 4 ft. Using the mean phosphorus concentration, determine the total mass of phosphorus in the pond.

2.2 During a hard rain event, the runoff from three streets comes together and flows into the storm drain. The runoff from A Street has a flow of 24 L/s and contains 87 mg/L TSS. That from B Street has 104 mg/L TSS at 18 L/s and the C Street runoff is 30 L/s at 45 mg/L TSS. (a) Find the flow rate and TSS concentration for the combined runoff from the three streets. (b) Find the TSS mass flow rate (mg/s) for the combined runoff streams.

2.4 Runoff from a parking area enters a treatment pond at 145 L/s containing 125 mg/L TSS. The effluent from the practice leaves also at 145 L/s, but at 33 mg/L TSS. What is the TSS mass accumulation rate (mg/s) in the pond?

2.5 A sample of runoff is collected for BOD measurement. The sample is diluted by a factor of 5 and the initial DO is 7.5 mg/L. After 5 days, the DO is 5.4 mg/L. What is the BOD of the runoff?

2.6 A 100-mL water sample is mixed with 200 mL of deionized water for BOD analysis. The DO of the mixture is 8.2 mg/L. The DO after 5 days is 4.9 mg/L. What is the BOD of the water sample?

2.7 A water sample contains 2.1×10^{-5} M phosphate. What is this concentration in mg/L as P?

2.8 The drinking water standard for nitrate is 10 mg/L as N. What is this concentration in M?

2.9 A river water has a TKN of 4.6 mg/L. The ammonium concentration is 9.3×10^{-5} M. What is the organic N concentration in mg/L as N?

2.10 During a six-month period (April 1 to September 30) four measurements of the lead concentration in a stream were taken, with the following results:

Date	4/12	5/4	7/13	9/3
C (mg/L)	0.183	0.322	0.148	0.266
Q (cfs)	46	29	13	17

Estimate the total lead load during the 183-day period.

2.11   At the confluence of two streams (A and B), TSS and flow measurements are made in stream A, one of the inflowing streams, with the results giving 238 mg/L TSS at a flow of 144 ft^3/s. The same measurements are made in the river, D, below the confluence of the streams, giving 286 mg/L TSS at 212 ft^3/s. Find the flow rate and TSS concentration in stream B.

2.12   The following data are the mean daily discharge (Q, cfs) and dissolved oxygen concentration (mg/L) for selected days on the Choptank River near Greensboro, MD.

Q	3.5	164	189	28	26	145	160	9	17
C	6.3	8.2	9.7	5.9	7.3	8.5	9.7	6.8	5.7

Fit the coefficients of Eq.(2.15) using least squares and graphically assess the goodness of fit.

2.13   Water quality measurements for a small river in an urbanized area near Washington, DC, are presented in the table below. Discuss the water quality with respect to individual water quality parameters and as an overall water quality rating.

Parameter	Value	Parameter	Value
pH	7.0	Nitrate Nitrogen (mg/L as N)	3.5
Dissolved Oxygen (mg/L)	7.3	Total Phosphorus (mg/L as P)	0.18
Suspended Solids (mg/L)	32	Chlorophyl-a (µg/L)	35
Total Organic Carbon (mg/L)	11	Fecal Coliform Bacteria (MPN/100 mL)	18,000

2.14   Referring to the water quality data of Problem 2.13, judge the overall water quality of the river based on the *worst* water quality parameter.

2.15   Referring to the water quality data of Problem 2.13, judge the overall water quality of the river using a geometric mean of the individual parameters instead of the sum. You will have to modify Table 2.5 to do this. Use your best judgment.

2.16   A nitrogen TMDL for a local river is 88000 lb/yr. Current point sources discharge 40000 lb/yr and non-point source discharges are estimated at 25000 lb/yr. A new factory wants to discharge 1 million gallons of treated wastewater per day into the river. What is the maximum allowable nitrogen concentration (mg/L as N) in the treated wastewater so that the TMDL is not exceeded? Use a 5% safety factor.

2.17   A TMDL for TSS equal to 10^7 lb/yr has been established for a river. Two point sources exist, one discharging 25 mg/L TSS at 12 mgd and the other discharging 38 mg/L at 1.5 mgd. For a 5% safety factor, what is the load currently allocated to nonpoint sources (lb/yr)?

2.8 REFERENCES

Evangelou, V.P. (1998) *Environmental Soil and Water Chemistry: Principles and Applications*, Wiley, New York.
Stumm, W. and Morgan, J.J. (1996) *Aquatic Chemistry*, Wiley, NY.

3

STATISTICAL METHODS FOR DATA ANALYSIS

NOTATION

C_v = coefficient of variation
F   = test statistic
H_0 = null hypothesis
H_A = alternative hypothesis
M_k = k^{th} moment about the mean

M'_k = k^{th} moment about the origin

n   = sample size
N   = total number of observations
P   = population parameter
R   = Dixon-Thompson test statistic
S   = sample standard deviation
S^2 = sample variance
S_e = standard error
t   = Student t statistic
x   = random variable
$\bar{x}$ = sample mean
z   = standard normal deviate
α = level of significance
α = location parameter of uniform distribution
α = probability of type I error
β = probability of type II error
β = scale parameter of uniform distribution
σ = population standard deviation
μ = population mean
λ = parameter of exponential distribution

3.1 INTRODUCTION

Measurements of water quality parameters are subject to considerable measurement error. Any error in measured values affects values computed with the data. For example, if four TKN measurements are made, 0.86, 0.47, 0.63, and 0.70 mg/L, the computed mean is 0.665 mg/L. If the 0.86 mg/L measurement was in error by 0.1, then the computed mean would be in error by 0.025 mg/L. The standard deviation would be affected even more. For example, if the 0.86 mg/L were really 0.76, then the standard deviation would be 0.125 mg/L rather than 0.162 mg/L. Such errors in computed statistics can significantly influence decisions based on the data.

The effect of measurement errors is compounded because it is often possible to only collect small samples. The constraint on sample size can be due to limited resources, equipment, time, or the availability of rainfall. The accuracy of computed statistics is inversely related to the sample size. Small samples often produce inaccurate statistics, which like measurement error affects the accuracy of decisions.

Measured water quality data represents information. To obtain the most from the information, it is important to use systematic methods of analysis, especially those methods that are based on the theory that underlies methods of data analysis. Statistical decision methods are systematic and have their basis in the theory of mathematical statistics. Therefore, an individual or organization involved in a legal case that involves measured data is more likely to be successful if they make a theoretical, systematic analysis of the data rather than subjective assessments of the data. Systematic, theoretical analyses enable measurement errors to be properly assessed and handled and the adverse effects of small samples to be minimized.

3.2 POPULATION AND SAMPLE MOMENTS

Whether summarizing a data set or attempting to find the population, one must characterize the sample. Moments are useful descriptors of data; for example, the mean, which is a moment, is an important characteristic of a set of observations on a random variable, such as rainfall volume or the concentration of a water pollutant. A moment can referenced to any point on the measurement axis; however, the origin (for example, zero point) and the mean are the two most common reference points. For statistical studies two moments are generally sufficient:

1. The mean is the first moment of values measured about the origin.
2. The variance is the second moment of values measured about the mean.

3.2.1 Mean

The *mean* is the first moment measured about the origin; it is also the average of all observations on a random variable. It is important to note that the population mean is most often denote as μ, while the sample mean is denoted by $\bar{x}$. For a discrete random variable, the mean is given by:

$$\left(\bar{x} \text{ or } \mu\right) = \sum_{i=1}^{n} x_i f(x_i) \tag{3.1}$$

If each sample point is given equal weight, then $f(x_i)$ equals $1/n$, and the mean of observations on a discrete random variable is given by:

$$\left(\bar{x} \text{ or } \mu\right) = \frac{1}{n} \sum_{i=1}^{n} x_i \tag{3.2}$$

Although the mean conveys certain information about either a sample or a population, it does not completely characterize a random variable.

Example 3.1. The following frequencies of total suspended solids measurements were collected from highway runoff along a major highway:

TSS (mg/L)	0-40	40-80	80-120	120-160	160-200	200-240
Frequency	2	14	23	19	12	6

The mean TSS concentration can be found. The centerpoint of each frequency range is used as the representative value for that histogram cell. Equation (3.1) is used to compute the sample mean of the 76 measured values:

$$\bar{X} = \sum_{i=1}^{6} xf(x) = [2(20) + 14(60) + 23(100) + 19(140) + 12(180) + 6(220)]/76$$
$$= 122.6 \text{ mg/L}$$

It is important to note that this value is not truly the sample mean because it is not based on the individual measurements. Instead, it is based on a histogram summary of the data. If all values fell near the lower limit of the histogram cell, then the mean might be as low as:

$$\bar{X}_L = [2(0) + 14(40) + 23(80) + 19(120) + 12(160) + 6(200)]/76$$
$$= 102.6 \text{ mg/L}$$

While it is best to use the mid-point of the histogram calls to compute a mean, it is better to use the actual measurements when they are available.

Example 3.2. Six measurements of lead (mg/L) were made from highway runoff: 43.6, 26.7, 63.2, 284, 51.1, and 38.7. Equation (3.2) is used to compute the mean of the six measurements:

$$\bar{X} = (43.6 + 26.7 + 63.2 + 284 + 51.1 + 38.7)/6 = 84.55 \text{ mg/L}$$

The computed mean may be a poor measure of central tendency because it is actually larger than 5 of the 6 values. The very large reading of 284 mg/L is an extreme event that may have been controlled by environmental conditions that did not exist when the other five measurements were made. The value could be checked as an outlier and censored if proven to be unrepresentative of the other data. This would yield a mean for the remaining five values of 40.3 mg/L, which is less than one-half of the mean when the extreme value is included.

3.2.2   Variance

The variance is the second moment about the mean. The variances of the population and sample are denoted by σ^2 and S^2, respectively. The units of the variance are the square of the units of the random variable; for example, if the random variable is measured in feet, the variance will have units of ft^2. For a discrete variable, the variance is computed by:

$$\left(S^2 \text{ or } \sigma^2\right) = \sum_{i=1}^{n} (x_i - \mu)^2 f(x_i) \tag{3.3}$$

If each of the n observations in a sample are given equal weight, $f(x) = 1/(n-1)$, then the variance is given by:

$$S^2 = \frac{1}{n-1} \sum_{i=1}^{n} (x_i - \mu)^2 \tag{3.4}$$

The variance of a sample can be computed using the following alternative equation:

$$S^2 = \frac{1}{n-1} \left[\sum_{i=1}^{n} x_i^2 - \frac{1}{n}\left(\sum_{i=1}^{n} x_i\right)^2 \right] \tag{3.5}$$

In Equations (3.3) and (3.4), the sample mean $(\bar{x})$ may be substituted for μ when μ is not known.

Variance is most important concept in statistics because almost all statistical methods require some measure of variance. Therefore, it is important to have a conceptual understanding of this moment. In general, it is an indicator of the closeness of the values of a sample or population to the mean. If all values in a sample equaled the mean, the sample variance would equal zero.

3.2.3   Standard Deviation

By definition, the standard deviation is the square root of the variance. It has the same units as both the random variable and the mean; therefore, it is a useful descriptor of the dispersion or spread of either a sample of data or a distribution function. The standard deviation of the population is denoted by σ, while the sample value is denoted by S.

Example 3.3. Five samples were taken to measure the dissolved oxygen in a stream, with the resulting values: 2.7, 2.9, 3.4, 3.1, and 2.9 mg/L. The computations for the mean, variance, and standard deviation of the five values are as follows:

Sample	x	x^2	$(x - \bar{x})$	$(x - \bar{x})^2$
1	2.7	7.29	-0.3	0.09
2	2.9	8.41	-0.1	0.01
3	3.4	11.56	0.4	0.16
4	3.1	9.61	0.1	0.01
5	2.9	8.41	-0.1	0.01
	15.0	45.28	0.0	0.28

From Eq. (3.2) the mean is:

$$\bar{x} = \frac{1}{5}(15) = 3.0 \text{ mg/L}$$

From Eq. (3.4) the sample variance is:

$$S^2 = \frac{1}{5-1}(0.28) = 0.07 \text{ (mg/L)}^2$$

Therefore, the sample estimate of the standard deviation is 0.26 mg/L. The small standard deviation indicates that the values fall close to the mean. It is common to express the data as the mean plus or minus one standard deviation. In this case, 3.0 ± 0.26. Note that 3 of 5 values fall within this range.

3.2.4   Coefficient of Variation

The coefficient of variation (C_v) is a dimensionless index of the relative spread of data. It is equal to the ratio of the standard deviation to the mean:

$$C_v = \frac{\sigma}{\mu} \text{ or } \frac{S}{\bar{\bar{X}}} \tag{3.6}$$

It is a useful characteristic because it is dimensionless and shows consistent values across independent samples, even when the means and variances are widely different.

The hypothetical data for Example 3.3 shows a coefficient of variation of 0.087. This is quite small. Actual measured water quality data generally have relatively large coefficients of variation. For example, Wu *et al.* (1998) presented results for samples from pollutant loadings for highway runoff. The coefficients of variation range from 0.32 to 1.84, although most were in the range from 0.6 to 1.0.

Example 3.4. In Example 3.2, it was shown that a single observation in a small sample can significantly influence the mean. For the sample of 6, the standard deviation is 98.46 mg/L. If the value of 284 mg/L is censored, the standard deviation drops to 13.64 mg/L. Again, the censoring of the extreme value had a significant effect on the standard deviation. In fact, the effect of the extreme event on the standard deviation is greater than the effect on the mean.

In addition to affecting the values of the first two moments, the extreme value has an effect on the coefficient of variation. The C_v is 1.16 when the extreme event is included and 0.34 when it is censored from the sample. Inclusion of the extreme event increases the relative variation as well as both the mean and standard deviation. This points out a significant problem associated with small samples; specifically, estimates made from small samples are subject to considerable scatter. In statistical terms, the scatter is referred to as *sampling variation*.

3.3 PROBABILITY DISTRIBUTIONS

While an exact value of a random variable cannot be predicted with certainty, its occurrence can be characterized by a probability distribution. As suggested in the previous section, the variability of values is referred to as sampling variation. But the likelihood of a value or range of values for a random variable is defined by a probability distribution and is called the sampling distribution. For example, the sampling distribution of the mean is $N(\mu, \sigma / \sqrt{n})$. Knowledge of the distribution that underlies a random variable is necessary to estimate the probability of values of the random variable occurring.

3.3.1 Probability

A student may be interested in determining the likelihood of his or her final grade. If the deadline for dropping courses is near, the likelihood of receiving a passing grade is of interest. In more formal terms, the student is interested in the probability of getting a passing grade. *Probability* is a scale of measurement used to describe the likelihood of an event, where an *event* is defined as the occurrence of a specific value of the random variable, which in this case is the "final grade." It is important to note that two random variables are of interest, the test score and the final grade. The student must use the value of one random variable to draw inferences about the other. The scale on which probability is measured extends from 0 to 1, inclusive, where a value of 1 indicates a certainty of occurrence of the event and a value of 0 indicates certainty of failure of the event to occur. Quite often, probability is specified as a percentage; for example, when the weather forecaster indicates that the chance of rain is 30 percent, past experience indicates that under similar meteorological conditions it has rained 3 out of 10 times. In this example, the probability was estimated empirically using the concept of relative frequency:

$$p(x = x_0) = \frac{n}{N} \qquad (3.7)$$

in which n is the number of observations on the random variable x that result in outcome x_0, and N is the total number of observations on x.

3.3.2 Types of Random Variables

Two types of random variables must be considered. First, a *discrete random variable* is one that may only take on distinct, usually integer values; for example, the outcome of a roll of a die may only take on the integer values from 1 to 6 and is therefore a discrete random variable. The number of samples and the number of days on which it rains in a year are discrete random variables. Second, the outcome of an event that can take on any value within a range is called *continuous*. For example, the average of all of the students' scores on a test has a maximum possible score of 100 and may take on any value, including nonintegers, between 0 and 100; thus the class average would be a continuous random variable. A distinction is made between these two types of random variables because the computation of probabilities is different.

The probability of a discrete random variable is given by a *mass function*. A mass function specifies the probability that the discrete random variable X equals some value X_k and is denoted by:

$$P(X_k) = P(X = X_k) \tag{3.8}$$

The probability of the random variable X has two important boundary conditions. First, the probability of an event X_k must be less than or equal to 1, and second, it must be greater than or equal to zero:

$$0 \leq P(X_k) \leq 1 \tag{3.9}$$

This property is valid for all possible values of k. Additionally, the sum of all possible probabilities must be equal to 1:

$$\sum_{k=1}^{N} P(X_k) = 1 \tag{3.10}$$

in which N is the total number of possible outcomes. It is often useful to present the likelihood of an outcome using the *cumulative mass function*, which is given by:

$$P(X \leq X_k) = \sum_{j=1}^{k} P(X_j) \tag{3.11}$$

The cumulative mass function is used to indicate the probability that the random variable X is less than or equal to X_k. It is inherent in the definition (Eq. 3.11) that the cumulative probability is defined as zero for all values less than the smallest X_k and 1 for all values greater than the largest value.

3.3.3 Uniform Distribution

A random variable x with a uniform distribution, which is bounded, has the following density function:

$$f(x) = \begin{cases} \dfrac{1}{\beta - \alpha} & \alpha \le x \le \beta \\ 0 & \text{otherwise} \end{cases} \tag{3.12}$$

in which α is the location parameter and represents the lower limit of the distribution, and β is the scale parameter and represents the upper bound of the distribution. Because the uniform, or rectangular, distribution is a continuous distribution, probabilities can be determined for any interval through integration of the density function.

The mean of the uniform distribution is:

$$\mu = \frac{\alpha + \beta}{2} \tag{3.13}$$

The variance is:

$$\sigma^2 = \frac{(\beta - \alpha)^2}{12} \tag{3.14}$$

with the standard deviation being the square root of the variance. Equations (3.13) and (3.14) use population moments and parameters. If sample statistics are available, $\overline{X}$ and S, the population parameters α and β can be approximated by:

$$\alpha = \overline{X} - S\sqrt{3} \tag{3.15a}$$
$$\beta = \overline{X} + S\sqrt{3} \tag{3.15b}$$

Example 3.5. Assume that the dissolved oxygen concentration in a stream is the value of a random variable for which all values are equally likely between the lower limit of 0 mg/L and upper limit of the saturation limit. On a particular day, the temperature is 8°C. The probability that the dissolved oxygen concentration is less than 4 mg/L can be found.

If all values are equally likely, the uniform distribution can be used to represent the likelihood of values of dissolved oxygen (DO). At 8°C, the saturation limit is 11.843 mg/L, with a lower limit of 0 mg/L. Therefore, in the population, $\alpha = 0$ and $\beta = 11.843$. Therefore, the probability of the DO being less than 4 mg/L is:

$$P(DO < 4) = \int_0^4 \frac{1}{\beta - \alpha} dx = \int_0^4 \frac{1}{11.843 - 0} dx = \frac{4 - 0}{11.843} = 0.338$$

3.3.4 Normal Distribution

The most frequently used density function is the normal or Gaussian distribution, which is a symmetric distribution with a bell-shaped appearance. The density function is given by:

$$f(x) = \frac{1}{\sqrt{2\pi}\sigma} \exp\left[-\frac{1}{2}\left(\frac{x-\mu}{\sigma}\right)^2\right] \quad \text{for } -\infty < x < \infty \qquad (3.16)$$

where μ is a location parameter and σ is a second parameter. Because an infinite number of values of μ and σ are possible, an infinite number of possible configurations of the normal distribution are possible. However, in spite of this, it is easy to define probabilities for selected portions of the distribution. The cumulative distribution function for the normal distribution can be determined by integrating Eq. (3.16).

3.3.4.1   Standard Normal Distribution

Because of the extensive use for the normal distribution and the infinite number of possible value of μ and σ, probabilities for a normal curve are usually evaluated using the standard normal distribution. The standard normal distribution is derived by transforming the random variable x to the random variable z using:

$$z = \frac{x-\mu}{\sigma} \qquad (3.17)$$

In practice, the sample values, $\overline{X}$ and S, may be used in place of μ and σ in Eq. (3.17). The density function for the standard normal curve is.

$$f(z) = \frac{1}{\sqrt{2\pi}} e^{-0.5z^2} \qquad (3.18)$$

The transformed variable z has a mean of 0 and a standard deviation of 1.

Probability estimates for the standard normal curve can be evaluated by integrating Eq. (3.18) between specific limits. For example, the probability that z is between 0.5 and 2.25 could be determined using:

$$P(0.5 \leq z \leq 2.25) = \int_{0.5}^{2.25} f(z)dz = \int_{0.5}^{2.25} \frac{1}{\sqrt{2\pi}} e^{-0.5z^2} dz \qquad (3.19)$$

Because probabilities for the standard normal distribution are frequently required, integrals, such as that of Eq. (3.19), have been computed and placed in tabular form. Table A.1 gives the probability that z is between $-\infty$ and any value of z from -3.40 to 3.40, in increments of z of 0.01. Probabilities for cases in which the lower limit is different from $-\infty$ can be determined using the following identity:

$$P(z_L < z < z_U) = P(-\infty < z < z_U) - P(-\infty < z < z_L) \qquad (3.20)$$

For example, the probability corresponding to the integral of Eq. (3.37) is given by:

$$P(0.5 < z < 2.25) = P(z < 2.25) - P(z < 0.5) \qquad (3.21a)$$
$$= 0.9878 - 0.6915 = 0.2963 \qquad (3.21b)$$

By transforming values using Eq. (3.17) the standard normal curve can be used to determine probabilities for all values of μ and σ. Since the population values of the parameters μ and σ are rarely known, estimates of μ and σ are the sample moments $\overline{X}$ and S_x, with $\mu = \overline{X}$ and $\sigma = S_x$. Of course, the accuracy of estimates of probabilities will depend on the accuracy the sample moments.

Example 3.6. Assume that the following ten samples from total phosphorus (TP) from a tidal estuary are collected:

$$0.71, 0.84, 0.33, 0.60, 0.54, 0.47, 0.79, 0.27, 0.58, 0.39 \quad \text{mg/L}$$

If they are assumed to be normally distributed, the probability of phosphorus readings above 1.0 mg/L, which is considered an indication of poor quality, can be found.

Since the population parameters are not known, they can be estimated using the sample moments. Therefore, the location and scale parameters are $\mu = \overline{X} = 0.552$ and $\sigma = S = 0.191$, i.e., TP ~ N(0.552, 0.191). Thus, the probability of interest is:

$$P(TPO_4 > 1 \, \text{mg/L}) = 1 - P(TPO_4 < 1 \, \text{mg/L})$$
$$= 1 - P\left(z < \frac{1 - 0.552}{0.191}\right) = 1 - P(z < 2.346)$$
$$= 1 - 0.9905 = 0.0095$$

Example 3.7. Monthly measurements of the pH of a small lake are collected for one year, with the following results:

$$8.3, 7.5, 6.8, 6.2, 7.0, 7.7, 8.4, 8.9, 7.9, 7.5, 8.3, 7.1$$

If the range of pH considered excellent is from 6.9 to 8.0 and if the values can assumed to be normally distributed, the percentage of the time that the pH is expected to fall in the excellent range is found.

The sample mean and standard deviation are 7.633 and 0.776, respectively. The probability that the pH falls in the excellent range is, therefore,

$$P(6.9 < pH < 8.0) = P(pH < 8.0) - P(pH < 6.9)$$
$$= P\left(z < \frac{8.0 - 7.633}{0.776}\right) - P\left(z < \frac{6.9 - 7.633}{0.776}\right)$$
$$= P(z < 0.4729) - P(z < -0.9446)$$
$$= 0.6818 - 0.1724 = 0.5094$$

Thus, the sample data suggest that the pH is in the 'excellent' range about 51% of the time.

3.3.4.2 Log-Normal Distribution

The normal distribution is symmetric. However, since many data sets are nonsymmetric with the long tail to the right, i.e., for large values, a common ploy is to take the logarithms of the data and fit the logarithms to a normal distribution. If the fit is good, then the data are assumed to be log-normally distributed.

To fit a log-normal distribution, the logarithms of the data are obtained and the mean and standard deviation of the logarithms are computed. Note that it is incorrect to compute the mean and standard deviation and then make a log transform of them, i.e., the mean of the logs does not equal the log of the mean.

Example 3.8. During an ecological study of a stream reach, 20 measurements of TKN (mg/L) were collected:

0.82	0.95	1.04	1.11	1.17	1.24	1.30	1.33
1.38	1.42	1.47	1.52	1.57	1.66	1.79	1.82
1.88	2.03	2.20	2.74				

A histogram of the data (see Fig. 3.1a) does not suggest a normal distribution; however, a histogram of the logarithms (see Fig. 3.1b) suggests that a log-normal distribution may underlie the TKN concentrations in the stream reach:

-0.086	-0.022	0.017	0.045	0.068	0.093	0.114	0.124
0.140	0.152	0.167	0.182	0.196	0.220	0.253	0.260
0.274	0.307	0.342	0.438				

The log mean $\bar{Y}$ and log standard deviation S_y are 0.1643 and 0.1284, respectively. If the TKN data are log-normally distributed, the probability that a TKN would be greater than 2 mg/L is desired.

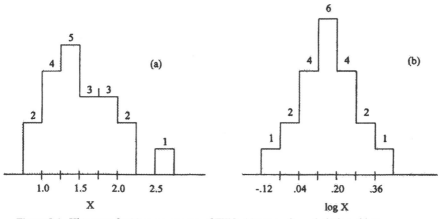

Figure 3.1. Histogram for 20 measurements of TKN: (a) untransformed; (b) logarithms

The probability must be computed in the log-space:

$$P(TKN > 2 \text{ mg/L}) = 1 - P(TKN < 2 \text{ mg/L})$$

$$= 1 - P\left(z < \frac{\log(2) - \bar{Y}}{S_y}\right) = 1 - P(z < 1.065)$$

$$= 1 - 0.8565 = 0.1435$$

3.3.5　*t* Distribution

The *t*, or Student's *t*, distribution is an unbounded distribution having a mean of zero and a variance that depends on the scale parameter v, which is sometimes termed the "degrees of freedom." As v increases toward infinity, the *t* distribution approaches the standard normal density function. In general, the standard normal distribution can be used instead of the *t* distribution for sample sizes greater than 30.

The computation of probabilities for the *t* distribution by the integration of the density function would be very tedious; therefore, values of the statistic are almost always obtained from tables. As a way of introducing the *t* table, it may be of interest to compare the structure of the *t* table with the table of the standard normal distribution. For the standard normal distribution, which is not a function of a parameter, the value of z appears on both the horizontal and vertical margins; the probabilities are located in the center of the table. The *t* distribution is a function of the parameter v and this is the argument located on the vertical margin of Table A.2. The probability is located on the horizontal margin and the value of the *t* statistic is located in the center of the table. When using *t* tables other than the one shown in Table A.2, it is very important to note just what probability is being specified by the table. In summary, the structures of the *t* and z tables are different in that the probability values for the z and *t* tables are located in the center and the horizontal margin, respectively. The *t* table in Table A.2 has *t* values for probabilities of 0.5, 0.4, 0.2, 0.1, 0.05, 0.025, 0.01, 0.005, and 0.001.

In most applications, decisions are based on critical values in one tail (Figure 3.2a). In some, applications, the probability of interest cuts off part of both tails. For example, for a probability of 5 percent divided equally between the two tails, the *t* value would be found by entering the table with values of $v = 5$ and $\alpha = 0.025$; this yields a *t* value of 2.571. Thus the *t* values of interest are all *t* values greater than 2.571 or less than −2.571 (see Fig. 3.2b).

3.4　A PROCEDURE FOR TESTING HYPOTHESES

How can a decision be made as to whether or not a sample statistic is likely to have been obtained from a specified population? Knowledge of the theoretical sampling distribution of a test statistic based on the statistic of interest can be used to make a test of a stated hypothesis. The test of hypothesis leads to a decision as to whether or not a stated hypothesis is valid. Tests are available for almost every statistic, with each test following the same basic steps.

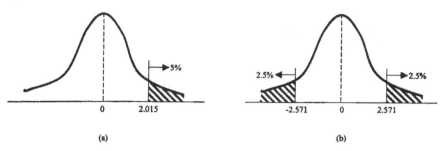

Figure 3.2. One-tailed and two-tailed probability computations: (a) one-tailed computations; (b) two-tailed computations

The following six steps can be used to perform a statistical analysis of a hypothesis:

1. Formulate hypotheses.
2. Select the appropriate statistical model (theorem) that identifies the test statistic and its distribution.
3. Specify the level of significance, which is a measure of risk.
4. Collect a sample of data and compute an estimate of the test statistic.
5. Obtain the critical value of the test statistic, which defines the region of rejection.
6. Compare the computed value of the test statistic (step 4) with the critical value (step 5) and make a decision by selecting the appropriate hypothesis.

Each of these six steps will be discussed in more detail in the following paragraphs.

3.4.1   Step 1: Formulation of Hypotheses

Hypothesis testing represents a class of statistical techniques that are designed to extrapolate information from samples of data to make inferences about populations.   The first step is to formulate two hypotheses for testing.   It is important to note that the hypotheses are composed of statements that involve either population distributions or parameters; hypotheses should not be expressed in terms of sample statistics.   The first hypothesis is called the *null hypothesis*, is denoted by H_0, and is always formulated to indicate that a difference does not exist.   The second hypothesis, which is called the *alternative hypothesis*, is formulated to indicate that a difference does exist.   Both are expressed in terms of the populations or population parameters.   The alternative hypothesis is denoted by either H_1 or H_A.

While the null hypothesis is always expressed as an equality, the alternative hypothesis can be a statement of inequality ($\neq$), less than ($<$), or greater than ($>$). The selection depends on the problem.   If standards for a water quality index indicated that a stream was polluted when the index was greater than some value, than the H_A would be expressed as a greater-than statement.   If the mean dissolved oxygen was not suppose to be lower than some standard, then the H_A would be a less-than statement.   If a direction is not physically meaningful, such as when the mean should not be significantly less than or significantly greater than some value, then a two-tailed inequality statement is used for the H_A. The statement of the alternative hypothesis is important in steps 5 and 6.

3.4.2   Step 2: The Test Statistic and its Sampling Distribution

The two hypotheses of step 1 provide for an equality or a difference between specified populations or parameters. To test the hypotheses, it is necessary to identify the test statistic that reflects the difference suggested by the alternative hypothesis. The specific test statistic is generally the result of known statistical theory. Theoretical models are available for all of the more frequently used hypothesis tests. In cases where theoretical models are not available, approximations have usually been developed.

3.4.3 Step 3: The Level of Significance

A pair of hypotheses were formulated in step 1; in step 2, a test statistic and its distribution were selected to reflect the problem for which the hypotheses were formulated. In step 4, data will be collected to test the hypotheses. Table 3.1 shows the situations that could exist in the population, but are unknown (i.e., H_0 is true or false) and the decisions that the data could suggest (i.e., accept or reject H_0). The decision table suggests two types of error:
  1.   Type I error: reject H_0 when, in fact, H_0 is true.
  2.   Type II error: accept H_0 when, in fact, H_0 is false.
These two types of incorrect decisions are not independent; for a given sample size, the magnitude of one type of error increases as the magnitude of the other type of error decreases. While both types of errors are important, the decision process most often considers only one of the errors, specifically the Type I error. The level of significance, which is usually the primary element of the decision process in hypothesis testing, represents the probability of making a type I error and is denoted by the Greek lowercase letter alpha, α. The level of significance should not be made exceptionally small, because the probability of making a type II error will then be increased.

Table 3.1. Decision table for hypothesis testing

Decision	Situation	
	H_0 is true	H_0 is false
Accept H_0	Correct decision	Incorrect decision: type II error
Reject H_0	Incorrect decision: type I error	Correct decision

3.4.4   Step 4: Data Analysis

After obtaining the necessary data, the sample is used to provide an estimate of the test statistic. In most cases, the data are also used to provide estimates of the parameters that are required to define the sampling distribution of the test statistic. Many tests require computing a statistic called the degrees of freedom, which is necessary to define the sampling distribution of the test statistic.

3.4.5   Step 5: The Region of Rejection

The region of rejection consists of those values of the test statistic that would be unlikely to occur when the null hypothesis is, in fact, true. Extreme values of the test statistic are least likely to occur when the null hypothesis is true. Thus the region of rejection usually lies in one or both tails of the distribution of the test statistic, with the location of the region of rejection depending on the statement of the alternative hypothesis. The region of acceptance consists of all values of the test statistic that are likely to occur if the null hypothesis is true.

Some computer programs avoid dealing with the level of significance as part of the output and instead compute and print the rejection probability. The *rejection probability* is the area in the tail of the distribution beyond the computed value of the test statistic. This concept is best illustrated by way of examples. Assume for a particular problem, a software package is used to analyze a set of data and prints out a computed value of the test statistic z of 1.92 and a rejection probability of 0.0274. This means that approximately 2.74% of the area under the probability distribution of z lies beyond a value of 1.92. To use this information for making a one-tailed upper test, the null hypothesis would be rejected for any level of significance larger than 2.74% and accepted for any level of significance smaller than 2.74%. For a 5% level, H_0 is rejected, while for a 1% level of significance, the H_0 is accepted. Printing the rejection probability places the decision in the hands of the reader of the output.

3.4.6   Step 6: Select the Appropriate Hypothesis

A decision on whether or not to accept the null hypothesis depends on a comparison of the computed value (step 4) of the test statistic and the critical value (step 5). The null hypothesis is rejected when the computed value lies in the region of rejection. Rejection of the null hypothesis implies acceptance of the alternative hypothesis. The decision for most hypothesis tests can be summarized in a table such as the following:

If H_A is	Then reject H_0 if
$P \neq P_0$	$S > S_{\alpha/2}$ or $S < S_{1-\alpha/2}$
$P < P_0$	$S < S_{1-\alpha}$
$P > P_0$	$S > S_\alpha$

where P is the parameter being tested against a standard value, P_0; S is the computed value of the test statistic; and $S_{\alpha/2}$ and $S_{1-\alpha/2}$ are the tabled values for the population and have an area of $\alpha/2$ in the respective tail.

Example 3.9.   Consider the comparison of runoff volumes from two watersheds that are similar in drainage area and other important watershed characteristics such as slope, but differ in the amount of land development. On one watershed, small pockets of land have been developed. The hydrologist is seeking an answer as to whether or not the small amount of land development has been sufficient to increase the volume of runoff from a storm. The watersheds are located near to each other and, therefore, most likely experience the same rainfall distributions. While rainfall characteristics are not measured, the total storm runoff volumes are measured.

The statement of the problem suggests that two means will be compared, one for a developed-watershed population μ_d and one for an undeveloped-watershed population μ_u. The hydrologist believes that the case where μ_d is less than μ_u is not rational and, therefore, prepares to test the following hypotheses:

$$H_O: \mu_d = \mu_u \qquad\qquad\qquad (3.22a)$$

$$H_A: \mu_d > \mu_u \qquad\qquad\qquad (3.22b)$$

Thus, a one-sided test of two means will be made, with the statement of the alternative hypothesis determined by the problem statement.

Several theorems are available for use in comparing two means, and the hydrologist will select the one theorem that is most appropriate for the data expected to be collected in step 4. For example, one theorem assumes equal variances that are unknown, while another theorem assumes variances that are known and do not have to be equal. A third theorem assumes unequal and unknown variances. The theorem should be specified before the data are collected.

In step 3, the level of significance needs to be specified. The implications of the two types of error are:

Type I: Conclude that H_O is false when it is true and wrongly assume that even spotty development can increase runoff volumes. This might lead to the requirement for unnecessary BMPs.

Type II: Conclude that H_O is true when it is not and wrongly assume that spotty development does not increase runoff volumes. This might allow increases in runoff volumes to enter small streams and ultimately cause erosion problems.

Assume that the local government concludes that the implications of a type II error are more significant than those of the type I error. They would, therefore, want to make β small, which may mean selecting a level of significance that is larger than the traditional 5%.

While the data have not been collected, the problem statement has been transformed into a research hypothesis (step 1), the relevant statistical theory has been identified (step 2), and the risk of errors of sampling have been considered (step 3). It is generally considered incorrect experimental practice to collect and peruse the data prior to establishing the first three steps of the test.

In step 4, the data are collected. Generally, the largest sample size that is practical to collect should be obtained. Accuracy is assumed to improve with increasing sample size. Once the data are collected and organized, the test statistic and parameters identified in the theorem are computed. It may also be necessary to check any assumptions that are specified in the theorem. For the case of the runoff volumes, the sample size may be limited by the number of storms that may occur during the period of time alloted to the experiment.

In step 5, the critical value would be obtained from the appropriate table. The value may depend on parameters, such as the sample size, from step 4. The critical value would also depend on the statement of the alternative hypothesis (step 1) and the level of significance (step 3). The critical value and the statement of the alternative hypothesis would define the region of rejection.

In step 6, the computed value of the test statistic is compared with the critical value. If it lies in the region of rejection, then the hydrologist might assume that the null hypothesis is not correct and that spotty development within a watershed causes increases in runoff volumes. The value of the level of significance would indicate the probability that the rejected null hypothesis is falsely being rejected.

3.4.7   Summary of Common Hypothesis Tests

It would not be possible to summarize all of the hypothesis tests that are available; in fact, if an entire textbook were written that did nothing more than summarize the available hypothesis tests, the book would be very large. Instead, a few of the more frequently used tests will be listed here just to show that all tests follow the same six steps and that it is not difficult to apply these tests. Table 3.2 provides a summary of frequently used tests. The table identifies the statistical characteristic involved, the null hypothesis for each test, the alternative hypothesis or hypotheses if more than one alternative is possible, the test statistic, and the decision criterion. The tests that are summarized in Table 3.2 are as follows:

(1) The test of a single mean with σ known, which is sometimes referred to as the z text
(2) The test of a single mean with σ unknown, which is sometimes referred to as the t test
(3) The test of two means with the random variables having identical population variances that are not known
(4) The test of a single variance to some standard value σ_0^2
(5) The chi-square goodness-of-fit test for the probability distribution of a random variable x.

Except for the chi-square goodness-of-fit test, all of the tests can be applied as either one-sided or two-sided tests. The chi-square goodness-of-fit test is always one-sided with the region of rejection in the upper tail of the χ^2 distribution. To apply the chi-square test the sample values of the random variable x are separated into k cells and the observed number of values (O_i) in each cell i is determined. The expected number of values (E_i) is computed from the probability function identified in the null hypothesis.

Table 3.2. Summary of hypothesis tests

Statistical characteristic	H_0	Test statistic	H_A	Reject H_0 if:
One mean (σ known)	$\mu = \mu_0$	$z = \dfrac{\bar{X} - \mu_0}{\sigma/\sqrt{n}}$ where $z = N(0,1)$	$\mu < \mu_0$ $\mu > \mu_0$ $\mu \neq \mu_0$	$z < -z_\alpha$ $z > z_\alpha$ $z < -z_{\alpha/2}$ or $z > z_{\alpha/2}$
One mean (σ unknown)	$\mu = \mu_0$	$t = \dfrac{\bar{X} - \mu_0}{S/\sqrt{n}}$ with $v = n-1$	$\mu < \mu_0$ $\mu > \mu_0$ $\mu \neq \mu_0$	$t < -t_\alpha$ $t > t_\alpha$ $t < -t_{\alpha/2}$ or $t > t_{\alpha/2}$
Two means ($\sigma_1 = \sigma_2$, but unknown)	$\mu_1 = \mu_2$	$t = \dfrac{\bar{X}_1 - \bar{X}_2}{S(1/n_1 + 1/n_2)^{0.5}}$ with $v = n_1 + n_2 - 2$ $$S^2 = \frac{(n_1-1)S_1^2 + (n_2-1)S_2^2}{n_1 + n_2 - 2}$$	$\mu_1 < \mu_2$ $\mu_1 > \mu_2$ $\mu_1 \neq \mu_2$	$t < -t_\alpha$ $t > t_\alpha$ $t < -t_{\alpha/2}$ or $t > t_{\alpha/2}$
One Variance	$\sigma^2 = \sigma_0^2$	$\chi^2 = \dfrac{(n-1)s^2}{\sigma_0^2}$ with $v = n-1$	$\sigma^2 < \sigma_0^2$ $\sigma^2 > \sigma_0^2$ $\sigma^2 \neq \sigma_0^2$	$\chi^2 < \chi_\alpha^2$ $\chi^2 > \chi_\alpha^2$ $\chi^2 < \chi_{\alpha/2}^2$ or $\chi^2 > \chi_{\alpha/2}^2$
Probability distribution	$x \sim f(x,p)$ where x = random variable p = parameters(s)	$\chi^2 = \sum_{i=1}^{k} \dfrac{(O_i - E_i)^2}{E_i}$ with $v = k-1$	$x \neq f(x,p)$	$\chi^2 > \chi_\alpha^2$

Example 3.10. In Example 3.7, 12 pH measurements produced a mean and standard deviation of 7.633 and 0.766, respectively. Can we conclude that the mean pH is significantly different from 7.0?

Since the population variance is not known, the *t* test should be used. The test of Table 3.2 has the test statistic:

$$t = \frac{\bar{X} - \mu_o}{S/\sqrt{n}} = \frac{7.633 - 7}{0.776/\sqrt{12}} = 2.826$$

For a sample size of 12, the test uses 11 degrees of freedom. For a two-tailed test at a 5% level of significance, the critical value from Table A.2 is 2.201. Therefore, the null hypothesis H_0: $\mu = 7$ can be rejected in favor of the alternative hypothesis H_A: $\mu \neq 7$.

Example 3.11. In Example 3.6, a sample of ten total phosphorus (TP) readings had a mean and a standard deviation of 0.552 and 0.191 mg/L, respectively. Can we conclude that the true mean is equal to 0.40 mg/L or is it greater than 0.40 mg/L?

For this case, the problem statement specifies a direction. Therefore, the null and alternative hypotheses are:

$$H_O: \mu = 0.40 \text{ mg/L}$$

$$H_A: \mu > 0.40 \text{ mg/L}$$

The alternative hypothesis is one-tailed, upper. If the null hypothesis is rejected, then it is assumed that the population mean is significantly greater than 0.40 mg/L. The t-statistic of Table 3.2 is used:

$$t = \frac{\overline{X} - \mu_o}{S/\sqrt{n}} = \frac{0.552 - 0.40}{0.191/\sqrt{10}} = 2.517$$

For nine degrees of freedom, the critical value is 1.833 for a 5% level of significance. The null hypothesis should be rejected, with the conclusion that the sample mean is significantly higher than the assumed population mean of 0.40 mg/L.

Note that 3 of the 10 measured TP values are less than 0.40 mg/L. However, recall that the sampling variation of the mean is smaller than the sampling distribution of the random variable. Thus, the fact that 30% of the sample values are less than the assumed population mean is not grounds for automatically accepting the null hypothesis.

Example 3.12. Wastewater from a battery manufacturing plant is discharged into a river. Eight random samples of the lead concentration were collected over a two-week period in the river reach above the plant's outfall:

$$17.4, 52.1, 38.3, 21.0, 12.7, 46.5, 62.1, 28.4 \text{ µg/L}$$

Over the same time period eleven samples of the lead concentration were collected in the stream reach immediately below the outfall:

$$162, 97.1, 143, 244, 51.8, 87.9, 138, 172, 63.0, 108, 152 \text{ µg/L}$$

Is it reasonable to conclude that the lead concentration in the reach below the outfall is significantly higher than the concentration above the outfall?

This is a case where the objective is to compare two samples rather than a sample against either a population or standard. Thus, the two-sample t test of Table 3.2 is used. It is a one-tailed test because the problem would dictate finding out if the concentration in the downstream reach were higher than in the upstream reach. This would dictate the following hypotheses: $H_O: \mu_u = \mu_d$ and $H_A: \mu_d > \mu_u$.

The means for the upstream ($\overline{X}_u$) and downstream ($\overline{X}_d$) reaches are 34.8 and 129.0 µg/L, respectively. The standard deviations for the upstream and downstream reaches are 17.79 and 55.17 µg/L, respectively. Therefore, the pooled standard deviation is:

$$S = \left[\frac{7(17.79)^2 + 10(55.17)^2}{8+11-2} \right]^{0.5} = 43.8\,\mu g/L$$

The computed value of the test statistic is:

$$t = \frac{\overline{X}_d - \overline{X}_u}{S\left(\dfrac{1}{n_d} + \dfrac{1}{n_u}\right)^{0.5}} = \frac{129 - 34.8}{43.8\left(\dfrac{1}{11} + \dfrac{1}{8}\right)^{0.5}} = 4.629$$

which has 17 degrees of freedom. Since the problem arises because of the concern that the plant discharge will increase the lead concentration, the computed test statistic should be positive if the alternative hypothesis were true. The critical value for a 5% level of significance is 1.74. Since the computed value exceeds the critical value, the null hypothesis is rejected. This suggests that the mean lead concentration downstream exceeds that for the upstream reach.

3.5   OUTLIER DETECTION

After plotting a frequency histogram, an event that is much larger or much smaller than the remainder of the sample may be evident. This will initially be labeled as an *extreme event*. Some data samples may contain more than one extreme event. Extreme events can create problems in data analysis and modeling. For example, an extremely large value can cause the sample mean and standard deviation to be much larger than the population values. In bivariate analysis (i.e., X vs. Y), an extreme point can adversely influence the sample value of a correlation coefficient; it could also distort the coefficients of the regression line, thus suggesting an effect that may not reflect the true relationship between the two variables.

Having subjectively decided that one or more values in a sample are extreme events, the values should be objectively evaluated. The objective would be to assess whether or not the extreme event(s) is likely to have occurred if the sample were correctly obtained from the assumed population. Statistical theory in the form of a hypothesis test can be used to make a decision. If the statistical test indicates that the observed extreme event is unlikely to have occurred when sampling from the assumed population, then the extreme event is called an *outlier*. That is, an outlier is a measured value that, according to a statistical test, is unlikely to have been drawn from the same population as the remainder of the sample data.

Having determined that an extreme event is an outlier, the question arises: What can be done with the value? If the value is kept in the sample, then it may distort any values or relationships computed from the sample. Eliminating a data point that has been proven to be an outlier should yield more accurate statistics and relationships with other variables. However, some individuals are against censoring (i.e., eliminating the statistically proven outlier from the sample). Their argument goes: If the value was actually measured, then such a value can occur and so it is incorrect to censor the value from the sample. Both are legitimate arguments.

While the general consensus is toward allowing proven outliers to be censored, every effort should be made to find the reason that such an extreme event occurred. If the data point is the result of sampling from another population, then it seems reasonable to analyze separately the samples from the two populations. Examples of multiple populations in hydrology are: hurricane versus nonhurricane floods; snowmelt floods versus non-snowmelt floods; channel erosion from cohesive and noncohesive beds.

While a number of methods have been proposed, the Dixon-Thompson method is commonly used. This method defines a test statistic as the ratio of two deviations and assumes that sampling is from a normal population. If one suspects that a log-normal population underlies the sample, the tests can be applied to the logarithms of the measured data. The Dixon-Thompson test has the advantage that it can be used for samples as small as 3. The hypotheses for this test are:

H_0: All sample points are from the same normal population   (3.23a)
H_A: The most extreme sample point in the sample is unlikely to
     have come from the normal population from which the
     remainder of the sample points were drawn   (3.23b)

It is only valid for testing one outlier. It can be used as either a one-tailed test or a two-tailed test. For example, if only low outliers are important to the physical problem, then the Dixon-Thompson test can be used for low outliers.

To conduct the Dixon-Thompson test, the data are ranked from smallest to largest, with the smallest value denoted as X_1 and the largest value denoted as X_n. Thus, the subscript indicates the rank of the value from smallest to largest. The test statistic R and critical value R_c depend on the sample size. Table 3.3 gives the equations used to compute both R and R_c. The null hypothesis is rejected if R is greater than R_c.

Example 3.13. The Dixon-Thompson test can be applied to the Floyd River flood series to test the largest value as an outlier. Consider the following annual maximum flood series for the Floyd River at James, Iowa, (06-6005) for the period from 1935 to 1973:

71,500	7440	4520	2260	1360
20,600	7170	4050	2060	1330
17,300	6280	3810	2000	1300
15,100	6250	3570	1920	970
13,900	5660	3240	1720	829
13,400	5320	2940	1460	726
8,320	4840	2870	1400	318
7,500	4740	2710	1390	

The mean ($\overline{X}$) and standard deviation (S_x) are 6771 cfs and 11,696 cfs, respectively. The log mean ($\overline{y}$) and log standard deviation (S_y) are 3.5553 and 0.46418, respectively. In this case, the value being examined is prespecified so it is appropriate to use a one-tailed upper test. The equation from Table 3.3 gives:

Table 3.3. Test Statistics and Critical Values (R_c) for the Dixon-Thompson Outlier Test as a function of sample size (n)

Sample Size	Low Outlier Test Statistic	High Outlier Test Statistic	Polynomial for Critical value, R_c	
3 to 7	$R=\dfrac{X_2-X_1}{X_n-X_1}$	$R=\dfrac{X_n-X_{n-1}}{X_n-X_1}$	$R_c = 1.975 - 0.4994n + 0.5895n^2 + 0.0025n^3$	(3.24)
8 to 10	$R=\dfrac{X_2-X_1}{X_{n-1}-X_1}$	$R=\dfrac{X_n-X_{n-1}}{X_n-X_2}$	$R_c = 1.23 - 0.125n + 0.005n^2$	(3.25)
11 to 13	$R=\dfrac{X_3-X_1}{X_{n-1}-X_1}$	$R=\dfrac{X_n-X_{n-2}}{X_n-X_2}$	$R_c = 0.90 - 0.03n$	(3.26)
14 to 25	$R=\dfrac{X_3-X_1}{X_{n-2}-X_1}$	$R=\dfrac{X_n-X_{n-2}}{X_n-X_3}$	$R_c = 0.9975 - 0.04268n + 0.000764n^2$	(3.27)
26 to 200	$R=\dfrac{X_n-\overline{X}}{S_X}$	$R=\dfrac{\overline{X}-X_1}{S_X}$	$R_c = 2.2795 + 0.025012n - 0.00018427n^2 + 4.61106x10^{-7}n^3$	(3.28)

$$R=\frac{X_n-\overline{X}}{S_x}=\frac{71,500-6771}{11,696}=5.53$$

The critical value is obtained from Eq. (3.28):

$$R_c = 2.2795+0.025012(39)-0.00018427(39)^2+4.611\times10^{-7}(39)^3$$
$$= 3.002$$

Thus, the largest flow is rejected and considered as an outlier by the Dixon-Thompson test.

Applying the Dixon-Thompson test for a log-normal assumption gives the following test statistic:

$$R=\frac{\log(71,500)-3.5553}{0.46418}=2.798$$

Thus, for the critical value of 3.002, the largest value would not be an outlier when assuming the data are from a log-normal population. The test statistic is only slightly below the critical value, which suggests that a less stringent level of significance would lead to the rejection of the null hypothesis.

It is important when applying this test to specify the hypotheses (one-tailed lower, one-tailed upper, or two-tailed) and the distribution (normal or log-normal) prior to collecting the data. The data should not be collected and examined prior to specifying the hypotheses and distribution.

3.6 PROBLEMS

3.1 What is the difference between the population and sample? An environmental engineer obtains five samples of dissolved oxygen from a stream. What is the corresponding population? How could the engineer use the sample to characterize the population?

3.2 Suspended solids concentrations were obtained for runoff from 60 sites. Convert the frequency histogram to a probability distribution. What is the probability that a site has a concentration of at least 90? What is the probability that the site had a value less than 70?

Range	0-59	60-69	70-74	75-79	80-84	85-89	90-94	95-100
Number	4	9	7	11	13	7	6	3

3.3 For the mass function of Problem 3.3, determine the following probabilities:
(a) $P(x = 4)$; (b) $P(x = 5)$; (c) $P(x \leq 5)$; (d) $P(x < 5)$; (e) $P(4 < x < 7)$;
(f) $P(x \geq 7)$.

3.4 Find the value of k that is necessary to make the following a legitimate density function:

$$f(x) = \begin{cases} kx^2 & \text{for } 0.2 \leq x \leq 0.5 \\ 0 & \text{otherwise} \end{cases}$$

Graph both the density and cumulative functions.

3.5 For the density function Problem 3.4, determine the following probabilities:
(a) $P(x < 0.3)$; (b) $P(0.25 < x < 0.4)$; (c) $P(x > 0.35)$; (d) $P(x = 0.42)$.

3.6 Compute the mean, variance, and standard deviation for the following density function:

$$f(x) = \frac{1}{\beta - \alpha} \quad \text{for } \alpha \leq x \leq \beta$$

3.7 Calculate the mean, variance, and standard deviation of the scores in the first row of the table below, assuming that (a) each has the frequency of 1; (b) each has the frequency shown in the second row of the table.

X_i	2	3	6	6	8
$f(X_i)$	3	4	6	2	2

3.8   Given the sample values (2, 3, 4, 6, 6, 8), compute the mean, variance, and standard deviation of the sample. What will the value of the mean, variance, and standard deviation be if (a) 2 points are added to each value; (b) each value is multiplied by 2?

3.9   Compute the mean and standard deviation of the uniform distribution in terms of the parameters α and β. Rearrange the two equations so that the parameters are a function of the moments.

3.10 Determine the following probabilities using the standard normal distribution:
      (a) $p(z > 1)$; (b) $p(z < 1)$; (c) $p(z < -1)$; (d) $p(0 < z < 2)$;
      (e) $p(-1 < z < 1)$; (f) $p(-1 < z < 2)$; (g) $p(-2 < z < -1)$;
      (h) $p(z < -3) + p(z > +3)$.

3.11 Given the sample (2, 3, 6, 6, 8) with a mean of 5 and a variance of 6, (a) calculate a z score for each observation; (b) transform all of the observations so that the new distribution has a mean of 10 and a variance of 9.

3.12 A random variable has a mean of 0.7 and a standard deviation of 0.1. Find the probability that the value of a sample point is (a) less than 0.55; (b) greater than 0.82; (c) between 0.61 and 0.88; (d) between 0.57 and 0.67. Assume a normal distribution.

3.13 A random variable has a mean of 25 and a standard deviation of 5. Find the probability that the value of a sample point is (a) less than 20, (b) greater than 32.5; (c) between 22.5 and 35.0. Assume a normal distribution.

3.14 A random variable has a mean of 0.7 and a standard deviation of 0.1. Find the value or values of the random variable such that (a) 5 percent of the area lies above it; (b) 2.5 percent of the area lies below it; (c) 0.5 percent of the area lies in each tail. Assume a normal distribution.

3.15 A random variable has a mean of 25 and a standard deviation of 5. Find the value of the random variable such that (a) 25 percent of the area lies above it; (b) 7 percent of the area lies below it; (c) 95 percent of the area lies above it. Assume a normal distribution.

3.16 The average annual precipitation for Washington, D.C., is 43 inches, with a standard deviation 6.5 inches. If the amounts can be assumed to have a normal distribution, find the probability that the precipitation in any one year is (a) greater than 55 inches; (b) less than 35 inches; (c) either less than 32 inches or greater than 51; (d) between 38 and 48 inches.

3.17 (a) Graph the right tail of the t distribution for values of v of 3, 10, 29, and ∞, (b) As v increases, what happens to the variance? (c) For the v values specified, approximately what area lies to the right of a t value of 2.5?

3.18 What are the characteristics of a null hypothesis and an alternative hypothesis?

3.19 Why is it necessary to state the null hypothesis as a finding of no significant difference (i.e., an equality) when the objective of the research may be to show a difference?

3.20 Distinguish between the sampling distribution of the random variable and the sampling distribution of the test statistic in the various steps of a hypothesis test.

3.21 Develop one-tailed upper, one-tailed lower, and two-tailed hypotheses related to the hydrologic effects of afforestation.

3.22 Develop one-tailed upper, one-tailed lower, and two-tailed hypotheses related to the hydrologic effects of clearing vegetation from a stream channel.

3.23 A sample of 20 yields a mean of 32.4. Test the two-sided hypothesis that the sample was drawn from a population with a mean of 35 (a) if the variance of the population is 33; (b) if the variance of the population is unknown, but the sample variance is 33. Use a level of significance of 5 percent.

3.24 A random sample of 10 has a mean of 110. Test the null hypothesis that $\mu = 120$ against the alternative hypothesis that $\mu < 120$ at the 5 percent level of significance. (a) Assume that the population standard deviation of 18 is known. (b) Assume that the population standard deviation is unknown and the sample value is 18.

3.25 A random sample of 25 has a mean of 4.8 with a standard deviation of 0.32. Test the null hypothesis that $\mu = 4.95$ against the alternative hypothesis that $\mu < 4.95$ at the 1 percent level of significance.

3.26 A random sample of 10 yields a mean and standard deviation, respectively, of 73.6 and 7.9. Assuming a two-sided test, test the hypothesis that the sample is from a population with a mean of 80. Use a level of significance of (a) 5 percent; (b) 1 percent.

3.27 A public water supply official claims that the average household water use is much less than 350 gallons per day. To test this assertion, a random sample of 200 water use records are complied. If the random sample showed an average use of 338 gallons per day and a standard deviation of 35 gallons per day, would it be safe to conclude at the 1 percent level of significance that μ is less than 350 gallons per day?

3.28 A random sample of 12 has a mean and standard deviation of 240 and 30, respectively. Test the null hypothesis that $\mu = 215$ against the alternatively hypothesis that $\mu > 215$ at a level of significance of (a) 5 percent; (b) 1 percent.

3.29 When using the Dixon-Thompson method to test for a low outlier with a sample size of 5, how much larger must X_2 be in comparison with X_1 for X_1 to be considered a low outlier? Assume $X_n - X_1 = 10$.

3.30 For a sample size of ten, with the Dixon-Thompson test, what minimum value of X_{10} is necessary to reject the largest sample value as a high outlier if $X_2 = 4.3$ and $X_9 = 11.4$?

3.31 Use the Dixon-Thompson method to test the following data for Seneca Creek, Maryland, for an outlier: 26100, 3020, 3160, 16000, 4900, 7850, 16000, 3260, 3010, 3620, 1070, 4950, 7410, 8250, and 2270.

3.32 The following data are the annual maximum discharges for the Grey River near Alpine, Wyoming. Test the data for outliers using the Dixon-Thompson method: 4210, 2010, 5010, 4290, 3720, 2920, 2500, 2110, 3110, 2420, 4280, 3860, 3150, 4050, 3260, 7230, 5170, 2550, 5220, 2550, 5220, 3650, 3590, and 650.

3.33 The following data are the annual maximum discharges for San Emigdio Creek, California: 6690, 538, 500, 340, 262, 234, 212, 208, 171, 159, 150, 145, 118, 94, 92, 60, 58, 54, 40, 38, 28, 27, 26, 25, 24, 17, 16, 8, and 1.23. Test the data for outliers using the Dixon-Thompson method.

4

STORMWATER HYDROLOGY

NOTATION

A   = drainage area
A_m = drainage area in mi^2
C   = runoff coefficient
C_w = weighted Rational coefficient
CN   = curve number
CN_c = composite curve number
CN_p = pervious area curve number
CN_w = weighted curve number
ΔE = change in elevation
f   = fraction of imperviousness
i   = rainfall intensity
I_a = initial abstraction
k   = velocity-slope coefficient
k_L = least squares estimate of k
L   = watershed length
n   = Manning's roughness coefficient
p   = exceedence probability
P   = rainfall depth
P_2 = 2-yr, 24-hr rainfall depth
q_p = peak discharge
q_u = unit peak discharge
R   = unconnected-to-total imperviousness
R_h = hydraulic radius
S   = retention for SCS method
S   = slope of flowpath
t_c = time of concentration
T   = return period
T_t = travel time

V   = velocity of flow
WP   = wetted perimeter

4.1 INTRODUCTION

As indicated in Chapter 2, many pollutants are transported from one location to another in the runoff. A pollutant load is computed using the concentration and the discharge rate.   To fully understand issues in water quality requires an understanding of the physical processes that control runoff. This includes the generation of pollutants, the transport of the pollutants to receiving waters, and the treatment of runoff that contains pollutants.   Watershed changes such as urbanization and deforestation complicate the task of making water quality estimates.

Designing for treatment of water quality pollutants requires an understanding of both modeling methods (see Chapter 5), data analysis methods (see Chapter 3), and the basic concepts of stormwater hydrology. Design methods usually require some measure of rainfall, such as a depth or an intensity, and indicators of watershed characteristics, such as the contributing drainage area and the land use. Storage of stormwater also plays a major role in several important water quality treatment methods. In addition to surface runoff, some treatment methods depend on water infiltrating into the ground-water system. Therefore, knowledge of the infiltration process is important. The sections of this chapter introduce important concepts related to the physical processes of stormwater hydrology.

4.2 THE HYDROLOGIC CYCLE

It is common to introduce the physical hydrologic processes as elements of the hydrologic cycle, which is a conceptual representation of the distribution and movement of water in the area, over the surface of the Earth, and through the ground. The hydrologic cycle shows that the processes are interdependent, with knowledge of each necessary to understand problems related to water quantity and quality as well as their solutions.

4.2.1 Water Quantity Perspective

The physical processes that control the distribution and movement of water are best understood in terms of the hydrologic cycle. Although the hydrologic cycle does not have a real beginning or ending point, we can begin the discussion with precipitation. For the purposes of this discussion, we will assume that precipitation consists of rainfall and snowfall. Rain falling on Earth may enter a water body directly, travel over the land surface from the point of impact to a watercourse, or infiltrate into the ground. Some rain is intercepted by vegetation; the intercepted water is temporarily stored on the vegetation until it evaporates back to the atmosphere. Some rain is stored in surface depressions, with almost all of the depression storage infiltrating into the ground. Water stored in depressions, water intercepted by vegetation, and water that infiltrates into the soil during the early part of a storm represent the initial losses. The loss is water that does not appear as runoff during or immediately following a rainfall event. Water entering an upland stream travels to increasingly larger rivers and then to the seas and oceans. The water that infiltrates into the ground may percolate to the water table or travel in the unsaturated zone until it reappears as surface flow. The amount of water stored in the soil determines, in part, the amount of rain that will infiltrate during the next storm event. Water stored in lakes, seas, and oceans evaporates back to the atmosphere, where it completes the cycle and is available for rainfall. Water also evaporates from soil devoid of vegetation. Rain that falls on vegetated surfaces may be intercepted; however, after the storage that is available for interception is filled, the water will immediately fall from the plant surfaces to the ground and infiltrate into the soil in a similar manner as the water that falls on bare ground infiltrates. Some of the water stored in the soil near plants is taken up by the roots of the vegetation, and subsequently passed back to the atmosphere from the leaves of the plants; this process is called *transpiration*.

Although it may appear that the hydrologic cycle of a natural environment is static, it is important to recognize that the landscape is constantly undergoing a transformation. High-intensity storms cause erosion of the land surface. Flood runoff from large-volume storms causes bankfull and high-velocity flows in streams with the potential for large amounts of channel erosion. During periods of extreme drought the perimeter of desert lands may increase. Forest fires caused either by natural means such as electrical storms or by the carelessness of human beings cause significant decreases in the available storage and decrease the surface roughness, both of which contribute to increases in surface runoff rates and volumes, as well as surface erosion. When mud flows are a potential problem, forest fires consume the vegetation, which contributes significantly to the increased production of debris. In many parts of the world, the largest floods result from the rapid melting of snow; such events can be as devastating as floods produced by large rainfall events. Flooding accompanying hurricanes and monsoons can also cause significant changes to the landscape, which by itself affects runoff rates and volumes from storms occurring long after the hurricane or monsoon. In summary, even in a natural environment, rainfall and runoff cause major changes in the watershed.

As the population of the world has increased, changes to the land have often been significant, with major changes to the runoff characteristics of a watershed as a result (Anderson 1970). Land clearing for agricultural development increases the amount of exposed soil, with obvious decreases in the protective covering of the natural vegetation. This loss of protective covering decreases the potential for infiltration, increases surface runoff, and can result in significant soil losses. Over the last two centuries, urbanization has caused significant changes to the landscape surrounding these urban centers. Urbanization has had significant effects on the processes of the hydrologic cycle for watersheds subject to the urban development. Clearing of the land has reduced the vegetation and therefore the availability of interception storage; grading of land surfaces reduces the available volume of depression storage. Impervious surfaces reduce the potential for infiltration and the resulting recharge of ground water storage. Impervious surfaces are also less rough than the natural surfaces and thus offer less resistance to the runoff; this change in roughness can increase runoff velocities and surface erosion. These changes to the processes of the hydrologic cycle cause significant changes in runoff characteristics. The reduced storage results in increased volumes of surface runoff (Sauer *et al.* 1981). The reduced surface roughness decreases the travel time of runoff. The reductions in both storage and travel time result in increased peak rates of runoff, which increase both flood damages due to overbank flows and channel erosion.

In an attempt to compensate for the lost natural storage, many localities require the replacement of the lost natural storage with human-made storage. While the stormwater detention basin is the most frequently used method of stormwater management, other methods are used, such as infiltration pits, rooftop and parking lot storage, and porous pavement. These engineering works do not always return the runoff characteristics to those that existed in the natural environment. In fact, poorly conceived methods of control have, in some cases, made flood runoff conditions worse.

Site development usually results in significant increases in impervious surfaces, which results in increased surface runoff rates and volumes. In many localities, stormwater control facilities are required. In the upper reaches of a site, swales can be used to move water away from buildings and transportation facilities. Concentrated runoff from swales may enter gutters and drainage ditches along roadways where the runoff may drain into highway inlets or small streams. At many sites where land development has resulted in large amounts of imperviousness, on-site detention basins may be required to control the increased runoff. The design of stormwater detention basins requires knowledge of routing of water through the hydraulic outlet structure, as well as knowledge about surface runoff into the detention basin. The design must consider meteorological factors, geomorphological factors, and the economic value of the land, as well as human value considerations such as aesthetic and public safety aspects of the design. The design of a stormwater detention basin should also consider the possible effects of inadequate maintenance of the facility.

4.2.2   Water Quality Perspective

The hydrologic cycle is also useful as a tool for understanding water quality processes. Pollution in the form of acid rain is a function of the atmospheric processes of the hydrologic cycle. The kinetic energy of the raindrops is responsible for dislodging soil particles. The runoff transports soil as well as pollutants attached for the soil. Watershed characteristics such as the slope and surface roughness affect the flow rate of the runoff. The watershed characteristics are also important in the treatment of water to reduce pollutants. The rate of runoff is a primary factor in setting the opportunity for the settlement of suspended materials.

Storage both within the watershed and in the stream network plays important water quality roles. Bioretention storage allows for infiltration, the settling of solid materials, the uptake of pollutants by vegetation, and the delay of the runoff. Small urban detention basins also allow for the settling of solids. Large dams can reduce settleable solids in the streamflow and provide time for nonconservative pollutants to be assimilated. Even small amounts of surface depression storage can have water quality benefits, as the depression may control water in the first flush of surface runoff.

4.3 PRECIPITATION

Storms differ in a number of characteristics, and the characteristics have a significant affect on water quality. The rainfall depth, duration, intensity, and frequency are the most important characteristics of rainfall events.

4.3.1   Depth-Duration-Frequency

The characteristics that must be identified in either assessing an actual storm or developing a design storm are:
1.   *Duration*: the length of time over which a precipitation event occurs
2.   *Depth*: the amount of precipitation that occurs over the storm duration
3.   *Frequency*: the frequency of occurrence of events that have the same depth and duration

Closely related to these definitions is the concept of intensity, which equals the depth divided by the duration. For example, a storm having a duration of 2 hr and a depth of 3 inches would have an intensity of 1.5 in./hr.

Volume and Depth. The volume of a storm is most often reported as a depth, with units of length such as inches or centimeters; in such cases, the depth is assumed to occur uniformly over the watershed. Thus the volume equals the depth times the watershed area. For example, if the 2-hr storm with a volume of 24 acre-in. occurred on a 6-acre watershed, the average depth of rainfall would be 4 in. and the intensity would be 2 in./hr. This interchanging use of units for storm volume often leads to confusion because the terms depth and volume are applied to a quantity having units of length. One might speak of the rainfall volume, but express it in inches. Such statements imply that the depth occurred uniformly over the entire watershed and the units are "area-inches," with the area of the watershed used to compute a volume in acre-inches or some similar set of units.

Just as each concept is important by itself, it is also important to recognize the interdependence of these terms. A specified depth of rainfall may occur from many different combinations of intensities and durations, and these different combinations of intensities and durations will have a significant effect on both runoff volumes and rates, as well as on engineering designs that require rainfall characteristics as input. For example, 3 inches of precipitation may result from any of the following combination of intensity and duration:

Intensity (in./hr)	Duration (hr)	Depth (in.)
12	0.25	3
6	0.50	3
3	1.00	3
1.5	2.00	3

Because the rainfall intensity is an important determinant of the hydrologic response, it is important to specify both the depth and duration (or intensity and duration) and not just the total volume (i.e., depth). Other definitions of importance are:

Frequency. Just as intensity, duration, and depth are interdependent, the fourth concept, frequency, is also a necessary determinant. Frequency can be discussed in terms of either the exceedance probability or the return period, which are defined as follows:

Exceedence probability: the probability that an event having a specified depth and duration will be exceeded in one time period, which is most often assumed to be one year.

Return period: the average length of time between events that have the same depth and duration.

The exceedance probability (p) and return period (T) are related by:

$$p = \frac{1}{T} \qquad (4.1)$$

The following tabulation gives selected combinations of Equation (4.1):

T (yr)	p	T (yr)	p
2	0.50	5	0.20
10	0.10	25	0.04
100	0.01	500	0.002

For example, if a storm of a specified duration and depth has a 1% chance of occurring in any one year, it has an exceedance probability of 0.01 and a return period of 100 years. The argument for not using the term return period to interpret the concept of frequency is that it is sometimes improperly interpreted. Specifically, some individuals believe that if a 100-yr rain (or flood) occurs in any one year, it cannot occur for another 100 years; this belief is false because it implies that storm events occur deterministically rather than randomly. Because storm events occur randomly, there is a finite probability that the 100-yr event could occur in two consecutive years, or not at all in a period of 500 years. Thus the exceedance probability concept is preferred by many. However, engineers commonly use the term return period, and its meaning should be properly understood.

4.3.2 Rainfall Maps

Rainfall depths are required for many design problems. Because of the frequent need for such information, the depth of rainfall for selected durations and frequencies are often provided in the form of maps that show lines of equal rainfall depths; the lines are called isohyets. Maps are available for the U.S. (see Figure 4.1 for the 100-yr, 24-hr map) and for individual states.

4.3.3 Intensity-Duration-Frequency

In addition to depth-duration-frequency curves, intensity-duration-frequency curves are readily available because a rainfall intensity is used as input to many hydrologic design methods. Because of the importance of the intensity-duration-frequency (IDF) relationship in hydrologic analyses, IDF curves have been compiled for most localities; the IDF curve for Baltimore, Maryland, is shown in Figure 4.2.

The intensity-duration-frequency (IDF) curve is most often used by entering with the duration and frequency to find the intensity. For example, the 10-yr, 2-hr rainfall intensity for Baltimore is found from Figure 4.2 by entering with a duration of 2 hr, moving vertically to the 10-yr frequency curve, and then moving horizontally to the intensity ordinate, which yields $i = 1.3$ in./hour. This corresponds to a storm depth of 2.6 in.

The IDF curve could also be used to find the frequency for a measured storm event. The predicted frequency is determined by finding the intersection of the lines defined by the measured intensity and the storm duration. If the volume rather than the intensity is measured, the intensity must be determined prior to determining the frequency. For example, if a 3-in. storm occurs in Baltimore during a period of 3 hr, the intensity is 1 in./hr. Using an intensity of 1 in./hr and a duration of 3 hr yields a return period of 10 yr or an exceedance probability of 0.1.

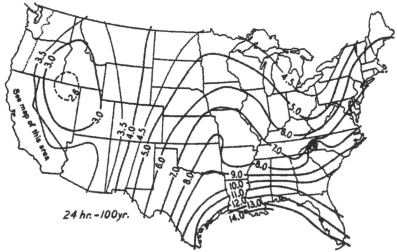

Figure 4.1. Twenty-four-hour rainfall, in inches, to be expected once in 100 years. (National Weather Service, 1961.)

4.3.4 Development of a Design Storm

To this point, the following characteristics of storms have been discussed: depth, duration, frequency, and intensity. Some design problems only require the total depth of rainfall for a specified duration and frequency. However, for many problems in hydrologic analysis, it is necessary to show the variation of the rainfall volume with time (SCS, 1986). A plot of rainfall depth (or intensity) versus time is referred to as a *hyetograph*. That is, some hydrologic problems require the storm input to the design method to be expressed as a hyetograph and not just as a total depth for the storm. Characteristics of a hyetograph that are important are the peak, the time to peak, and the distribution, as well as the depth, duration, and frequency. In developing a design storm for any region, empirical analyses of measured rainfall records are made to determine the most likely arrangement of the ordinates of the hyetograph. For example, some storm events will have an early peak (front loaded), some a late peak (rear loaded), some will peak in the center of the storm (center loaded), and some will have more than one peak. The empirical analysis of measured rainfall hyetographs at a location will show the most likely of these possibilities, and this finding can be used to develop the design storm.

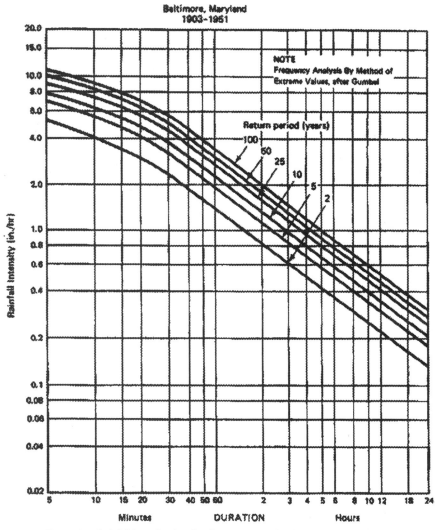

Figure 4.2. Rainfall intensity-duration-frequency. (National Weather Service, 1961.)

Figure 4.3. Approximate geographic areas for SCS rainfall distributions. (SCS, 1986.)

The SCS developed four dimensionless rainfall distributions using the Weather Bureau's Rainfall Frequency Atlases (NWS, 1961). The rainfall-frequency data for areas less than 400 mi^2, for durations to 24 hr. and for frequencies from 1 yr to 100 yr were used. Data analyses indicated four major regions, and the resulting rainfall distributions were labeled type I, IA, II, and III. The locations where these design storms are applicable are shown in Figure 4.3.

The distributions are based on the generalized rainfall depth-duration-frequency relationships shown in technical publications of the Weather Bureau. Rainfall depths for durations from 6 min to 24-hr were obtained from the depth-duration-frequency information in these publications and used to derive the storm distributions. Using increments of 6 min, incremental rainfall depths were determined. For example, the maximum 6-min depth was subtracted from the maximum 12-min depth and this 12-min depth was subtracted from the maximum 18-min depth, and so on to 24 hours. The distributions were formed by arranging these 6-min incremental depths such that for any duration from 6 min to 24 hr, the rainfall depth for that duration and frequency is represented as a continuous sequence of 6-min depths. The location of the peak was found from the analysis of measured storm events to be location dependent. For the regions with type I and IA storms, the peak intensity occurred at a storm time of about 8 hr, while for the regions with type II and III storms, the peak was found to occur at the center of the storm. Therefore for type II and III storm events, the greatest depth is assumed to occur at about the middle of the 24-hr period, the second largest incremental depth during the next time period and the third largest in the interval preceding the maximum intensity. This after-before cycling continues with each incremental rainfall depth in decreasing order of magnitude. Thus the smaller increments fall at the beginning and end of the 24-hr storm. This procedure results in the maximum depth being contained within the maximum 1-hr depth, the maximum 1-hr depth is contained with the maximum 6-hr depth, and so on. Because all of the critical storm depths are contained within the storm distributions, the distributions are appropriate for designs on both small and large watersheds.

This procedure can be used to show the values of the cumulative rainfall distribution of the SCS type II storm for a time increment of 0.25 hr; the values are given in Table 4.1 for hyetograph times from 10.00 to 14.00 hr. The largest incremental rainfall (27.6%) occurred between hyetograph times 11.75 and 12.00 hr. The next largest (10.4%) was placed just prior to the largest incremental amount. The third largest (4.4%) was placed just after the largest incremental amount. This procedure is followed approximately for the remainder of the storm. The resulting distributions are most often presented with the ordinates given on a dimensionless scale. The ordinates are given in Table 4.2 for all four SCS synthetic design storms.

Example 4.1. The procedure of forming a design storm can be illustrated with a simplified example. The design storm will have the following characteristics: duration, 6 hr; frequency, 50 yr; time increment, 1 hr; location, Baltimore. The intensity-duration-frequency curve of Figure 4.2 can be used to obtain the rainfall intensities for durations of 1 to 6 hr in increments of 1 hr; these intensities are shown in column 3 of Table 4.3. The depth (that is, duration times intensity) is shown in column 4, with the incremental depth (column 5) taken as the difference between the depths for durations 1 hr apart. The incremental depths were used to form the 50-yr design storm, which is shown in column 6, by placing the largest incremental depth in hour 3 and the second largest incremental depth in hour 4. The remaining incremental depths were placed by alternating their location before and after the maximum incremental depth. It is important to notice that the maximum three hours of the design storm has a volume of 3.90 in., which is the depth for a 3-hr duration from Figure 4.2; this will be true for any storm duration from 1 to 6 hr. The cumulative form of the design storm is given in column 7 of Table 4.3.

A dimensionless design storm can be developed by transforming the cumulative design storm of column 7 of Table 4.3 by dividing each ordinate by the total depth of 4.62 in. The dimensionless cumulative design storm derived from the 50-yr intensities is shown in column 8 of Table 4.3. The calculation of a dimensionless design storm for a 2-yr frequency is shown in the lower part of Table 4.3. In comparing the 50-yr and 2-yr dimensionless design storms, it should be apparent that a cumulative design storm could be developed for any design frequency by multiplying the 6-hr rainfall depth for that frequency by the average ordinates of the dimensionless cumulative design storms of Table 4.3, which is approximately [0.05, 0.14, 0.77, 0.89, 0.97, 1.00]. Based on this dimensionless cumulative design storm (see Figure 4.4a), the 10-yr cumulative design storm, which has a 6-hr depth of 3.48 in. (0.58 in./hr from Figure 4.2 multiplied by 6 hr), would be [0.17, 0.49, 2.68, 3.10, 3.38, 3.48 in.], which is shown using the ordinate on the right-hand side in Figure 4.4a. Thus the 10-yr design storm would be [0.17, 0.32, 2.19, 0.42, 0.28, 0.10 in.]. The 10-yr, 6-hr design storm is shown in Figure 4.4b with ordinates expressed as an intensity.

Table 4.1. SCS 24-hour Type II design Storm for hyetograph times from 10 to 14 hours and an increment of 0.25 Hour

Time (hr)	Cumulative Rainfall (in.)	Incremental Rainfall (in.)
10.00	0.181	0.010
10.25	0.191	0.012
10.50	0.203	0.015
10.75	0.218	0.018
11.00	0.236	0.021
11.25	0.257	0.026
11.50	0.283	0.104
11.75	0.387	0.276
12.00	0.663	0.044
12.25	0.707	0.028
12.50	0.735	0.023
12.75	0.758	0.018
13.00	0.776	0.015
13.25	0.791	0.013
13.50	0.804	0.011
13.75	0.815	0.010
14.00	0.825	

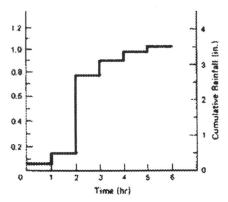

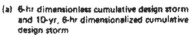

(a) 6-hr dimensionless cumulative design storm and 10-yr, 6-hr dimensionalized cumulative design storm

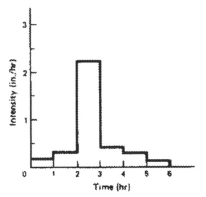

(b) 10-yr, 6-hr design storm for Baltimore

Figure 4.4. Formation of dimensionless and dimensionalized design storms

Table 4.2. SCS cumulative, dimensionless one-day storms

Time (hrs)	Type I	Type IA	Type II	Type III
0	0	0	0	0
0.5	0.008	0.010	0.0053	0.0050
1.0	0.017	0.020	0.0108	0.0100
1.5	0.026	0.035	0.0164	0.0150
2.0	0.035	0.050	0.0223	0.0200
2.5	0.045	0.067	0.0284	0.0252
3.0	0.055	0.082	0.0347	0.0308
3.5	0.065	0.098	0.0414	0.0367
4.0	0.076	0.116	0.0483	0.0430
4.5	0.087	0.135	0.0555	0.0497
5.0	0.099	0.156	0.0632	0.0568
5.5	0.112	0.180	0.0712	0.0642
6.0	0.126	0.206	0.0797	0.0720
6.5	0.140	0.237	0.0887	0.0806
7.0	0.156	0.268	0.0984	0.0905
7.5	0.174	0.310	0.1089	0.1016
8.0	0.194	0.425	0.1203	0.1140
8.5	0.219	0.480	0.1328	0.1284
9.0	0.254	0.520	0.1467	0.1458
9.5	0.303	0.550	0.1625	0.1659
10.0	0.515	0.577	0.1808	0.1890
10.5	0.583	0.601	0.2042	0.2165
11.0	0.624	0.624	0.2351	0.2500
11.5	0.655	0.645	0.2833	0.2980
12.0	0.682	0.664	0.6632	0.5000
12.5	0.706	0.683	0.7351	0.7020
13.0	0.728	0.701	0.7724	0.7500
13.5	0.748	0.719	0.7989	0.7835
14.0	0.766	0.736	0.8197	0.8110
14.5	0.783	0.753	0.8380	0.8341
15.0	0.799	0.769	0.8538	0.8542
15.5	0.815	0.785	0.8676	0.8716
16.0	0.830	0.800	0.8801	0.8860
16.5	0.844	0.815	0.8914	0.8984
17.0	0.857	0.830	0.9019	0.9095
17.5	0.870	0.844	0.9115	0.9194
18.0	0.882	0.858	0.9206	0.9280
18.5	0.893	0.871	0.9291	0.9358
19.0	0.905	0.844	0.9371	0.9432
19.5	0.916	0.896	0.9446	0.9503
20.0	0.926	0.908	0.9519	0.9570
20.5	0.936	0.920	0.9588	0.9634
21.0	0.946	0.932	0.9653	0.9694
21.5	0.956	0.944	0.9717	0.9752
22.0	0.965	0.956	0.9777	0.9808
22.5	0.974	0.967	0.9836	0.9860
23.0	0.983	0.978	0.9892	0.9909
23.5	0.992	0.989	0.9947	0.9956
24.0	1.000	1.000	1.0000	1.0000

Table 4.3. Development of 6-hour dimensionless cumulative design storms for Baltimore

(1) T (yr)	(2) Duration (hr)	(3) Intensity (in./hr)	(4) Depth (in.)	(5) Incremental Depth (in.)	(6) Design Storm (in.)	(7) Cumulative Design Storm (in.)	(8) Dimensionless Cumulative Design Storm
50	1	3.00	3.00	3.00	0.25	0.25	0.0054
	2	1.75	3.50	0.50	0.40	0.65	0.141
	3	1.30	3.90	0.40	3.00	3.65	0.790
	4	1.05	4.20	0.30	0.50	4.15	0.898
	5	0.89	4.45	0.25	0.30	4.45	0.963
	6	0.77	4.62	0.17	0.17	4.62	1.000
2	1	1.35	1.35	1.35	0.10	0.10	0.046
	2	0.82	1.64	0.29	0.19	0.29	0.134
	3	0.61	1.83	0.19	1.35	1.64	0.759
	4	0.50	2.00	0.17	0.29	1.93	0.894
	5	0.42	2.10	0.10	0.17	2.10	0.972
	6	0.36	2.16	0.06	0.06	2.16	1.000

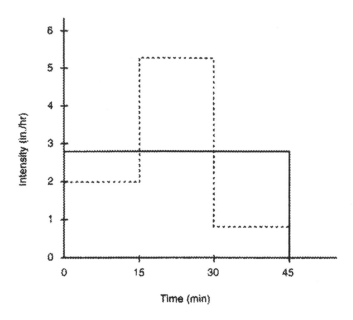

Figure 4.5.   Comparison of a 45-minute constant-intensity design storm with the maximum three
ordinates of the SCS type II

    Concern about the 24-hr duration of the SCS design storms has been expressed.
The argument is that flooding on small watersheds results from short duration
storms, not 24-hr storms. Many prefer the constant-intensity design storm even
though such a storm is unlikely to occur because storms usually begin and end with
a period of low intensity. In this regard, the SCS design storms are more rational.
    Example 4.2. Assume that a small watershed has a time of concentration of 45
min and that a 10-yr return period is used. From Figure 4.2, the design intensity is
2.8 in./hr, which yields a depth of 2.1 in. Using the maximum 45-min period of the
SCS 24-hr design storm (see Table 4.2), the corresponding proportions are 0.104,
0.276, and 0.044. Multiplying these by the 24-hr depth of 4.8 in. for a 10-yr storm
yields ordinates of 0.50, 1.32, and 0.21 in., respectively, for a total depth of 2.03 in.
This is close to the 2.1-in. depth for the constant-intensity storm. The intensity
hyetographs for the two storms are shown in Figure 4.5. The three intensities are
2.00, 5.30, and 0.84 in./hr. This example illustrates that the 24-hr design storm is
capable of reflecting intensities of small-duration storms.

4.4 WATERSHED CHARACTERISTICS

    In making assessments of water quality, it is often necessary to provide
estimates of watershed characteristics. For example, low flow studies relate the low
flow rates to the watershed area. Estimates of soil erosion from land surfaces

require the slope and length of the plot. The ability of vegetated buffer zones to trap sediment depend on the slope, length, and roughness of the buffer strip. The parameters most commonly used in water quality studies are the drainage area, the length of the stream or watershed, the slope, indicators of land use such as the runoff curve number, and the time of travel of water through a certain portion of a watershed. Thus inputs to water quality methods are discussed in this section.

4.4.1 Watershed: Definition and Delineation

The concept of a watershed is basic to many analyses. Since large watersheds are made up of many smaller watersheds, it is necessary to define the watershed in terms of a point; this point is usually the location at which the analysis is being made and is referred to as the watershed "outlet." With respect to the outlet, the watershed consists of all land area that sheds water to the outlet during a rainstorm. Using the concept that "water runs downhill," a watershed is defined by all points enclosed within an area from which rain falling at these points will contribute water to the outlet. This is best shown pictorially, as in Figure 4.6. The shaded area of Figure 4.6 represents the watershed for the outlet at point A. Water is contributed to the outlet from many smaller areas, which are also watersheds. For example, if a design were being made at point B rather than point A, the watershed would be the small area enclosed within the dashed lines. The watershed for point B is made up of smaller watersheds, with the two stream tributaries reflecting the collecting areas for water that results from rain on the watershed.

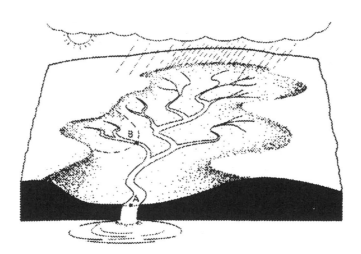

Figure 4.6. Delineation of watershed boundary.

4.4.2 Drainage Area

The drainage area (A) is probably the single most important watershed characteristic for hydrologic design. It reflects the volume of water that can be generated from rainfall. It is common in hydrologic design to assume a constant depth of rainfall occurring uniformly over the watershed. Under this assumption, the volume of water available for runoff would be the product of the rainfall depth and the drainage area. Thus the drainage area is required as input to models ranging from simple linear prediction equations to complex computer models.

4.4.3 Watershed Length

The length (L) of a watershed is the second watershed characteristic of interest. While the length increases as the drainage area increases, the length of a watershed is important in hydrologic computations; for example, it is used in time-of-concentration calculations. The watershed length is also highly correlated with channel length.

Watershed length is usually defined as the distance measured along the main channel from the watershed outlet to the basin divide. Since the channel does not extend to the basin divide, it is necessary to extend a line from the end of the channel to the basin divide following a path where the greatest volume of water would travel. The straight-line distance from the outlet to the farthest point on the watershed divide is not usually used to compute L because the travel distance of flood waters is conceptually the length of interest. Thus, the length is measured along the principal flow path. Since it will be used for hydrologic calculations, this length is more appropriately labeled the hydrologic length.

While the drainage area and length are both measures of watershed size, they may reflect different aspects of size. The drainage area is used to indicate the potential for rainfall to provide a volume of water. The length is usually used in computing a time parameter, which is a measure of the travel time of water through the watershed.

4.4.4 Watershed Slope

Flood magnitudes reflect the momentum of the runoff. Slope is an important factor in the momentum. Both watershed and channel slope may be of interest. Watershed slope reflects the rate of change of elevation with respect to distance along the principal flow path. Typically, the principal flow path is delineated, and the watershed slope (S) is computed as the difference in elevation (ΔE) between the end points of the principal flow path divided by the hydrologic length of the flow path (L):

$$S = \frac{\Delta E}{L} \qquad\qquad (4.2)$$

The elevation difference ΔE may not necessarily be the maximum elevation difference within the watershed since the point of highest elevation may occur along a side boundary of the watershed rather than at the end of the principal flow path.

Where the design work requires the watershed to be subdivided, it will be necessary to compute the slopes of each subarea. It may also be necessary to compute the channel slopes for the individual sections of the steams that flow through the subareas. When computing the slope of a subarea, the principal flow path for that subarea must be delineated. It should reflect flow only for that subarea rather than flow that enters the subarea in a channel. The stream-reach slope may also be necessary for computing reach travel times.

4.4.5 Land Cover and Use

During a brief rain shower, people will often seek cover under a nearby tree; they recognize that the tree provides a temporary shelter since it intercepts rain during the initial part of the storm. It would seem to follow that a forested watershed would have less flood runoff than a watershed devoid of tree cover.

Rooftop runoff provides another example of the effect of land cover on runoff rates and volumes. During a rainstorm it is obvious that flow from the downspouts on houses starts very shortly after the start of rain. Rooftops are impervious, steeply sloped, planar surfaces, so there is little to retard the flow. Flow down a grassy hill of the same size as the rooftop will begin long after similar flow over a rooftop. The grassy hill sheds water at a slower rate and has a smaller volume because some of the water infiltrates into the topsoil and the grass is hydraulically rougher than the shingles on the roof. It would seem to follow from this example that flow from impervious surfaces would have greater volumes and smaller travel times than flow over pervious surfaces that have similar size, shape, and slope characteristics.

These two conceptual examples should illustrate that the land cover significantly affects the runoff characteristics of a watershed. When watershed characteristics other than the land cover are held constant, the runoff characteristics of a watershed, which includes the volume of runoff, the timing of runoff, and maximum flood flow rates, can differ significantly. Therefore, land cover and use serve as inputs for problems in hydrologic analysis and design.

Many descriptors of land cover/use are used in hydrologic design. Most often, a qualitative description of land cover is transformed into a quantitative index of runoff potential. For example, the Rational Method (see Section 4.5) uses a runoff coefficient C to reflect the runoff potential of a watershed. Larger values of C reflect increased runoff potential. The value for commercial property ($C = 0.75$) indicates a greater runoff potential than the value for a residential area ($C = 0.3$), which in turn has a greater runoff potential than a forested area ($C = 0.15$). From this it should be evident that the runoff potential increases as a watershed is transformed from a forested cover to an urban land cover.

The Natural Resources Conservation Service (NRCS, formerly SCS) uses a different land cover/use index in their models than is used in the Rational Method. The runoff curve number (CN) is the NRCS cover type index. It is an integration of the hydrologic effects of land use, soil type, and antecedent moisture. The runoff curve number is discussed in Section 4.4.9.

Urban land covers are especially important in hydrology. Many hydrologic design problems result from urban expansion. The percentage of imperviousness is a commonly used index of the level of urban development. High-density residential areas characteristically have percentages of imperviousness from 40% to 70%. Commercial and industrial areas are characterized by impervious cover often from 70% to 90%. Impervious covers in urban areas are not confined to the watershed surface. Channels are often lined with concrete to increase the flow capacity of the channel cross section and to quickly remove flood waters. Channel lining is often criticized because it can transfer the flooding problem from an upstream reach to a reach downstream.

4.4.6   Surface Roughness

Roughness implies unevenness of texture. Sandpaper is thought of as being rough. Yet degrees of roughness are evidenced by the different grades of sand paper. Surface roughness is also important in hydrologic design. As indicated in the previous section, a grass surface is hydrologically rougher than a shingled roof. The grass retards the flow to a greater extent than does the shingles.

Manning's roughness coefficient (n) is the most frequently used index of surface roughness. While the units of n are rarely shown, its units are dependent on the dimension system used. Values of n for hydrologic overland flow surfaces are given in Table 4.4. Quite frequently, tables of n values give a range of values for each surface. This reflects the uncertainty in the value. Unless a specific reason for selecting a value within the stated range is obvious, the mean value of the endpoints of the range should be used.

4.4.7   Channel Cross Sections

Channel cross sections are a very important part of hydrologic analysis and design for a number of reasons. Many types of design problems require cross-section information, including the cross-sectional area. The wetted perimeter, slope, roughness, and average velocity are other important characteristics.

It is only necessary to walk beside a river or stream to recognize that, even over short distances, stream cross sections can take on a wide variety of shapes and sizes. Small streams in upland areas often have V-shaped cross sections. As the stream drains larger areas, past floods have sculptured out larger, often rectangular or trapezoidal sections. Many rivers have cross sections that include a relatively small V- or U-shaped channel that carries the runoff from small storms and the runoff during periods between storm events. This small channel area is often part of a much larger flat area on one or both sides of the channel; this flat area is called the *floodplain*, which is the area that is covered with water during times of higher discharges. The floodplain generally has low side slopes and could easily be represented as a rectangular cross section. Because of channel instability and erosional processes, cross sections can change shape during flood events, and where such changes take place, they should be accounted for in developing a discharge rating table. Channel cross sections can also change because of development within the floodplain.

Table 4.4. Manning's roughness coefficient (n) for overland flow surface

Surface	Recommend n
Fallow	0.010
Bare land	0.010
Smooth concrete	0.011
Graveled surface	0.012
Asphalt	0.012
Bare clay	0.013
Good wood	0.014
Cast iron	0.018
Smooth earth	0.023
Corrugated metal pipes	0.024
Conventional tillage	
no residue	0.09
with residue	0.19
Grass	
Short	0.15
Dense	0.24
Bermuda grass	0.41
Woods	
No underbrush	0.20
Light underbrush	0.40
Dense underbrush	0.80
Rangeland	0.13

4.4.8   Channel Roughness

The roughness of a surface affects the characteristics of runoff, whether the water is on the surface of the watershed or in the channel. With respect to the hydrologic cycle, the roughness of a surface retards flow over the surface. For overland flow, increased roughness delays the runoff and should increase the potential for infiltration. Reduced velocities associated with increased roughness should also decrease the amount of erosion. The general affects of roughness on flow in a channel are similar to those for overland flow.

Manning's roughness coefficient (n) is required for a number of hydraulic computations. It is a necessary input in floodplain delineation. Also, a number of methods for estimating the timing of runoff use n as an input. It is also used in the design of stable channel systems. Thus it is an important input.

A number of methods exist for estimating the roughness coefficient. First, books that provide a series of pictures of stream channels with a recommended n value for each picture are available; thus for any natural channel a value of n can be obtained by comparing the roughness characteristics of the channel with the pictures and using the n value from the picture that appears most similar (Hicks and Mason, 1991). Second, tables (for example, see Table 4.5) that give typical or average values of n for various channel conditions are available. However, the accuracy of

n values from such tables are highly dependent on the degree of homogeneity of channel conditions and the degree of specificity of the table. The picture comparison and tabular look-up methods are used frequently because of their simplicity and because studies have not shown them to be highly inaccurate.

4.4.9   Runoff Curve Numbers

The NRCS runoff *curve number* (*CN*) was developed as an index that represents the combination of a hydrologic soil group and a land use and treatment class (McCuen 2002). Empirical analyses suggested that the *CN* was a function of three factors: soil group, the cover complex, and antecedent moisture conditions.

Table 4.5. Recommended design values of Manning roughness coefficients, n[a]

	Manning n Range[b]
I. Unlined open channels[c]	
A. Earth, uniform section	
1. Clean, recently completed	0.016–0.018
2. Clean, after weathering	0.018–0.020
3. With short grass, few weeds	0.022–0.027
4. In graveled soil, uniform section, clean	0.022–0.025
B. Earth, fairly uniform section	
1. No vegetation	0.022–0.025
2. Grass, some weeds	0.025–0.030
3. Dense weeds or aquatic plants in deep channels	0.030–0.035
4. Sides, clean gravel bottom	0.025–0.030
5. Sides, clean, cobble bottom	0.030–0.040
C. Dragline excavated or dredged	
1. No vegetation	0.028–0.033
2. Light brush on banks	0.035–0.050
D. Rock	
1. Based on design section	0.035
2. Based on actual mean section	
a. Smooth and uniform	0.035–0.040
b. Jagged and irregular	0.040–0.045
E. Channels not maintained, weeds and brush uncut	
1. Dense weeds, high as flow depth	0.08–0.12
2. Clean bottom, brush on sides	0.05–0.08
3. Clean bottom, brush on sides, highest stage of flow	0.07–0.11
4. Dense brush, high-stage	0.10–0.14
II. Roadside channels and swales with maintained vegetation[d,e]	
(values shown are for velocities of 2 and 6 ft/sec):	
A. Depth of flow up to 0.7 ft	
1. Bermuda grass, Kentucky bluegrass, buffalo grass	
a. Mowed to 2 in.	0.07–0.045
b. Length 4 to 6 in.	0.09–0.05
2. Good stand, any grass	
a. Length about 12 in.	0.18–0.09
b. Length about 24 in.	0.30–0.15
3. Fair stand, any grass	
a. Length about 12 in.	0.14–0.08
b. Length about 24 in.	0.25–0.13

[a]Estimates are from the Bureau of Public Roads for straight alignments

(continued)

Table 4.5. Recommended design values of Manning roughness coefficients, n^a (*Continued*)

	Manning n Range[b]
B. Depth of flow 0.7–1.5 ft	
1. Bermuda grass, Kentucky bluegrass, buffalo grass	
a. Mowed to 2 in.	0.05–0.035
b. Length 4 to 6 in.	0.06–0.04
2. Good stand, any grass	
a. Length about 12 in.	0.12–0.07
b. Length about 24 in.	0.20–0.10
3. Fair stand, any grass	
a. Length about 12 in.	0.10–0.06
b. Length about 24 in.	0.17–0.09
III. Natural stream channels[f]	
A. Minor streams[g] (surface width at flood stage less than 100 ft)	
1. Fairly regular section	
a. Some grass and weeds, little or no brush	0.030–0.035
b. Dense growth of weeds, depth of flow materially greater than weed height	0.035–0.05
c. Some weeds, light brush on banks	0.04–0.05
d. Some weeds, heavy brush on banks	0.05–0.07
e. Some weeds, dense willows on banks	0.06–0.08
f. For trees within channel, with branches submerged at high stage, increase all above values by	0.01–0.10
2. Irregular sections, with pools, slight channel meander; increase value in 1a-e by	0.01–0.02
3. Mountain streams, no vegetation in channel, banks usually steep, trees and brush along banks submerged at high stage	
a. Bottom of gravel, cobbles, and few boulders	0.04–0.05
b. Bottom of cobbles, with large boulders	0.05–0.07
B. Floodplains (adjacent to natural streams)	
1. Pasture, no brush	
a. Short grass	0.030–0.035
b. High grass	0.035–0.05
2. Cultivated areas	
a. No crop	0.03–0.04
b. Mature row crops	0.035–0.045
c. Mature field crops	0.04–0.05
3. Heavy weeds, scattered brush	0.05–0.07
4. Light brush and trees[h]	
a. Winter	0.05–0.06
b. Summer	0.06–0.08

[b]Use larger value for poor construction

4.4.9.1 Soil Group Classification

The soil scientists of the U.S. Natural Resources Conservation Service (NRCS) classified more than 4000 soils on the basis of their runoff potential and grouped them into four hydrologic soil groups that are identified by the letters A, B, C, and D. Soil characteristics that are associated with each group are given in Table 4.6. The NRCS soil group can be identified at a site using one of three ways: (1) soil characteristics, (2) county soil surveys, and (3) minimum infiltration rate.

County soil surveys, which are available from Soil Conservation Districts, give detailed descriptions of the soils at locations within a county; these surveys are usually the best means of identifying the soil group. The published soil survey is the most common method for identifying the soil group. All of the soils in the county are shown on a detailed map. The map consists of many sheets that were made from aerial photographs. On each sheet of the detailed map, areas with the same soil characteristics are outlined and identified by symbols. In addition to the soil maps, a guide is provided. The guide lists all of the soils in the county in alphabetical order by map symbol. The hydrologic soil group is also given in the county soil survey. Additional details about soils in the county are provided in the survey.

Table 4.6. Soil group classifications

Group A	Deep sand; loess; aggregated silts
Group B	Shallow loess; sandy loam
Group C	Clay loams; shallow sandy loam
Group D	Heavy plastic clays; soils that swell when wet

4.4.9.2   Hydrologic Condition

The type of vegetation or ground cover on a watershed, and the quality or density of that cover, have a major impact on the infiltration capacity of a given soil. Further refinement in the cover type is provided by the definition of cover quality as follows:

Poor: Heavily grazed or regularly burned areas. Less than 50% of the ground surface is protected by plant cover or brush and tree canopy.

Fair: Moderate cover with 50 to 75% of the ground surface protected by vegetation.

Good: Heavy or dense cover with more than 75% of the ground surface protected by vegetation.

In most cases, the cover type and quality of a watershed in existing conditions can be readily determined by a field review of a watershed. In ultimate-planned open spaces, the soil cover condition shall be considered as "good."

4.4.9.3   Curve Number Tables

Table 4.7 shows the *CN* values for the different land uses, treatment, and hydrologic condition; separate values are given for each soil group. For example, the *CN* for a wooded area (good cover) and soil group B is 55; for soil group C, the *CN* would increase to 70. If the cover on soil group B is poor, the *CN* will be 66.

4.4.9.4   Estimation of CN Values for Urban Land Uses

The *CN* table (Table 4.7) provides *CN* values for a number of urban land uses. For each of these, the *CN* is based on a specific percent of imperviousness. For example, the *CN* values for commercial land use are based on an imperviousness of 85%. For urban land uses with percentages of imperviousness different than those shown in Table 4.7, curve numbers can be computed using a weighted CN approach, with a CN of 98 used for the impervious areas and the CN for open space (good condition) used for the pervious portion of the area. Thus CN values of 39, 61, 74, and 80 are used for hydrologic soil groups A, B, C, and D, respectively. These are the same CN values for pasture in good condition. The following equation can be used to compute a weighted CN (CN_w):

Table 4.7. Runoff curve numbers (average watershed condition, $I_a = 0.2S$)

Land Use Description			Curve Numbers for Hydrologic Soil Group			
			A	B	C	D
Fully developed urban areas[a] (vegetation established)						
Lawns, open spaces, parks, golf courses, cemeteries, etc.						
Good condition; grass cover on 75% or more of the area			39	61	74	80
Fair condition; grass cover on 50% to 75% of the area			49	69	79	84
Poor condition; grass cover on 50% or less of the area			68	79	86	89
Paved parking lots, roofs, driveways, etc.			98	98	98	98
Streets and roads						
Paved with curbs and storm sewers			98	98	98	98
Gravel			76	85	89	91
Dirt			72	82	87	89
Paved with open ditches			83	89	92	93
		Average % impervious[b]				
Commercial and business areas		85	89	92	94	95
Industrial districts		72	81	88	91	93
Row houses, town houses, and residential with lots sizes 1/8 acre or less		65	77	85	90	92
Residential: average lot size						
1/4 acre		38	61	75	83	87
1/3 acre		30	57	72	81	86
1/2 acre		25	54	70	80	85
1 acre		20	51	68	79	84
2 acre		12	46	65	77	82
Developing urban areas[c] (no vegetation established)						
Newly graded area			77	86	91	94
Western desert urban areas						
Natural desert landscaping (pervious area only)[f]			63	77	85	88
Artificial desert landscaping			96	96	96	96

Land Use Description	Treatment or Practice[d]	Hydrologic Condition	Curve Numbers for Hydrologic Soil Group			
			A	B	C	D
Cultivated agricultural land						
Fallow	Straight row or bare soil		77	86	91	94
	Conservation tillage	Poor	76	85	90	93
	Conservation tillage	Good	74	83	88	90
Row crops	Straight row	Poor	72	81	88	91
	Straight row	Good	67	78	85	89
	Conservation tillage	Poor	71	80	87	90
	Conservation tillage	Good	64	75	82	85
	Contoured	Poor	70	79	84	88
	Contoured	Good	65	75	82	86
	Contoured and	Poor	69	78	83	87
	conservation tillage	Good	64	74	81	85

(continued)

STORMWATER MANAGEMENT FOR SMART GROWTH

Table 4.7. Runoff curve numbers (average watershed condition, $I_a = 0.2S$) (*Continued*)

Land Use Description	Treatment or Practice[d]	Hydrologic Condition	Curve Numbers for Hydrologic Soil Group			
			A	B	C	D
	Contoured and terraces	Poor	66	74	80	82
	Contoured and terraces	Good	62	71	78	81
	Contoured and terraces	Poor	65	73	79	81
	and conservation tillage	Good	61	70	77	80
Small grain	Straight row	Poor	65	76	84	88
	Straight row	Good	63	75	83	87
	Conservation tillage	Poor	64	75	83	86
	Conservation tillage	Good	60	72	80	84
	Contoured	Poor	63	74	82	85
	Contoured	Good	61	73	81	84
	Contoured and	Poor	62	73	81	84
	conservation tillage	Good	60	72	80	83
	Contoured and terraces	Poor	61	72	79	82
	Contoured and terraces	Good	59	70	78	81
	Contoured and terraces	Poor	60	71	78	81
	and conservation tillage	Good	58	69	77	80
Close-seeded	Straight row	Poor	66	77	85	89
legumes	Straight row	Good	58	72	81	85
rotations	Contoured	Poor	64	75	83	85
meadows[e]	Contoured	Good	55	69	78	83
	Contoured and terraces	Poor	63	73	80	83
	Contoured and terraces	Good	51	67	76	80
Noncultivated agricultural land						
Pasture or range	No mechanical treatment	Poor	68	79	86	89
	No mechanical treatment	Fair	49	69	79	84
	No mechanical treatment	Good	39	61	74	80
	Contoured	Poor	47	67	81	88
	Contoured	Fair	25	59	75	83
	Contoured	Good	6	35	70	79
Meadow		—	30	58	71	78
Forestland—grass or		Poor	55	73	82	86
orchards—evergreen		Fair	44	65	76	82
deciduous		Good	32	58	72	79
Brush		Poor	48	67	77	83
		Fair	35	56	70	77
		Good	30	48	65	73
Woods		Poor	45	66	77	83
		Fair	36	60	73	79
		Good	25	55	70	77
Farmsteads		—	59	74	82	86
Forest-range						
Herbaceous		Poor	[f]	80	87	93
		Fair		71	81	89
		Good		62	74	85
Oak-aspen		Poor		66	74	79
		Fair		48	57	63
		Good		30	41	48

87

Table 4.7. Runoff curve Numbers (average watershed condition, $I_a = 0.2S$) (*Continued*)

Land Use Description	Treatment or Practice[d]	Hydrologic Condition	Curve Numbers for Hydrologic Soil Group			
			A	B	C	D
Juniper		Poor	[g]	75	85	89
		Fair		58	73	80
		Good		41	61	71
Sage-grass		Poor		67	80	85
		Fair		51	63	70
		Good		35	47	55

[a] For land uses with impervious areas, curve numbers are computed assuming that 100% of runoff impervious areas is directly connected to the drainage system. Pervious areas (lawn) are considered to be equivalent to lawns in good condition. The impervious areas have a CN of 98.
[b] Includes paved streets.
[c] Use for the design of temporary measures during grading and construction. Impervious area percent for urban areas under development vary considerably. The user will determine the percent impervious. Then using the newly graded area CN, the composite CN van be computed for any degree of development.
[d] For conservation tillage poor hydrologic condition, 5 to 20% of the surface is covered with residue (less than 750-lb/acre row crops or 300-lb/acre small grain). For conservation tillage good hydrologic condition, more than 20% of the surface is covered with residue (greater than 750-lb/acre row crops or 300-lb/acre small grain).
[e] Close-drilled or broadcast.
  For noncultivated agricultural land:
    Poor hydrologic condition has less than 25% ground cover density.
    Fair hydrologic condition has between 25 and 50% ground cover density.
    Good hydrologic condition has more than 50% ground cover density.
  For forest-range:
    Poor hydrologic condition has less than 30% ground cover density.
    Fair hydrologic condition has between 30 and 70% ground cover density.
    Good hydrologic condition has more than 70% ground cover density.
[f] Composite CN's for natural desert landscaping should be computed based on the impervious area percentage ($CN = 98$) and the pervious area CN. The pervious area CN's are assumed equivalent to desert shrub in poor hydrologic condition.
[g] Curve numbers for group A have been developed only for desert shrub

$$CN_w = CN_p(1-f) + f(98) \qquad (4.3)$$

in which f is the fraction (not percentage) of imperviousness and CN_p is the curve number for the pervious portion (39, 61, 74, or 80). To show the use of Equation (4.3), the CN values for commercial land use with 85% imperviousness are as follows:

A soil: $39(0.15) + 98(0.85) = 89$
B soil: $61(0.15) + 98(0.85) = 92$
C soil: $74(0.15) + 98(0.85) = 94$
D soil: $80(0.15) + 98(0.85) = 95$

These are the same values shown in Table 4.7. If only 70% were impervious, the CNs would be 80, 87, 91, and 93 for A, B, C, and D soils, respectively.

4.4.9.5   Effect of Unconnected Impervious Area on Curve Numbers

Many local drainage policies require runoff from certain types of impervious land cover (for example, rooftops, driveways, patios) to be directed to pervious surfaces rather than connected to storm drain systems. Such a policy is based on the belief that disconnecting these impervious areas will require smaller and less costly drainage systems and lead both to increased ground water recharge and to improvements in water quality. In the belief that disconnecting some impervious surfaces will reduce both the peak runoff rates and volumes of direct flood runoff, some developers believe that they should be given credit for the reductions in peak rates and volumes in the design of drainage systems. Thus, a need to account for the effect of disconnecting impervious surfaces on runoff rates and volumes exists.

Estimating CNs for areas with some unconnected imperviousness requires three important variables: the pervious area CN, the percentage of impervious area, and the percentage of the imperviousness that is unconnected. The existing method (Eq. 4.3) for computing composite CN values is based on the pervious area CN and the percentage of imperviousness. A correction factor to the composite CN values is based on the pervious area CN and the percentage of imperviousness. A correction factor to the composite CN is a function of the percentage of unconnected imperviousness. The correction can only be applied when the percentage of impervious is less than 30%. A correction is not made if the total imperviousness is greater than 30%. The adjusted CN (CN_c), can be computed with the following equation:

$$CN_c = CN_p + f(98 - CN_p)(1 - 0.5R) \qquad (4.4)$$

in which f is the fraction of impervious cover, CN_p is the pervious area CN, and R is the ratio of the unconnected impervious area to the total impervious area.

Example 4.3. To illustrate the use of Eq. (4.4), consider the case of a drainage area having 25% imperviousness, a pervious area CN of 61, and 50% of the imperviousness unconnected. Using 25% imperviousness with 50% unconnected, the composite CN would be 68:

$$CN_c = 61 + 0.25(98 - 61)[1 - 0.5(0.5)] = 68 \qquad (4.5)$$

If credit were not given for unconnected imperviousness, the percentage of imperviousness and the pervious area CN could be used with Eq. (4.3) to find the composite CN. For 25% imperviousness and a pervious area CN of 61 the composite CN would equal 70, which means that disconnecting 50% of the impervious cover allows for the CN to be reduced by 2. This lower CN will lead to a reduction in peak discharge and runoff volume.

4.4.10 Time of Concentration

The velocity of water across a lawn, down a street gutter, in a swale, flowing in a stream, or in a storm flow pipe has important implications in both water quantity and water quality. Where runoff velocities are relatively high, the flood peaks are also high. High runoff velocities are responsible for greater erosion rates, and pollutants will travel greater distances when suspended in water flowing at higher velocities. Therefore, the control of runoff velocities is beneficial to both flow rates and pollution control.

The *time of concentration* (t_c) is defined as the time required for a particle of water to flow hydraulically from the most distant point in the watershed to the outlet or design point; methods of estimation based on this definition use watershed characteristics and sometimes a precipitation index such as the 2-yr, 2-hr rainfall intensity.

In practice, terminology is not always applied consistently; therefore, it is worthwhile clarifying some common terms associated with the time of concentration. The term overland flow could be applied to a number of flow regimes. It is, however, worthwhile separating this into sheet flow and concentrated flow. Sheet flow occurs in the upper reaches of a watershed, usually over very short flow paths. Typically, this is evident on steeply sloped paved surfaces, where it appears as shallow layers, often with small waves. A kinematic wave equation is usually used to compute travel times for sheet flow. After some distance, topography usually causes the flow to concentrate in rills, then swales or gutters, and then gullies. Flow on these paths is referred to as concentrated flow.

Manning's equation is used to estimate velocities of overland flow, both sheet flow and concentrated flow. It can be shown that the kinematic-wave equation for estimating sheet-flow travel time is based on Manning's equation with the assumption that the hydraulic radius equals the product of the rainfall intensity and the travel time. The velocity method for concentrated flow uses Manning's equation with an assumed depth and Manning's n. Frequently used curves of velocity versus slope are valid only as long as the assumed depth and n are appropriate.

4.4.10.1 Velocity Method

The velocity method is based on the concept that the travel time (T_t) for a particular flow path is a function of the length of flow (L) and the velocity (V):

$$T_t = \frac{L}{60V} \qquad (4.6)$$

in which T_t, L, and V have units of min, ft, and ft/sec, respectively. The travel time is computed for the principal flow path. Where the principal flow path consists of segments that have different slopes or land covers, the principal flow path should be divided into segments and Equation (4.6) should be used for each flow segment. The time of concentration is then the sum of the travel times:

$$t_c = \sum_{i=1}^{k} T_{ti} = \sum_{i=1}^{k} \left(\frac{L_i}{60V_i} \right) \qquad (4.7)$$

in which k is the number of segments and the subscript i refers to the flow segment.

The velocity of Eq. (4.6) is a function of the type of flow (sheet, concentrated flow, gully flow, channel flow, pipe flow), the roughness of the flow path, and the slope of the flow path. A number of methods have been developed for estimating the velocity. Flow velocities in pipes and open channels can be computed using Manning's equation:

$$V = \frac{1.486}{n} R_h^{2/3} S^{1/2} \qquad (4.8)$$

in which V is the velocity (ft/sec), n is the roughness coefficient, R_h is the hydraulic radius (ft), and S is the slope (ft/ft). Values of n for overland-flow surfaces are given in Table 4.4. Values of n for channels can be obtained from Table 4.5.

Equation (4.8) can be simplified so that V is only a function of the slope by assuming values for n and R_h. This gives a relationship between the velocity and the average slope of the surface:

$$V = kS^{0.5} \qquad (4.9)$$

in which V is the velocity (ft/sec) and S is the slope (ft/ft). Thus, k equals $1.486 R_h^{2/3}/n$. For Eq. (4.9), the value of k is a function of the land cover with the effect measured by the value of n and R_h. Values for selected land covers are given in Table 4.8. These are shown in Figure 4.7. Equation 4.9 and Figure 4.7 are a simplification of Manning's equation (Equation 4.8). Values were assumed for n and R_h and used to compute the values for k for Eq. (4.9). Other curves could be developed for different flow regimes by selecting values for n and R_h. The assumed values of n and R_h are given in Table 4.8.

After short distances, runoff tends to concentrate in rills and then gullies of increasing proportions. Where roughness coefficients exist for such flow, Manning's equation can be used. Alternatively, the average velocity can be obtained using Eq. (4.9) or Figure 4.7, where values are given for grassed waterways, unpaved gullies, and paved areas.

4.4.10.2 Sheet-Flow Travel Time

At the upper reaches of a watershed, runoff does not concentrate into well-defined flow paths, such as gullies or swales. Instead it flows over the surface at reasonably uniform, shallow depths. Hydrologically speaking, this is sheet flow. It is evident on long, sloping streets during rainstorms. After some distance, the flow begins to converge into concentrated flow paths that have depths noticeably greater than that of the sheet flow. The distance from the upper end of the watershed or flow surface to the point where significant concentrated flow begins is termed the sheet-flow length. For steeply sloped, impervious surfaces the sheet-flow length can be several hundred feet. For shallow slopes or for pervious surfaces concentrated flow will begin after relatively short sheet-flow lengths.

In the upper reaches of a watershed, sheet-flow runoff during the intense part of the storm will flow as a shallow layer with a reasonably constant depth. An equation, referred to as the kinematic wave equation for the equilibrium time, can be developed using Manning's equation with the assumption that the hydraulic radius equals the product of the rainfall intensity and the travel time, i.e., $R_h = i\, T_t$.

Using the velocity equation of Eq. (4.6) with the travel time equal to the time of concentration, Manning's equation becomes:

$$V = \frac{L}{T_t/60} = \frac{1.486(60)}{n} R_h^{2/3} S^{1/2} = \frac{1.486(60)}{n} \left(\frac{iT_t}{60(12)} \right)^{2/3} S^{1/2} \quad (4.10)$$

in which i [=] in./hr, T_t [=] min, S [=] ft/ft, and L [=] ft. Solving for the travel time yields:

$$T_t = \frac{0.938}{i^{0.4}} \left(\frac{nL}{\sqrt{S}} \right)^{0.6} \quad (4.11)$$

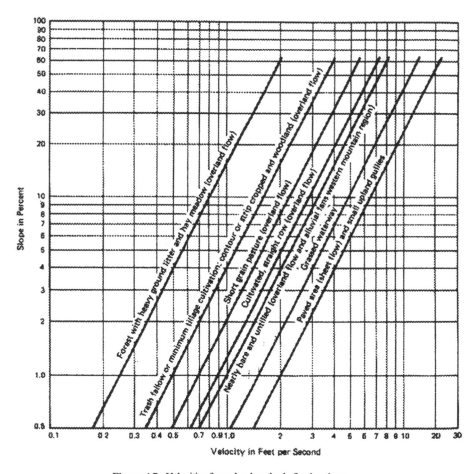

Figure 4.7. Velocities for upland method of estimating t_c.

Equation (4.11) requires the rainfall intensity i for the time of concentration. Since T_t is not initially known, it is necessary to assume a value of T_t to obtain i from a rainfall IDF curve (Ragan and Duru 1972). As an example, Figure 4.2 is the IDF curve for Baltimore, MD. Then T_t is computed. If the initial assumption for T_t was incorrect, then a new estimate of i is obtained from the IDF curve using the computed value of T_t. The iterative process should be repeated until the value of T_t does not change. Generally only two or three trials are necessary for convergence.

To by-pass the necessity to solve Eq. (4.11) iteratively, Welle and Woodward (1986) assumed a power-model relationship between rainfall intensity and rainfall duration. Using a 2-yr return period, they substituted the 2-yr, 24-hr rainfall depth for the rainfall intensity i and derived the following alternative model for Eq. (4.11):

$$ T_t = \frac{0.42}{P_2^{0.5}} \left(\frac{nL}{S^{0.5}} \right)^{0.8} \qquad (4.12) $$

in which L is the flow length (ft), S is the average slope (ft/ft), P_2 is the 2-yr, 24-hr rainfall depth (in.) and T_t [=] min. Eq. (4.12) has the advantage that an iterative solution is not necessary.

The n values for use with Eqs. (4.11) and (4.12) are given in Table 4.8 and are for very shallow flow depths, 0.1 ft or so; these values reflect the effects of raindrop impact; drag over plane surfaces; obstacles such as litter, crop ridges, and rocks; and erosion and transportation of sediment. The rainfall depth P_2 for Eq. (4.12) can be obtained from the local IDF curve, such as Figure 4.2 for Baltimore.

In addition to the previously mentioned assumptions, these kinematic equations make the following assumptions: (1) no local inflow; (2) no backwater effects; (3) no storage effects; (4) the discharge is only a function of depth, for example, $q = ay^b$; and (5) planar, nonconverging flow. These assumptions become less realistic for any of the following conditions: the slope decreases, the surface roughness increases, or the length of the flow path increases. Recognizing that Eqs. (4.11) and (4.12) can yield unusually long times of concentration, limits are often, somewhat arbitrarily, placed on flow lengths. Common length limits are from 100 ft to 300 ft. TR-55 (SCS, 1986) places a limit of 300 ft on Eq. (4.12); however, many feel that this is too long and a shorter length of 100 ft should be used. This limit would not necessarily apply to Eq. (4.11). However, it should be recognized that the five assumptions given above can be violated for the other inputs, n and S. Empirical evidence (McCuen and Speiss, 1995) suggests that Eq. (4.11) (but not Eq. 4.12) should only be used for flow conditions where the term $nL/\sqrt{S}$ equals approximately 100. For a given n and S, the allowable length can be estimated as $100\, S^{0.5} n^{-1}$.

Example 4.4. At four locations, a local public works department measures the velocity of flow in a gutter when the gutter is flowing full. The slope of the gutters cover the range of slopes found in the jurisdiction. The best estimate of k for Eq. (4.9) is found for the measured values of velocity (V, ft/sec) and slope (S, ft/ft):

S	0.009	0.023	0.014	0.006
V	3.4	6.8	3.9	3.2

Table 4.8. Coefficients of velocity versus slope relationship for estimating travel times with the velocity method

Land Use/Flow Regime	n	R_h (ft)	k
Forest			
Dense underbrush	0.8	0.25	0.7
Light underbrush	0.4	0.22	1.4
Heavy ground litter	0.2	0.20	2.5
Grass			
Bermudagrass	0.41	0.15	1.0
Dense	0.24	0.12	1.5
Short	0.15	0.10	2.1
Short grass pasture	0.025	0.04	7.0
Conventional tillage			
With residue	0.19	0.06	1.2
No residue	0.09	0.05	2.2
Agricultural			
Cultivated straight row	0.04	0.12	9.1
Contour or strip cropped	0.05	0.06	4.6
Trash fallow	0.045	0.05	4.5
Rangeland	0.13	0.04	1.3
Alluvial fans	0.017	0.04	10.3
Grassed waterway	0.095	1.0	15.7
Small upland gullies	0.04	0.5	23.5
Paved area (sheet flow)	0.011	0.06	20.8
Paved area (sheet flow)	0.025	0.2	20.4
Paved gutter	0.011	0.2	46.3

Each pair (V, S) can be used to estimate the k value by rearranging Eq. (4.9):

$$\hat{k} = V / S^{0.5} \tag{4.13}$$

Then the average of the four computed values of k could be used as the estimated $\bar{k}$. Another estimate could be made using least squares (see Chapter 5). The least-squares estimator is:

$$k_L = \sum_{i=1}^{4} \left(VS^{0.5}\right) / \sum_{i=1}^{4} S \tag{4.14}$$

The calculations for these estimates are given in Table 4.9.

    The average of the computed $\hat{k}$ values is 38.7. The least squares estimate is:

$$k_L = 2.0633 / 0.052 = 39.7 \tag{4.15}$$

Statistically, the least squares estimate of 39.7 is the better of the two. This is in reasonable agreement with the value for k of 46 given in Table 4.8. The difference is within the bounds of sampling variation.

This example illustrates analysis, or the derivation of a k value from measured data. While the four sets of measurements are hypothetical, the example does point out that k values are subject to variation from site to site. It is important to recognize this when applying Eq. (4.9) in design. Both R_h and n are subject to considerable error variation, which affects the accuracy of the value of k.

Example 4.5. Consider a short-grass surface of 120 ft in length and at a slope of 3%, or 0.03 ft/ft. Assume that the local drainage policy requires use of a 10-yr return period for design. Manning's n for short grass is 0.15 (see Table 4.4). When using Eq. (4.11), an estimate of T_t is used to obtain i from Figure 4.2. In this example, a T_t of 5 min is assumed; thus, the initial estimate of the intensity is 8 in./hr, which gives the following estimate of the sheet-flow travel time:

$$T_t = \frac{0.938}{i^{0.4}} \left[\frac{0.15(120)}{(0.03)^{0.5}} \right]^{0.6} = \frac{15.21}{i^{0.4}} = \frac{15.21}{(8)^{0.4}} = 6.62 \text{ min.} \qquad (4.16)$$

Since this is greater than the initial estimate of 5 min, the revised travel time of 6.6 min is entered into Figure 4.2 to obtain a new estimate of the intensity, which is 7.3 in./hr. This produces a new estimate of T_t:

$$T_t = \frac{15.21}{(7.3)^{0.4}} = 6.87 \text{ min.} \qquad (4.17)$$

Thus, a travel time of 7 minutes is reasonable for this portion of the flowpath.

As an alternative, the NRCS kinematic wave equation, Eq. (4.12), could be used. The 2-yr, 24-hr intensity is 0.13 in./hr, which yields a P of 3.12 in. Thus, Eq. (4.12) yields:

$$T_t = \frac{0.42}{(3.12)^{0.5}} \left[\frac{0.15(120)}{0.03^{0.5}} \right]^{0.8} = 9.8 \text{ min.} \qquad (4.18)$$

Table 4.9. Estimation of the velocity method coefficient

S	V	$\hat{k}$	$VS^{0.5}$
0.006	3.2	41.3	0.2479
0.009	3.4	35.8	0.3226
0.014	3.9	33.0	0.4615
0.023	6.8	44.8	1.0313
0.052		154.9	2.0633
		$\bar{k} = 38.7$	

This is slightly different from the 7-min estimate obtained with Eq. (4.12). The difference between the two estimates is due in part to the return period used and the structural coefficients of the two equations.

The two kinematic wave equations can yield quite different estimates of travel time. No one, to date, has compared the two equations with measured data, so a preference based on empirical evidence cannot be made. One author (RHM) believes that the exponent of 0.8 in Eq. (4.12) is too large and makes computed travel times too sensitive to $nL / \sqrt{S}$. Thus, he prefers the theoretical model of Eq. (4.11) over Eq. (4.12).

4.5   RATIONAL FORMULA

The most widely used uncalibrated equation is the Rational Method, which relates the peak discharge (q_p, ft^3/sec) to the drainage area (A, acres), the rainfall intensity (i, in./hr), and the runoff coefficient (C):

$$q_p = CiA \qquad (4.19)$$

The rainfall intensity is obtained from an IDF curve (see Figure 4.2) using both the return period and a duration equal to the time of concentration as input. The value of the runoff coefficient is a function of the land use, cover condition, soil group, and watershed slope. Table 4.10 is an example of a table of C values. A primary use of the Rational Method has been for design problems for small urban areas such as the sizing of inlets and culverts, which are characterized by small drainage areas and short times of concentration. For such designs, short duration storms are critical, which is why the time of concentration is used as the input duration for obtaining i from the IDF curve.

Example 4.6. Consider the design problem where a peak discharge is required to size a storm-drain inlet for a 2.1-acre parking area in Baltimore, with a time of concentration of 0.1 hr and a slope of 1.8%. For a 25-yr design return period, the rainfall intensity (see Figure 4.2) is 8.6 in./hr, and the runoff coefficient (see Table 4.10) is 0.85. Therefore, the design discharge is:

$$q_p = 0.85(8.6)(2.1) = 15.4 \, \text{ft}^3/\text{sec} \qquad (4.20)$$

Some drainage policies provide for a minimum time of concentration, with 15 to 20 min often being specified. If the preceding design were for a project where the minimum t_c was 15 min, the design intensity would be 6.5 in./hr, and the peak discharge would be 11.6 ft^3/sec.

The use of Eq. (4.19) assumes that the watershed is homogeneous in these characteristics so that the runoff coefficient used provides unbiased estimates. Where a drainage area is characterized by distinct subareas that have different runoff potentials, the watershed should be subdivided, and the equation should be applied separately to each area. Where a watershed is not homogeneous, but is characterized by highly dispersed areas that can be characterized by different runoff coefficients, a weighted runoff coefficient should be determined. The weighting is based on the area of each land use and is found by the equation:

$$C_w = \frac{\sum\limits_{j=1}^{n} C_j A_j}{\sum\limits_{j=i}^{n} A_j} \qquad (4.21)$$

in which A_j is the area for landcover j, C_j is the runoff coefficient for area j, n is the number of distinct landcovers within the watershed, and C_w is the weighted runoff coefficient. The weighted coefficient can be used with Eq. (4.19).

The discussion to this point concerning the Rational Method has used the method only to compute the peak discharge for a contributing area. The method can also be used for nonhomogeneous watersheds in which the watershed is divided into homogeneous subareas and where multiple inlets and pipe systems are involved. Where a watershed has distinct areas of nonhomogeneity, every attempt should be made to subdivide the watershed into homogeneous subareas and then use the Rational Method for each subarea or group of subareas.

The method described here is an attempt to provide an equal level of protection to each structural element of the total drainage area. It is based on the following two rules for using Equation (4.19):

1. For each inlet area at the headwater of a drainage area, the Rational Method (Eq. 4.19) is used to compute the peak discharge.
2. For locations were drainage is arriving from two or more inlet areas, the longest time of concentration is used to find the design intensity, a weighted runoff coefficient is computed, and the total drainage area to that point is used with Eq. (4.19).

Table 4.10   Coefficients (C) for the Rational Formula

Description of area	C
Business	
Downtown	0.85
Neighborhood	0.60
Residential	
Single-family	0.40
Multiunits, detached	0.50
Multiunits, attached	0.70
Industrial	
Light	0.65
Heavy	0.75
Parks, Golf Courses	0.20
Playgrounds	0.30
Parking lots	0.85

4.6   TR-55 GRAPHICAL PEAK-DISCHARGE METHOD

For many peak-discharge-estimation methods, the input includes variables to reflect the size of the contributing area, the amount of rainfall, the potential watershed storage, and the time-area distribution of the watershed. These variables are often translated into input variables such as the drainage area, the depth of rainfall, an index reflecting land use and soil type, and the time of concentration. The TR-55 Graphical Method is typical of many peak-discharge methods that are based on inputs such as that described. The Rational Method was based on these same four inputs.

The Graphical method uses the drainage area A_m in sq. mi., not in acres like the Rational method. The Graphical method uses the 24-hour rainfall depth in inches for the design return period, not the rainfall intensity like the Rational method. The peak discharge q_p is computed by:

$$q_p = q_u A_m Q \qquad (4.22)$$

in which q_u is the unit peak discharge (ft^3/s/mi^2/in.), and Q is the runoff depth (inches) for the design return period. For a SCS type II storm the unit peak discharge can be computed by:

$$\log(q_u) = C_0 + C_1 \log(t_c) + C_2 [\log(t_c)]^2 \ \ for\ 0.3 \le t_c \le 10\ hr \qquad (4.23a)$$

and

$$\log(q_u) = \left[C_0 + C_1 \log(0.3) + C_2 (\log(0.3))^2 \right] e^{C_3 [\log(t_c) - \log(0.3)]}$$

$$\times \left(\frac{\log(t_c)}{\log(0.3)} \right)^{0.2} \ \ for\ 0.02 \le t_c < 0.3\ hr \qquad (4.23b)$$

in which the coefficients C_0, C_1, C_2, and C_3 are given in Table 4.11 for the Type II storm. The runoff depth Q (in.) is computed using the runoff curve number CN and the 24-hr rainfall depth P (in.) using the following:

$$Q = \begin{cases} \dfrac{(P - 0.2\ S)^2}{P + 0.8\ S} & \text{for } P \ge 0.2\ S \\[2mm] 0 & \text{for } P < 0.2\ S \end{cases} \qquad (4.24)$$

where

$$S = \frac{1000}{CN} - 10$$

To obtain the values of C_0, C_1, C_2, and C_3 it is necessary to compute the I_a/P ratio where:

$$I_a = 0.2 \, S \tag{4.25}$$

Table 4.11 Coefficients C_0, C_1, C_2, and C_3 for the unit peak discharge of the SCS Graphical Method for the Type II storm distribution as a function of the initial abstraction-to-rainfall depth (I_a/P) ratio

I_a/P	C_0	C_1	C_2	C_3
0.10	2.55323	-0.61512	-0.16403	0.090
0.30	2.46532	-0.62257	-0.11657	0.081
0.35	2.41896	-0.61594	-0.08820	0.078
0.40	2.36409	-0.59857	-0.05621	0.075
0.45	2.29238	-0.57005	-0.02281	0.072
0.50	2.20282	-0.51599	-0.01259	0.070

4.7   PROBLEMS

4.1   Discuss the elements of the hydrologic cycle that are important in the hydrologic processes relevant to runoff from a 1-acre lot in a residential area.

4.2   Discuss the elements of the hydrologic cycle that are important in the hydrologic processes relevant to runoff from a 120-acre commercial development.

4.3   Rainfall is the water input to watersheds. Discuss the water outputs from a 1-ac watershed that is (a) forested; (b) residential; (c) an impervious parking lot.

4.4   What hydrologic role does watershed vegetation play in the infiltration process?

4.5   Stream and river channel are sources of concentrated flows. Discuss the water inputs to and outputs from a 500-ft section of a channel.

4.6   From the perspective of the continuity of mass, discuss the hydrologic processes of a watershed undergoing urbanization.

4.7   From the standpoint of the hydrologic cycle discuss the effect of a dam built into a river.

4.8   What is the exceedence probability for: (a) a 15-yr storm? (b) a 160-yr storm? (c) a 6-month storm?

4.9   In terms of a major hurricane, discuss the concepts of rainfall duration, depth, intensity, and frequency.

4.10 Why is the term *exceedence probability* preferred by hydrologists even though the public continues to use the term *return period*?

4.11 What is the return period of a storm with an exceedence probability of: (a) 0.14? (b) 0.003? (c) 0.992?

4.12 A storm with a constant intensity rainfall of 2.73 in./hr occurs for 25 minutes over 56-acre watershed. Compute the average depth (in.) and the volume (ac-ft) of rainfall.

4.13 Using Figure 4.1 estimate the 100-yr, 24-hr rainfall intensity for (a) Miami, (b) NYC, (c) Minneapolis, (d) St. Louis.

4.14 Figure 4.2 shows a typical intensity-duration-frequency curve. If it were transformed to a volume-duration-frequency curve (i.e., depth (in.) vs. duration (hr)) for each return period, how would the graph appear?

4.15 From the intensity-duration-frequency curve of Figure 4.2 determine the missing values of intensity, depth, duration, or frequency for the following table:

Intensity (in./hr)	Depth (in.)	Duration (hr)	Frequency (yr)
–	–	0.75	5
–	–	6	25
8.0	–	0.25	–
–	2.5	0.5	–
0.3	–	24	–
4.0	–	–	50

4.16 From the intensity-duration-frequency curve of Figure 4.2 determine the missing values of intensity, depth, duration, or frequency for the following table:

Intensity (in./hr)	Depth (in.)	Duration (hr)	Frequency (yr)
2.0	4.0	–	–
–	4.8	24	–
–	–	6	75
–	0.58	0.083	–
0.3	–	12	–

4.17 Determine the average depth of rainfall for a 25-yr, 6-hr storm over a 120-mi^2 watershed in Baltimore (Figure 4.2). Also, compute the total rainfall volume in acre-ft.

4.18 Using the procedure described in Section 4.3.4 develop a 10-yr, 1-hr design storm for Baltimore using increments of 10 min. Also place the design storm in dimensionless form.

4.19 Using the procedure described in Section 4.3.4 develop a 100-yr, 24-hr design storm for Baltimore using a time increment of 3 hr. Also, place the design storm in dimensionless form.

4.20 Using the 25-yr, 24-hr rainfall volume for Baltimore (Figure 4.2), convert the dimensionless NRCS type II storm (Table 4.1) to a dimensioned design storm. Use a 3-hr time increment.

4.21 Using the resulting design storm of Example 4.1 (Table 4.3), compute the maximum 2-hr intensity from the 6-hr storm and compare it to the value from Figure 4.2. Discuss the result.

4.22 Using the 10-yr, 24-hr rainfall volume for Baltimore (Figure 4.2), convert the dimensionless NRCS type II storm (Table 4.1) to a dimensioned design storm. Use a 2-hr time increment.

4.23 Using the 50-yr, 24-hr rainfall volume for Baltimore (Figure 4.2), convert the dimensionless NRCS type II storm (Table 4.1) to a dimensioned design storm. Use a 4-hr time increment.

4.24 Points on an overland flowpath have the following elevations and are separated by the distances shown:

Point	1	2	3	4	5	6
Elevation (ft)	247	231	210	202	188	181
Distance (ft)		365	640	725	560	875

Compute the slope of each section of the flowpath and report the values in ft/ft, %, ft per 100 ft, and ft/mi.

4.25 Two points on a flowpath have elevations of 432 and 397 ft. The points are separated by 2.4 miles along a flowpath. Compute the slope in ft/ft, %, ft per 100 ft, and ft/mile.

4.26 Two adjacent subwatersheds have characteristics as follows: Subarea 1: length = 8600 ft; elevation drop = 170 ft. Subarea 2: length = 3300 ft; elevation drop = 250 ft. Compute the slope of each subarea. Can you legitimately conclude that the average watershed slope for the entire area is the average of the two subarea slopes? Explain.

4.27 Compute the channel slope between sections 1 and 6 and for each of the five reaches. Compute the average of the computed slopes for the five reaches and compare it to the estimate for the entire reach. Discuss the results.

Section	Elevation (ft)	Distance from outlet (ft)
1	134	0
2	141	4300
3	146	5600
4	180	7800
5	235	10,400
6	242	11,300

4.28 If errors in measures of elevation are ±0.5 ft and measurements of channel length have an accuracy of ±6% of the measured length, what is the expected accuracy of computed slopes?

4.29 Does the length of grass affect the value of Manning's roughness? Discuss.

4.30 A large roof is covered by slate. Propose and justify a value for Manning's roughness coefficient.

4.31 Estimate the fraction increase in channel roughness due to vegetation in the channel.

4.32 Assume that channel roughness (n) and the velocity of flow (V, fps) are related by $V = 0.22/n$. Assume n has an accuracy of ±20%. What is the affect of the inaccuracy of n on V for (a) an unlined open channel (earth, uniform section); (b) a Bermuda grass covered grass lined roadside channel with a flow depth of 6 inches; (c) a minor stream with a gravel bottom with no vegetation.

4.33 From the standpoint of the hydrologic processes, why does a poor hydrologic condition produce more surface runoff than good hydrologic condition?

4.34 A watershed varies in land use and soil, as follows:

Sector	Soil group	Land use	Treatment	Condition	Area (ac)
1	C	Row crop	Straight row	Poor	24
2	C	Row crop	Contoured	Good	35
3	B	Small grain	Straight row	Good	16
4	A	Meadow	–	–	29

Compute the average watershed curve number.

4.35 The following summarizes the soil and land use for a 58-acre watershed:

Sector	Area (ac)	Soil group	Land use
1	12	A	Residential (2-ac lots)
2	21	B	Commercial
3	16	B	Parks, golf courses (fair)
4	9	C	Residential (1/2-ac lots)

Compute the average curve number (CN).

4.36 A residential subwatershed is 50% impervious on an A soil. Compute the curve number for the watershed.

4.37 A residential subwatershed is 35% impervious on a B soil. Compute the curve number.

4.38 An urban watershed has areas with 62% imperviousness on an A soil and 37% imperviousness on B soil. The area with A soil is 43% of the total watershed area. Compute the watershed curve number.

4.39 A suburban watershed has the following characteristics:

Part	Area (ac)	Soil	Imperviousness
1	23	B	12%
2	14	B	41%
3	32	C	28%

Compute the watershed CN.

4.40 A watershed is 28% impervious with 80% of the imperviousness unconnected. Compute the weighted curve number for an A soil.

4.41 A 15-ac watershed is 23% impervious on a B soil. Compute the weighted curve number for 0%, 25%, 50%, 75%, and 100% unconnected.

4.42 From the standpoint of hydrologic processes discuss why unconnecting impervious areas from the direct runoff drainage system reduces runoff volumes.

4.43 In a housing development, the contractor directed all rooftop runoff onto grassed surfaces. After the hours are occupied, several homeowners connect the rooftop runoff with a buried pipe to the asphalt-lined drainage depth that parallels the roadway. This reduces wet lawns following storms. What is the hydrologic effect of this on the drainage system?

4.44 For the case of Example 4.3, what is the sensitivity of the weighted CN to the percentage of imperviousness unconnected?

4.45 From the perspective of hydrologic processes explain how the slope, the watershed size, the roughness of the flowpath, and the intensity of the rainfall affect the time of concentration.

4.46 Discuss the relative importance of flowpath roughness and slope on (a) an upland grass-covered swale and (b) a small natural stream.

4.47 Compute the hydraulic radius of a channel that has a trapezoidal cross-section (width 24 ft, side slopes of 6:1 (h:v), with a 1-ft diameter semicircular low-flow channel at the center-bottom of the trapezoidal section. Assume a flow depth of 3.5 ft.

4.48 A small agricultural watershed (straight row crops) has an elevation drop of 21 ft and a principal flow path of 780 ft. Compute the time of concentration for the watershed using the velocity method.

4.49 A paved highway section has a slope of 4% and a length of 1800 ft. Determine the time of concentration using the velocity method.

4.50 A hydrologist expects to make time-of-concentration estimates for a land use not covered by the coefficients of Table 4.9. The hydrologist makes the following measurements:

Site	Elevation Drop (ft)	Length (ft)	Travel Time (min)
1	6	180	3.9
2	2	230	7.5
3	1	90	3.8
4	22	310	4.7
5	9	150	2.7

Estimate the value of k for Eq. (4.9).

4.51 A hydrologist expects to make time-of-concentration estimates for a land use not covered by the coefficients of Table 4.9. The hydrologist makes the following measurements:

Site	Elevation Drop (ft)	Length (ft)	Travel Time (min)
1	4	460	6.2
2	4	390	5.3
3	10	520	4.7
4	12	470	4.1
5	12	360	3.0
6	20	480	2.4
7	39	550	2.8

Estimate the value of k for Eq. (4.9).

4.52 A flow path consists of the following sections:

Section	Slope (%)	Length (ft)	Land Use
1	3.5	260	Forest (light underbrush)
2	4.1	490	Short grass
3	3.4	370	Upland gulley
4	3.1	420	Grassed waterway

Estimate the time of concentration using the velocity method.

4.53 A flow path consists of the following sections:

Section	Slope	Length	Land Use
1	2.8	190	Forest (dense underbrush)
2	2.5	310	Paved gutter
3	1.9	440	Grassed waterway

Estimate the time of concentration using the velocity method.

4.54 The principal flow path for a 95-acre watershed is a 3100-ft stream with a base width of 7 ft, side slopes of 2.5:1 (h:v), and a depth of about 1 ft at bankfull flow. The stream is a regular section with heavy brush on the banks. The elevation drop over the length of the stream is 56 ft. Estimate the time of concentration.

4.55 The principal flow path for a 210-acre watershed is a 5300-ft stream with a base width of 10 ft, side stops of 3:1 (h:v), and a depth of 1.4 ft at bankfull flow. The stream is an unlined open channel, earth, uniform section, with short grass. The elevations of the upper and lower sections of the stream are 258 ft and 146 ft, respectively. Estimate the time of concentration.

4.8 REFERENCES

Anderson, D.C. (1970) *Effects of Urban Development on Floods in Northern VA.* U.S.G.S. Water Supply Paper 2001-C.

Hicks, D.M. and Mason, P.D. (1991) *Roughness Characteristics of New Zealand Rivers*, Water Resources Survey, DSIR Marine and Freshwater, Wellington, New Zealand.

McCuen, R.H. (2002) "Approach to Confidence Interval Estimation for Curve Numbers," *J. Hydrologic Engineering, ASCE*, 7(1): 43-48.

McCuen, R.H. and Spiess, J.M. (1995) "Assessment of Kinematic Water Time of Concentration," *J. Hydraulic Engineering, ASCE*, **121**(3): 256-266.

National Weather Service (NWS) (1961) *Rainfall Frequency Atlas of the United States*, Tech. Paper #40. U.S. Dept. Commerce, Washington, DC.

Ragan, R.M. and Duru, J.O. (1972) "Kinematic Wave Nomograph for Time of Concentration," *J. Hydraulic Div., ASCE*, **98**(HY10): 1765-1771.

Sauer, V.B. and others (1981) *Magnitude and Frequency of Urban Floods in the United States.* U.S.G.S. Reston, VA.

Soil Conservation Service (SCS) (1984) *Computer Program for Project Formulation*, Tech Release No. 20. Washington, DC.

Soil Conservation Service (SCS) (1986) *Urban Hydrology for Small Watersheds*, Tech Release No. 55. Washington, DC.

Welle, P.I. and Woodard, D. (1986) "Time of Concentration," Tech. Note: Hydrol. No. N4, Soil Conservation Service, NENTC, Chester, PA

5

INTRODUCTION TO MODELING

NOTATION

b	=	regression coefficient
b_0	=	regression coefficient (intercept)
b_1	=	regression coefficient (slope)
d	=	particle diameter
e_i	=	residual or error of observation i
F	=	least squares objective function
K	=	frequency factor
n	=	sample size
P_i	=	exceedence probability
Q	=	discharge rate
R	=	correlation coefficient
R	=	removal percentage
r_{ij}	=	correlation between variables i and j
s	=	days since last street sweeping
S	=	sample standard deviation
S_e	=	standard error of estimate
S_x	=	sample standard deviation
T	=	return period
V	=	velocity of street sweeper
$\overline{X}$	=	sample mean
$\overline{Y}$	=	sample mean
$\hat{Y}$	=	predicted value of a random variable
z	=	standard normal deviate
ν	=	degrees of freedom

5.1   INTRODUCTION

Models are an accepted tool for making many environmental decisions. They allow estimates to be made at locations where measured data are not available. Models can be used to make projections into the future for conditions that are expected to exist at that time. For example, land use policies may depend on projected water quality loads at ultimate watershed development. Since the local watersheds are not at ultimate levels of development, estimates are needed for the policy makers to set bounds on development. Thus, models that can provide reasonable estimates of the effect of land development practices on the water quality parameters are needed as input to the decisions.

Often times, calibrated models are available that are useful in solving standard problems. In other cases, models are not available, and one must be developed before the parties involved in the decision will accept the policy. Therefore, knowledge of modeling methods are essential if the model is to be accepted by the involved parties. Acceptance may depend on the expected accuracy of projections and the ability of the model to provide extrapolations for expected future conditions.

Modeling is a complex topic about which entire textbooks are devoted. However, an understanding of the basic principles of model development is important for those who use models in decision making. This chapter is devoted to those principles.

5.2   UNIVARIATE FREQUENCY ANALYSIS

The concept of frequency applies to runoff characteristics as well as to rainfall characteristics. The peak of the discharge hydrograph is an important design variable, so the frequency of a peak discharge plays a central role in hydrology. Engineers commonly estimate the 100-yr peak discharge in their design work. Many water quality projects use the more frequent streamflow magnitude as the standard for design. The frequency concept for runoff can be discussed in terms of either the return period or the exceedence probability.

Since discharge rates are central to many design problems in hydrology, the process of analysis as it applies to discharge warrants special treatment. Frequency analysis is a statistical tool that can be applied to any random variable, not just runoff. We could, for example, make a frequency analysis of rainfall characteristics, water quality or toxic waste parameters, or the magnitude of earthquakes.

Design problems such as the delineation of flood profiles require estimates of discharge rates. A number of methods of estimating peak discharge rates are available. The methods can be divided into two basic groups: those intended for use at sites where gaged streamflow records are available and those intended for use at sites where such records are not available; those two groups will referred to as methods for gaged and ungaged sites, respectively.

5.2.1 Population versus Sample

In frequency modeling, it is important to distinguish between the population and the sample. Frequency modeling is a statistical method that deals with a single random variable and is, thus, classified as a univariate method. The goal of univariate prediction is to make estimates of either probabilities or magnitudes of random variables. The objective of univariate data analysis is to use sample information to determine the appropriate population density function. The input requirements for frequency modeling include a data series and a probability distribution that is assumed to describe the occurrence of the random variable. The data series could include the largest instantaneous peak discharge to occur each year of the record or values obtained from systematic water quality samples. The probability distribution could be either the normal or lognormal distribution. The population consists of a mathematical model that is a function of one or more parameters. For example, the normal distribution is a function of two parameters: the mean μ and standard deviation σ. In addition to identifying the correct function, it is necessary to quantify the parameters.

A frequently used procedure called the method of moments (Ang and Tang, 1975) equates characteristics of the sample (for example, sample moments) to characteristics of the population (for example, population parameters). It is important to note that estimates of probabilities and magnitudes are made using the assumed population and not the sample of data; the sample is used only in identifying the population.

5.2.2 Regionalization

It is important to point out that frequency analysis may actually be part of a more elaborate problem of analysis. Specifically, separate frequency analyses can be performed at a large number of sites within a region and the value of the random variable X for a selected exceedence probability determined for each site; these values can then be used to develop a regression model using the random variable Y as the criterion or dependent variable. As an example, regression equations that relate a peak discharge of a selected exceedence probability to watershed characteristics are widely used in hydrologic design (Jennings et al., 1994). These equations are derived by (1) making a frequency analysis of annual maximum discharges at a number (n) of stream gage stations in a region, (2) selecting the value of the peak discharge from each of the n frequency curves for a selected exceedence probability, say the 100-yr flood, and (3) developing the regression equation relating the n values of peak discharge to watershed characteristics for the same n watersheds. This process is referred to *regionalization*.

5.2.3 Probability Paper

Since frequency analyses are often presented graphically, a special type of graph paper, which is called probability paper, is required. The paper has two axes, with the ordinate being used to plot the value of the random variable, and the probability of its occurrence is given on the abscissa. The probability scale will vary depending on the probability distribution being used. The example of Figure

5.1 uses normal probability paper. The probability scale represents the cumulative normal distribution. The scale at the top of the graph is the exceedence probability; the probability that the random variable will be equaled or exceeded in one time period. It ranges from 99.99% to 0.01%. The ordinate of probability paper is used for the random variable, such as the peak discharge. For example, for the frequency curve of Figure 5.1, the probability of Y being greater than 7 in one time period is 0.023; therefore, the probability of Y not being greater than 7 in one time period is 0.977. If the sample of data is from the distribution function that was used to scale the probability paper, the data will follow the pattern of the population line when properly plotted on the paper. If the data do not follow the population line, then either (1) the sample is from a different population or (2) sampling variation has produced a nonrepresentative sample. In most cases, the former reason is assumed to be the cause, especially when the sample size is reasonably large.

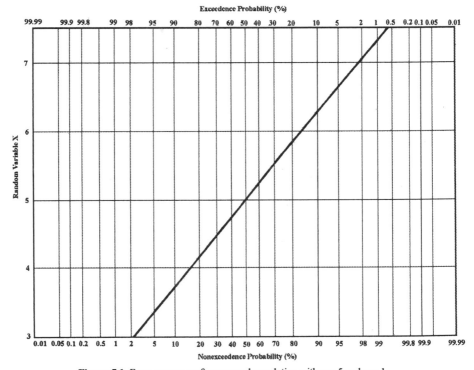

Figure 5.1 Frequency curve for a normal population with $\mu = 5$ and $\sigma = 1$

5.2.4   Mathematical Model

As an alternative to a graphical solution using probability paper, a frequency analysis may be conducted using a mathematical model. Actually, the mathematical model is preferred to the graphical model when making estimates of either a probability or a magnitude. A model that is commonly used in hydrology for normal, log-normal, and log-Pearson Type III analyses has the form:

$$Y = \overline{Y} + KS \qquad (5.1)$$

in which Y is the value of the random variable having mean $\overline{Y}$ and standard deviation S, and K is a frequency factor. Depending on the underlying population, the specific value of K reflects the probability of occurrence of the value Y. Equation (5.1) can be rearranged to solve for K when Y, $\overline{Y}$, and S are known and we wish to estimate the probability of X occurring:

$$K = \frac{Y - \overline{Y}}{S} \qquad (5.2)$$

In summary, Eq. (5.1) is used for the synthesis case when the probability is known and an estimation of the magnitude is needed, while Eq. (5.2) is used for the synthesis case where the magnitude is known and the probability is needed.

5.2.5 Procedure

In a broad sense, frequency analysis can be divided into two phases: deriving the population curve and plotting the data to evaluate the goodness of fit. The following procedure is often used to derive the frequency curve to represent the population:
1. Hypothesize the underlying density function.
2. Obtain a sample and compute the sample moments.
3. Equate the sample moments and the parameters of the proposed density function.
4. Construct a frequency curve that represents the underlying population.
This procedure is referred to as method-of-moments estimation because the sample moments are used to provide numerical values for the parameters of the assumed population. The computed frequency curve, which represents the population, can then be used for estimation of magnitudes for a given return period or probabilities for specified values of the random variable.

It is important to recognize that it is not necessary to plot the data points in order to make probability statements about the random variable. While the four steps listed above lead to an estimate of the population frequency curve, the data should be plotted to ensure that the population curve is a good representation of the data. The plotting of the data is a somewhat separate part of a frequency analysis; its purpose is to assess the quality of the fit rather than act as a part of the estimation process.

5.2.6 Sample Moments

The sample mean $(\overline{Y})$ and the standard deviation (S) can be computed using Eqs. (3.2) and (3.5), respectively. The equations can also be used when the data are transformed by taking the logarithms; in this case, the log transformation should be made prior to computing the moments.

5.2.7   Plotting Position Formulas

It is important to note from the discussed steps that it is not necessary to plot the data before probability statements can be made using the frequency curve; however, the data should be plotted to determine how well the measured data agree with the fitted curve of the assumed population. A rank-order method is used to plot the data. This involves ordering the data from the largest event to the smallest event, assigning a rank of 1 to the largest event and a rank of n to the smallest event, and using the rank (i) of the event to obtain a probability plotting position. Numerous formulas have been proposed; however, the Weibull model is widely used:

$$P_i = \frac{i}{n+1} \tag{5.3}$$

in which i is the rank of the event, n is the sample size, and P_i values give the exceedence probabilities for an event with rank i. The data are plotted by placing a point for each value of the random variable at the intersection of the value of the random variable and the value of the exceedence probability at the top of the graph. The plotted data should fall near the population line if the assumed population model is a reasonable assumption.

5.2.8   Return Period

The concept of return period is used to describe the likelihood of flood magnitudes. The return period is the reciprocal of the exceedence probability:

$$T = \frac{1.0}{P} \tag{5.4}$$

Just as a 25-yr rainfall has a probability of 0.04 of occurring in anyone year, a 25-yr flood has a probability of 0.04 of occurring in anyone year. It is incorrect to believe that once a 25-yr event occurs that it will not occur again for another 25 years. Two 25-yr events can occur in consecutive years. Then again, a period of 100 years may pass before a second 25-yr event occurs following the passage of one 25-yr event.

5.2.9   The Normal Distribution

The normal distribution is commonly used for frequency analyses. Following the general procedure outlined above, the specific steps used to develop a curve for a normal population are as follows:
1.   Assume that the random variable has a normal distribution with population parameters μ and σ.
2.   Compute the sample moments $\overline{Y}$ and S.

3. For the normal distribution, the parameters and sample moments are related by $\mu = \overline{Y}$ and $\sigma = S$.
4. A curve is fitted as a straight line with $(\overline{Y} - S)$ plotted at an exceedence probability of 0.8413 and $(\overline{Y} + S)$ at an exceedence probability of 0.1587.

The frequency curve of Figure 5.1 is an example for a normal distribution with a mean of 5 and a standard deviation of 1. It is important to note that the curve passes through the two points: $(\overline{Y} - S, 0.841)$ and $(\overline{Y} + S, 0.1587)$.

5.2.10   The Log-Normal Distribution

When a poor fit to observed data is obtained, a different distribution function should be considered. For example, when the data demonstrate a concave-upward curve, it is reasonable to try a log-normal distribution or an extreme-value distribution.

The same procedure that is used for fitting the normal distribution can be used to fit the log-normal distribution. The underlying population is assumed to be log normal. The data must first be transformed to logarithms, $X = \log Y$. This transformation creates a new random variable X. The mean and standard deviation of the logarithms are computed and used as the parameters of the population; it is important to recognize that the logarithm of the mean does not equal the mean of the logarithms, which is also true for the standard deviation. Thus one should not use the logarithms of the mean and standard deviation as parameters; the mean and standard deviation of the logarithms should be computed and used as the parameters.

When using a frequency curve for a log-normal distribution, the value of the random variable Y and the moments of the logarithms ($\overline{X}$ and S_y) are related by the equation

$$X = \overline{X} + zS_x \qquad\qquad (5.5)$$

in which z is the value of the standardized normal variate; values of z and the corresponding probabilities can be taken from Appendix A.1. Equation (5.5) can be used to estimate either flood magnitudes for a given exceedence probability or an exceedence probability for a specific discharge. To find a discharge for a specific exceedence probability the standard normal deviate z is obtained from Appendix A.1 and used in Eq. (5.5) to compute the discharge. To find the exceedence probability for a given discharge X, Eq. (5.5) is rearranged by solving for z. With values of X, $\overline{X}$, and S_x, a value of z is computed and used with Appendix A.1 to compute the exceedence probability. Of course, the same values of both X and the probability can be obtained directly from the frequency curve.

5.3   BIVARIATE MODELING

To this point, the discussion has been concerned with a single random variable. In such analyses, it is important to identify the underlying distribution of the random variable and make estimates of its parameters. In many cases, variation in the value of a random variable is associated with variation in one or more additional variables. By establishing the relationship between two or more variables, it may be possible to reduce the uncertainty in an estimate of the random variable of interest.

5.3.1   Correlation Analysis

A correlation analysis provides a means of drawing inferences about the strength of a relationship between two or more variables. That is, it is a measure of the degree to which the values of these variables vary in a systematic manner. Thus it provides a quantitative index of the degree to which one or more variables can be used to predict the value of another variable.

It is just as important to understand what correlation analysis cannot do as it is to know what it can do. Correlation analysis does not provide an equation for predicting the value of a variable; fitting an equation is the task of regression analysis. Also, correlation analysis does not indicate whether or not a relationship is causal; that is, it is necessary for the investigator to determine whether or not there is a cause-and-effect relationship between the variables. Correlation analysis only indicates whether or not the degree of common variation is significant (McNemar, 1969).

5.3.1.1   Graphical Analysis

The first step in examining the relationship between two variables is to perform a graphical analysis. Visual inspection of the data can provide the following information:
1. Identify the degree of common variation, which is an indication of the degree to which the two variables are related.
2. Identify the range and distribution of the sample data points.
3. Identify the presence of extreme events.
4. Identify the form of the relationship between the two variables.
5. Identify the type of the relationship.
Each of these factors is of importance in the statistical analysis of sample data and decision making.

When variables show a high degree of association, one assumes that a causal relationship exists. If a causal relationship is expected, the association demonstrated by the sample data provides empirical support for the assumed relationship. Common variation implies that when the value of one of the random variables is changed, the value of the other variable will change in a systematic manner. For example, an increase in the peak discharge occurs when the amount of impervious cover is increased. If the change in the one variable is highly predictable from a given change in the other variable, a high degree of common variation exists.

It is important to use a graphical analysis to identify the range and distribution of the sample data points so that the stability of the relationship can be assessed and so that one can assess the ability of the data sample to represent the distribution of the population. If the range of the data is limited, a computed relationship may not be stable; that is, it may not apply to the distribution of the population. If an attempt is made to use the sample to project the relationship between the two random variables, a small change in the slope of the relationship will cause a large change in the predicted estimate of Y for values of X at the extremes of the range of the population.

Relationships can be linear or nonlinear. Since the statistical methods to be used for the two forms of a relationship differ, it is important to identify the form. Additionally, the most frequently used correlation coefficient depends on a linear relationship existing between the two random variables; thus the correlation of a nonlinear relationship may be low even when the relationship shows a strong systematic relationship.

5.3.1.2 Bivariate Correlation

Correlation is the degree of association between the elements of two samples of data, that is, between observations on two variables. Correlation coefficients provide a quantitative index of the degree of association. Examples of variables that are assumed to have a causal relationship and significant correlation are the cost of living and wages, examination grades and IQ, and the volumes of rainfall and streamflow. However, examples that lack a cause-and-effect relationship but may have significant correlation are (1) the crime rate and the sale of chewing gum over the past two decades, and (2) annual population growth rates in nineteenth-century France and annual cancer death rates in the United States in the twentieth century. The correlation coefficient is computed by:

$$R = \frac{\sum_{i=1}^{n} X_i Y_i - \left(\sum_{i=1}^{n} X_i\right)\left(\sum_{i=1}^{n} Y_i\right)/n}{\left[\sum_{i=1}^{n} X_i^2 - \left(\sum_{i=1}^{n} X_i\right)^2 /n\right]^{0.5} \left[\sum_{i=1}^{n} Y_i^2 - \left(\sum_{i=1}^{n} Y_i\right)^2 /n\right]^{0.5}}$$

(5.6)

5.3.2   Regression Analysis

The commonly used linear bivariate model has the form:

$$\hat{Y} = b_0 + b_1 X$$

(5.7)

in which $\hat{Y}$ is the predicted value of the criterion variable Y, X is the value of the predictor variable, and b_0 and b_1 are the intercept and slope coefficients, respectively. The intercept coefficient is the value of Y where X is equal to zero. The slope coefficient represents the rate of change of Y with respect to change in X.

5.3.2.1 Principle of Least Squares

The values of the slope and intercept coefficients of Eq. (5.7) can be computed using the principle of least squares, which is a process of obtaining "best" estimates of the coefficients. Regression is the tendency for the expected value of one of two jointly correlated random variables to approach more closely the mean value of its set than any other. The principle of least squares is used to regress Y on X of Eq. (5.22). The procedure is described by numerous authors (Mendenhall and Sincich, 1999; McCuen, 2003; Roscoe, 1975). To express the principle of least squares, it is important to define the error, e, or residual, as the difference between the predicted and measured value of the criterion variable:

$$e_i = \hat{Y}_i - Y_i \qquad (5.8)$$

in which $\hat{Y}_i$ is the ith predicted value of the criterion variable, Y_i is the ith measured value of Y, and e_i is the ith error. It is important to note that the error is defined as the measured value of Y subtracted from the predicted value. With Eq. (5.8) a positive residual indicates overprediction, while a negative residual indicates underprediction. The objective function for the principle of least squares is to minimize the sum of the squares of the errors:

$$F = \min \sum_{i=1}^{n} \left(\hat{Y}_i - Y_i \right)^2 \qquad (5.9)$$

in which n is the number of observations on the criterion variable (i.e., the sample size).

5.3.2.2   Zero-Intercept Model

A linear model that forces Y to equal zero when X is equal to zero is often necessary for reasons of rationality. The following linear model satisfies this constraint:

$$\hat{Y} = bX \qquad (5.10)$$

in which b represent the slope of the relationship. The first task is to provide a least squares estimator of b. The objective function, F, of Eq. (5.9) is still used. To derive a value for b using least squares optimization, the solution procedure is to differentiate F with respect to the unknown (*note*: we do *not* find the derivative of Y with respect to X), set the derivative equal to zero, and solve for the unknown. Equation (5.9) is rewritten as:

$$F = \sum_{i=1}^{n} (Y - Y)^2 = \sum_{i=1}^{n} (bX - Y)^2 \qquad (5.11)$$

Taking the derivative, setting it to zero, and solving for b yields:

$$b = \frac{\sum\limits_{i=1}^{n} X_i Y_i}{\sum\limits_{i=1}^{n} X_i^2} \qquad (5.12)$$

Equation (5.12) is the least squares estimator for the coefficient b.

Example 5.1. Consider the case for a sample of 3. Table 5.1 shows the necessary computations. Therefore, the value of b equals 45/70 or 0.64286. The following equation could then be used to derive a future estimate of Y for a future value of X:

$$\hat{Y} = 0.64286X \qquad (5.13)$$

It is important to note that Eq. (5.12) is valid only for the linear model of Eq. (5.10) that does not include an intercept; Eq. (5.12) would not be valid for Eq. (5.7) since Eq. (5.7) includes an intercept. It is important to note that Eq. (5.12) is the least squares estimator, but it can provide biased estimates of Y. The sum of the errors in the last column of Table 5.1 is not zero and yields a bias of 0.000013. While in this case, the bias is essentially zero, in other cases the bias for Eq. (5.12) can be substantial. It should always be computed.

Table 5.1. Computation for Example 5.1

i	x_i	x_i^2	y_i	$x_i y_i$	$\hat{y}$	e
1	3	9	2	6	1.92858	–0.07142
2	5	25	3	15	3.21430	0.21430
3	6	36	4	24	3.85716	–0.14284
Sum		70		45		0.00004

5.3.2.3   Reliability of the Regression Equation

Having evaluated the coefficients of the regression equation, it is of interest to evaluate the reliability of the regression equation. The following criteria should be assessed in evaluating the model: (1) the correlation coefficient; (2) the standard error of estimate; (3) the rationality of the coefficients; and (4) the degree to which the underlying assumptions of the regression model are met.

Correlation Coefficient. As suggested previously, the correlation coefficient (R) is an index of the degree of linear association between two random variables. It must be recognized that, if the measured data are not representative of the population (i.e., data that will be observed in the future), the correlation coefficient will not be indicative of the accuracy of future predictions. The square of the correlation coefficient (R^2) equals the percentage of the variance in the criterion variable that is explained by the predictor variable.

Standard Error of Estimate. In the absence of additional information, the mean is the best estimate of the criterion variable; the standard deviation S_y of Y is an indication of the accuracy of prediction. If Y is related to one or more predictor variables, the error of prediction is reduced from S_y to the standard error of estimate, S_e. Mathematically, the standard error of estimate equals the standard deviation of the errors and has the same units as Y:

$$S_e = \left[\frac{1}{v} \sum_{i=1}^{n} \left(\hat{Y}_i - Y_i \right)^2 \right]^{0.5}$$

(5.14)

in which v is the degree of freedom, which equals the sample size minus the number of unknowns. For the bivariate model of Eq. (5.7), $p = 1$ and $v = n - 2$. For the zero intercept model of Eq. (5.10), v is equal to $n - 1$. In terms of the separation of variation concept discussed previously, the standard error of estimate equals the square root of the ratio of the unexplained variation to the degrees of freedom. It is important to note that S_e is based on $(n - p - 1)$ degrees of freedom, while the error, S_y, is based on $(n - 1)$ degrees of freedom. Thus in some cases S_e may be greater than S_y. To assess the reliability of the regression equation, S_e should be compared with the bounds of zero and S_y. If S_e is near S_y, the regression has not been successful. If S_e is much smaller than S_y and is near zero, the regression analysis has improved the reliability of prediction. The standard error of estimate is often preferred to the correlation coefficient because S_e has the same units as the criterion variable and its magnitude is a physical indicator of the expected errors.

Example 5.2. A sampling program for a small wetland involves measuring the total sediment removal at one-half hour intervals. The data are given in Table 5.2. The slope coefficient is:

$$b_1 = \frac{1135.5 - 18(455)/8}{51 - (18)^2 / 8} = 10.643$$

(5.15)

The intercept is computed by $\bar{Y} - b_1 \bar{X}$:

$$b_0 = \frac{455}{8} - 10.64 \left(\frac{18}{8} \right) = 32.929$$

(5.16)

This coefficient indicates that, at the retention time of zero, the percentage removal is 33%. This is irrational as a value of 0 would be expected. The slope coefficient of 10.6 indicates that the removal percentage increases by 10.6% for each one-hour increase in retention time. If the data are plotted, it is evident that the relationship is nonlinear. Therefore, the slope coefficient for a linear model should not be interpreted as a reflection of the change of Y with respect to change in X.

The standard error of estimate is a useful indicator of goodness of fit. For the errors shown in Table 5.2, the standard error is:

Table 5.2. Regression of sediment removal (Y, %) on retention time (X, hrs)

X	Y	X^2	Y^2	XY	$\hat{Y}$	e
0.5	28	0.25	784	14	38.25	10.25
1.0	47	1.00	2209	47	43.57	-3.43
1.5	54	2.25	2916	81	48.89	-5.11
2.0	58	4.00	3364	116	54.21	-3.79
2.5	63	6.25	3969	157.5	59.54	-3.46
3.0	66	9.00	4356	198	64.86	-1.14
3.5	68	12.25	4624	238	70.18	2.18
4.0	71	16.00	5041	284	75.50	4.50
18.0	455	51.00	27263	1135.5		0

$$S_e = \left[\frac{1}{8-2}(195.58)\right]^{0.5} = 5.709\% \qquad (5.17)$$

This value indicates that a predicted removal percentage has an expected error of ±5.7%. If this is compared with the standard deviation of Y (S_y), in the form S_e/S_y, the ratio indicates that the error is 41% of the error in the mean. This is considered reasonably good. Equation (5.6) can be used to compute the correlation coefficient:

$$R = \frac{1135.5 - 18(455)/8}{\left[51 - (18)^2/8\right]^{0.5}\left[27263 - (455)^2/8\right]^{0.5}} = 0.9267 \qquad (5.18)$$

This indicates good agreement between the predicted and measured values of Y. The square of R indicates that the linear relationship explains 85.9% of the variation in Y. Since the sum of the errors is zero, the model is unbiased.

While the goodness-of-fit statistics suggest a reasonably accurate model, the linear model does not follow the trend of the data. This illustrates that goodness-of-fit statistics can suggest an accurate model even when the model is structurally irrational. Also, the problem of the irrational intercept is significant. Therefore, the model should probably not be used either for prediction or to show the effect of X on Y.

5.4   MULTIPLE REGRESSION ANALYSIS

Most of the concepts introduced in the discussion of bivariate regression are also valid when discussing multiple regression. For example, the principle of least squares is used to calibrate the coefficients, and the correlation coefficient is used to indicate the goodness of fit. The main difference is the intercorrelation between predictor variables and its effect on the rationality of the coefficients.

5.4.1   Correlation Matrix

After a graphical analysis, the bivariate correlation coefficients should be computed for each pair of variables; this includes the correlation between the criterion variable and each predictor variable, as well as the correlation between each pair of predictor variables. The correlations are best presented in matrix form. The correlation matrix is a means of presenting in an organized manner the correlations between pairs of variables in a data set; it appears as:

$$
\begin{array}{c c}
& \begin{array}{c c c c c c} X_1 & X_2 & X_3 & \cdots & X_p & Y \end{array} \\
\begin{array}{c} X_1 \\ X_2 \\ X_3 \\ \vdots \\ X_p \\ Y \end{array} &
\begin{bmatrix}
1.0 & r_{12} & r_{13} & \cdots & r_{1p} & r_{1Y} \\
    & 1.0 & r_{23} & & r_{2p} & r_{2Y} \\
    & & 1.0 & & r_{3p} & r_{3Y} \\
    & & & & & \\
    & & & & 1.0 & r_{pY} \\
    & & & & & 1.0
\end{bmatrix}
\end{array}
$$

Note that the correlation matrix, which includes p predictor variables (X_i, $i = 1, 2, ..., p$) and the criterion variable (Y), is shown in triangular form because the matrix is symmetric (i.e., $r_{ij} = r_{ji}$); also, the elements on the principal diagonal equal 1.0 because the correlation between a variable with itself is unity. The matrix is $(p + 1) \times (p + 1)$. Whereas the correlation matrix shown above uses the sample correlation coefficients (r_{ij}), the correlation matrix for the population can be indicate using ρ_{ij} as the elements.

For the purposes of developing a regression equation, it is important to recognize that the correlation matrix actually consists of two parts, a $p \times p$ matrix of correlations between the predictor variables and a $p \times 1$ vector of correlations between the criterion variable and each predictor variable. It must be emphasized that the two types of correlations, which may be referred to as intercorrelations or multicolinearity (Draper and Smith, 1966) and predictor-criterion correlations, are both Pearson product-moment correlations computed with Eq. (5.16); their difference is only important when calibrating regression equations. In general, one hopes for low intercorrelations and high predictor-criterion correlations; unfortunately, the opposite occurs all too often.

Understanding the potential effects of intercorrelation is important because intercorrelation can lead to an irrational model. The least squares solution requires matrix inversion, and the higher the intercorrelation in the $p \times p$ predictor R matrix, the greater the likelihood of irrational coefficients. Near-singular matrices are subject to round-off problems. High intercorrelation can result from physical conditions or from problems related to the experimental design. For example, the watershed length and area are highly correlated from a physical standpoint, so including both these variables can cause irrational effects. An experimental design that leads to an outlier or statistical correlation of predictor variables can lead to the same result as intercorrelation associated with physically related variables. As the inclusion of correlated predictors does not generally lead to improved accuracy, it is better to delete one of the variables rather than generate a model that includes irrational coefficients.

5.4.2   Calibration of the Multiple Linear Model

The components of a regression analysis include the model, the objective function, and the data set. The data set consists of a set of n observations on p predictor variables and one criterion variable, where n should be, if possible, at least four times greater than p. The data set can be viewed as a matrix having dimensions of n by $(p + 1)$. The principle of least squares is used as the objective function (Hirsch et al., 1993). The model, in raw-score form, is:

$$\hat{y} = b_0 + b_1 X_1 + b_2 X_2 + \cdots + b_p X_p \tag{5.19}$$

in which X_j $(j = 1, 2, \ldots, p)$ are the predictor variables, b_j $(j = 1, 2, \ldots, p)$ are the partial regression coefficients, b_0 is the intercept coefficient, and $\hat{y}$ is the predicted value of the criterion variable y. Using the least squares principle and the model of Eq. (5.19), the objective function becomes:

$$F = \min \sum_{i=1}^{n} e_i^2 = \min \sum_{i=1}^{n} \left(b_0 + \sum_{j=1}^{p} b_j X_{ij} - y_i \right)^2 \tag{5.20}$$

in which F is the value of the objective function. It should be noted that the predictor variables include two subscripts, with i indicating the observation and j the specific predictor variable. The method of solution is to take the $(p + 1)$ derivatives of the objective function, Eq. (5.20), with respect to the unknowns, b_j $(j = 0, 1, \ldots, p)$ setting the derivatives equal to zero, and solving for the unknowns.

5.4.3   Evaluating a Multiple Regression Model

After a multiple regression model has been calibrated, we may ask: How well does the linear model represent the observed data? The following criteria should be used in answering this question: (1) the rationality of the coefficients; (2) the coefficient of multiple determination; (3) the standard error of estimate, which is usually compared with S_y; (4) the relative importance of the predictor variables; and (5) the characteristics of the residuals. The relative importance of each of these five criteria may vary with the problem as well as with the person.

Example 5.3. Street sweeping is one means of collecting particulates before they are washed from the roadways into receiving water bodies. The effectiveness of street sweeping is measured by the percentage removed (R, %) of the deposited material.   The effectiveness depends on the size of the particles (d, mm), the velocity (V, mph) of the sweeper, and the number of days (S, days) since the last sweeping or rainfall. The percentage removed is expected to decrease as each of the three predictors increases.

Measurements for eight street sections in an urban area are given in Table 5.3, with the removal ranging from 28% to 45%. A multiple linear regression model of the data was performed, with the following result:

$$\hat{R} = 52.39 - 20.18d - 2.948V - 1.124S \qquad (5.21)$$

The signs of the regression coefficients agree with those of the correlation coefficients and are rational in effect. However, the intercept coefficient of 52.4% would suggest that an efficiency of more than 52.4% cannot be achieved. This magnitude is the result of the data and the linear structure of Eq. (5.21). It may be preferable to set the intercept to 100% and use a nonlinear model structure that still accurately fits the data.

The goodness-of-fit statistics indicate a moderately accurate model. The correlation coefficient is high ($R = 0.899$, $R^2 = 0.809$) but this is not unexpected when the number of degrees of freedom is small. The correlation coefficient does not account for the degrees of freedom, which hinders its ability to be a reliable measure of accuracy. The standard error ratio (S_e/S_y) of 0.578 is a better measure of prediction accuracy, as it accounts for the loss of degrees of freedom for the fitting of four coefficients. In this case, the standard error ratio indicates only moderate accuracy. The standard error of estimate of 3.12% indicates moderate accuracy.

The standardized partial regression coefficients (t) are given a Table 5.4. They indicate that the sweeper velocity and the dry deposition period are about equally important and more important than the particle size. This differs from the indication given by the predictor-criterion correlation coefficients given in Table 5.4. The three R values are essentially the same. The smaller t value for the mean particle diameter occurs because d is intercorrelated with V (0.35) and S (0.49).

Table 5.3. Results for Example 5.3, including particle size (d), sweeper velocity (V), street slope (S), removal efficiency (R), predicted efficiency ($\hat{R}$), and error (e)

D	V	S	R	$\hat{R}$	e
(mm)	(mph)	(days)	(%)	(%)	(%)
0.015	3.5	3	34	38.4	4.40
0.010	2.0	3	45	42.9	-2.08
0.075	5.0	4	33	31.6	-1.36
0.045	4.0	6	36	33.0	-3.05
0.055	4.0	7	30	31.6	1.63
0.042	5.0	8	28	27.8	-0.18
0.018	2.5	9	35	34.5	-0.45
0.110	3.0	10	29	30.1	1.09
0.046	3.62	6.2	33.75	mean	
0.034	1.09	2.7	5.392	st. dev	

Table 5.4. Results for Example 5.3, including correlation matrix, regression coefficients (b), and standardized partial regression coefficients (t) with $|R_u| = 0.633$

	d	V	S	R	b	t
d	1.00	0.35	0.49	-.614	-20.18	-.127
V		1.00	-0.01	-.636	-2.948	-.598
S			1.00	-.620	-1.124	-.565
R				1.00	52.39 = intercept	

5.5 NONLINEAR MODELS

In most empirical analyses, linear models are attempted first because of the relative simplicity of linear analysis. Also, linear models are easily applied, and the statistical reliability is easily assessed. Linear models may be rejected because of either theoretical considerations or empirical evidence. Specifically, theory may suggest a nonlinear relationship between a criterion variable and one or more predictor variables; for example, many biological growth curves are characterized by nonlinear forms. Where a model structure cannot be identified by theoretical considerations, empirical evidence can be used in model formulation and may suggest a nonlinear form. For example, the hydrologic relationship between peak discharge and the drainage area of a watershed has been found to be best represented by a log-log equation, which is frequently referred to as a power model. A nonlinear form may also be suggested by the residuals that result from a linear analysis; that is, if a linear model produces nonrandomly distributed residuals (i.e., local biases are apparent), the underlying assumptions are violated; a nonlinear functional form may produce residuals that satisfy the assumptions that underlie the principle of least squares.

5.5.1 The Power Model

Linear models were separated into bivariate and multivariate; the same separation is applicable to nonlinear models. It is also necessary to separate nonlinear models on the basis of the functional form. Although polynomial and power models are the most frequently used nonlinear forms, it is important to recognize that other model structures are available such as square root, exponential, and logarithmic models. Since the power form is widely used, it is of importance to identify the structure. The multivariate model is:

$$\hat{y} = b_0 x_1^{b_1} x_2^{b_2} \cdots x_p^{b_p} \qquad (5.22)$$

where $\hat{y}$ is the predicted value of the criterion variable, and x_i is the ith predictor variable in the multivariate case, b_j ($j = 0, 1, \ldots, p$) is the jth regression coefficient, and p is the number of predictor variables. The bivariate form is just a special case of the multivariate form.

5.5.2   Transformation and Calibration

The multivariate power model of Eq. (5.22) can be evaluated by making a logarithmic transformation of both the criterion and the predictor variables:

$$z = \log y \qquad (5.23)$$

$$c = \log b_0 \qquad (5.24)$$

$$w_i = \log x_i \qquad (5.25)$$

The resulting model has the form:

$$\hat{z} = c + \sum_{i=1}^{p} b_i w_i \qquad (5.26)$$

The coefficients of Eq. (5.26) can be evaluated using a multiple regression analysis. The value of b_0 can be determined by making the following transformation:

$$b_o = 10^c \qquad (5.27)$$

Example 5.4. Small samples are, unfortunately, common because systematic data collection programs are not often funded. Small samples present a number of problems, most noticeably extreme events that may be statistical outliers. Also the goodness-of-fit statistics may be difficult to interpret.

Table 5.5 includes seven measurements of turbidity (T, NTU) and discharge (Q, ft³/s) from the Choptank River near Greensboro, MD. If the data are plotted, it is evident that one of the observations is much larger than the other six. One problem with extreme events is that they can control the coefficients of a fitted model (see Figure 5.2). The extreme point in this sample was tested using the Dixon-Thompson test (see Section 3.5) to determine if it is a statistical outlier. The turbidity value of 160 NTU is almost four times larger than the mean of the other six measurements, The Dixon-Thompson statistic is:

$$R = \frac{X_n - X_{n-1}}{X_n - X_1} = \frac{160 - 63}{160 - 17} = 0.678 \qquad (5.28)$$

The critical values are 0.503, 0.562, and 0.630 for 5%, 2 ½ %, and 1% levels of significance, respectively. Therefore, the computed value of 0.678 indicates a rejection probability of less than 1%, which indicates that the largest value is an outlier. Therefore, the data will be analyzed both with and without the extreme event.

Using the seven paired measurements, the following log-linear model was fitted with the data:

Table 5.5. Measured discharge (Q) and turbidity (T) for the Choptank River, and predicted turbidities ($\hat{T}_1$, $\hat{T}_2$, and $\hat{T}_3$) and errors (e_1, e_2, and e_3) for the power models of Eqs. (5.30), (5.31), and (5.32), respectively.

Q (cfs)	T (NTU)	$\hat{T}_1$	e_1	$\hat{T}_2$	e_2	$\hat{T}_3$	e_3
17	5.0	2.01	-2.99	3.00	-2.00	6.04	1.04
37	3.1	3.02	-0.08	4.50	1.40	2.75	-0.35
41	2.0	3.18	1.18	4.75	2.75	2.48	0.48
43	3.5	3.26	-0.24	4.87	1.37	2.36	-1.14
53	3.9	3.64	-0.26	5.44	1.54	1.91	-1.99
63	0.7	3.99	3.29	5.95	5.25	1.60	0.90
160	20.0	6.49	-13.51	9.69	-10.31	---	---

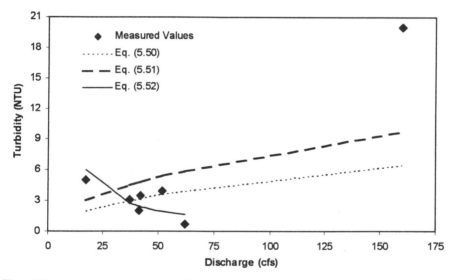

Figure 5.2. Relationship between turbidity and discharge for (a) power model; (b) unbiased power model; and (c) power model after censoring outlier..

$$\log \hat{T} = -0.3413 + 0.5234\,Q \tag{5.29}$$

This yields a correlation coefficient between $\log T$ and $\log Q$ of 0.349, which is not statistically significant. Transforming the intercept using Eq. (5.27) yields the following power model:

$$\hat{T_1} = 0.4557Q^{0.5234} \qquad (5.30)$$

This model yields the following goodness-of-fit statistics for estimating values of T: bias = -1.80 NTU, relative bias = -33.0%, S_e = 6.38 NTU, S_e/S_y = 0.973, R = 0.459, and R^2 = 0.211. The bias is especially significant, as it indicates that the average prediction will be underestimated by 33%. The standard error ratio of 0.973 indicates that the model of Eq. (5.30) is not any better than just using the mean turbidity of 5.46 NTU. The correlation coefficient of 0.459 is not statistically significant. All of the statistics indicate that the model is not reliable.

It is of interest to note that the correlation coefficient computed in the log space for Eq. (5.29) differs from the one computed for Eq. (5.30). The value of 0.349 for Eq. (5.29) reflects the accuracy of estimates of log T, not T. Therefore, the correlation coefficient for the log space should not be used as a measure of the accuracy of predictions.

This analysis is of interest for three reasons: the effects of small samples, outliers, and bias. These factors may not be unrelated. The bias of Eq. (5.30) can be eliminated by changing the intercept such that the sum of the errors is zero. This produces the following model:

$$\hat{T_2} = 0.6804Q^{0.5234} \qquad (5.31)$$

In addition to eliminating the bias, this improves the accuracy of the model, as the S_e/S_y drops to 0.840 and the correlation coefficient increases to 0.642, which in statistically significant at the 6.1% level of significance. However, the model does not provide accurate predictions as the standard error of 5.51 NTU is larger than all of the measured turbidities except for the outlier.

To examine the effect of the outlier, a least squares analysis was made on the other six sample pairs. The resulting power model is:

$$\hat{T_3} = 106.4Q^{-1.013} \qquad (5.32)$$

The most noticeable issue is the negative exponent, which indicates that, as the discharge increases, the turbidity decreases. The model underpredicts with a bias of -0.178 or -5.88% of the mean. The standard error is 1.37 NTU, which is 45% of the mean. The correlation coefficient is -0.582, or R^2 = 0.339. The log-space correlation coefficient is -0.649. While this has a larger magnitude than the correlation coefficient of -0.582, it is valid only for predicting log T, not T. The errors are shown in Table 5.5.

This example is intended to illustrate several problems that frequently arise in modeling: outliers, small samples, and poor correlation. Two alternative analyses were made to try to understand the data and produce a reliable prediction equation. First, the model was unbiased to remove the effect of a consistent underprediction. While this improved the prediction accuracy slightly, the equation was still relatively inaccurate. Second, in the presence of an outlier, the effect of the outlier often dominates the values of the fitted coefficients, as the least squares criterion establishes the coefficients to keep large errors to a minimum. It was evident in this case, as the removal of the outlier led to a very different model (Eq. 5.32) than either of the models based on the data set with the outlier (Eqs. 5.30 and 5.31). If Eq. (5.32) is used to predict the censored measurement for $Q = 160$, the estimated value is 0.623 NTU, which is not close to the measured value of 20 NTU. This illustrates the problem of extrapolating a model (Klinesmith and McCuen, 2004). While some small sample models may provide reasonable goodness-of-fit statistics, they may still not accurately reflect the processes being modeled.

5.6 PROBLEMS

5.1   Measurements of sediment inflow and outflow are made at 15 detention basins for the purpose of predicting the trap efficiency of basins. What is the sample? The population?

5.2   Random samples of a water quality parameter are collected periodically from the outflow of a bioretention facility. What is the population that the sample data will represent?

5.3   Briefly describe the role of analysis and synthesis in water quality modeling.

5.4   Discuss the relationship between the line drawn on a frequency graph and the mathematical equations, Eqs. (5.1) and (5.2).

5.5   With respect to a normal distribution frequency analysis, what is the method of moments?

5.6   For a sample of 17 measurements compute the plotting position probabilities using the Weibull formula.

5.7   Discuss the following statement: If a 100-yr flood occurs this year, it is safe to conclude that a similar flood will not occur for another 100 years.

5.8   When plotting a normal frequency curve, why is $\overline{X} - S$ plotted at a probability of 0.8413 and the magnitude $\overline{X} + S$ at a probability of 0.1587?

5.9   If sample values from a normal population yield a mean of 4.76 and a standard deviation of 0.62, compute: (a) the 75-yr magnitude; (b) the return period for an event of magnitude 5.6; (c) the probability that a magnitude of 3.4 will be exceeded in any one time period.

5.10 If sample values from a normal population yield a mean of 237 and a standard deviation of 56, compute (a) the 40-yr magnitude; (b) the return period for an event with a magnitude of 325; (c) the probability that a magnitude of 185 will not be exceeded in any one time period.

5.11 Using the following sample of phosphorus concentration (mg/L) for the last six years, estimate (a) the expected 50-yr concentration; (b) the probability in any year of a concentration of 4.5. Assume a normal distribution. $X = \{2.61, 0.80, 1.46, 3.72, 2.07, 3.11\}$.

5.12 Using the following sample data, show that the log mean is not equal to the mean of the logs; also show that the log standard deviation is not equal to the standard deviation of the logs. $X = \{7, 15, 11, 14\}$.

5.13 Compute the log mean and log standard deviation of the following organic nitrogen (mg/L) measurements: $\{0.62, 4.3, 2.1, 1.3, 1.5, 0.86\}$

5.14 Assume the log mean of fecal coliforms is 4.1 and the log standard deviation is 0.44 for storm runoff samples from an urban area. Assuming a log-normal distribution, compute (a) the probability of a sample value exceeding 60,000; (b) the value below which 80% of the samples will lie.

5.15 Annual average baseflow discharges for a particular watershed are as follows: $\{12, 31, 20, 17, 28, 15, 9, 19, 11\}$. Assuming that they are log-normally distributed, compute (a) the probability that the annual average baseflow in a year will be less than 6 cfs; (b) the magnitude above which 95% of the annual values will lie.

5.16 Measurements are made on 11 streams of the sum of the percentage agricultural and urban land covers (X) and the mean total nitrogen $(Y\ mg/L)$:

Y	0.13	0.68	0.93	0.81	1.21	0.93	0.67	0.89	0.59	0.82	1.02
X	74	19	28	16	12	24	15	20	8	19	35

Graph the data, and discuss the characteristics of the sample data evident from the graph.

5.17 Measurements are made of the channel cross-sectional area (C, ft^2), the contributing drainage area (A, mi^2), and the percentage of impervious cover $(I, \%)$ for 14 watersheds in a region:

C	28	32	28	31	22	27	23	26	19	22	24	15	20	14
A	1.82	1.51	1.37	1.24	1.17	1.12	0.94	0.93	0.84	0.82	0.70	0.62	0.47	0.39
I	3	29	11	36	6	21	12	28	5	19	31	7	22	10

Using a graphical analysis, assess the contributions of drainage area and imperviousness on the channel cross-sectional area.

5.18 Using the data of Problem 5.17, compute the total, explained, and unexplained variations for regression of C on A and C on I. Also, compute the values of R and R^2.

5.19 Derive least squares estimators for b_1 and b_2 for the equation $\hat{y} = b_1 X + b_2 X^2$.

5.20 Using the linear model of Eq. (5.7) and the data of Problem 5.17, regress C on A to compute the intercept and slope coefficients. Compute the residuals using Eq. (5.8) and graph them versus the values of the imperviousness given in Problem 5.17. Discuss the results.

5.21 Regress C on A using the data of Problem 5.17. Then regress A on C. Using the latter equation, transform it to predict C from A. Compare the coefficients of the transformed equation to the coefficients from the regression of C on A. Discuss the reason for the disparity.

5.22 Using the principle of least squares, derive the normal equations for the model $\hat{y} = a + bX^2$.

5.23 Regress the nitrogen data Y of Problem 5.16 on the land use X. Compute the correlation coefficient. Discuss the expected prediction accuracy.

5.24 Derive a relationship between the standard error of estimate S_e, the standard deviation of the criterion variable S_y, the correlation coefficient R, and the sample size.

5.25 Compute the correlation matrix for the data of Problem 5.17.

5.26 Using the principle of least squares derive the normal equations for the two-predictor model $\hat{y} = b_1 X_1 + b_2 X_2$.

5.27 The stream reaeration coefficient (r) is a necessary parameter for computing the oxygen deficit sag curve for a stream reach. The coefficient is a function of the mean stream velocity in feet per second (X_1), water depth in feet (X_2), and the water temperature in degrees Celsius (X_3).

X_1	X_2	X_3	r
1.4	1.3	14	2.89
1.7	1.3	23	4.20
2.3	1.5	17	4.17
2.5	1.6	11	3.69
2.5	1.8	20	3.78
2.8	1.9	15	3.56
3.2	2.3	18	3.35
3.3	2.5	11	2.69
3.5	2.6	25	3.58
3.9	2.9	19	3.06
4.1	3.3	13	2.41
4.7	3.4	24	3.30
5.3	3.5	16	3.05
6.8	3.7	23	4.18
6.9	3.7	21	4.08

(a) Plot each pair of the variables and discuss (1) the degree of correlation between each pair (low, moderate, high), (2) the possible effect of intercorrelation between the predictor variables, and (3) the apparent relative importance of the predictor variables.
(b) Compute the correlation matrix. Discuss the results.
(c) Compute the goodness-of-fit statistics (R, S_e) and discuss the implications.

5.28 The mean annual discharge (Q) of a river is an important parameter for many engineering design problems, such as the design of dam spillways and levees. Accurate estimates can be obtained by relating Q to the watershed area (A) and the watershed slope (S).

A (mi^2)	S (ft/ft)	Mean annual discharge (ft^3/sec)
36	0.005	50
37	0.040	40
45	0.004	45
87	0.002	110
450	0.004	490
550	0.001	400
1200	0.002	650
4000	0.0005	1550

(a) Plot each pair of the variables and discuss (1) the degree of correlation between each pair (low, moderate, high), (2) the possible effect of intercorrelation between the predictor variables, and (3) the apparent relative importance of the predictor variables.
(b) Compute the correlation matrix. Discuss the results.
(c) Compute the goodness-of-fit statistics (R, S_e) and discuss the implications.

5.29 Make a complete multiple regression analysis of the following data:

A	6	36	37	45	58	87	130	470
S	0.006	0.005	0.040	0.004	0.002	0.002	0.001	0.004
D	0.8	1.0	1.0	0.8	1.0	1.2	2.3	1.7
Q	14	50	39	43	56	107	195	492

where A = drainage area (mi^2), S = channel slope (ft/ft), D = average depth (ft), Q = average discharge (ft^3/s). In addition to computing the regression coefficients, the goodness-of-fit statistics, and the residuals, discuss the relative importance of the three predictor variables.

5.30 Using the data of Problem 5.29 fit a bivariate power model between the discharge Q and the drainage area A: $\hat{Q} = aA^b$. Assess its accuracy using the bias (i.e., the average error) and the standard error of estimate.

5.31 Using the data of Problem 5.17, fit a two-predictor power model $\hat{C} = aA^b I^d$. Assess its accuracy using the bias and standard error of estimate.

5.7 REFERENCES

Ang, A. and Tang, W.H. (1975). *Probability in Engineering Planning and Design*, Vol. 1, John Wiley & Sons, Inc., New York.

Draper, N. and Smith, H. (1966). *Applied Regression Analysis*. John Wiley and Sons, Inc., New York.

Hirsch, R.M., Helsel, D.R., Cohn, T.A., and Gilroy E.J. (1993). "Statistical Treatment of Hydrologic Data." Chap. 17 of *Handbook of Hydrology*. (D.R. Maidment, Ed.). McGraw-Hill Book, CO., New York.

Interagency Advisory Common Water Data. (1982). *Guidelines for Determining Flood Flow Frequency*. Bulletin 17B, U.S. Dept. Interior, Reston, VA.

Jennings, M.E., Thomas, Jr., W.O., and Riggs, H.C. (1994). *Nationwide Summary of U.S. Geological Survey Regional Regression Equations for Estimating Magnitude and Frequency of Floods for Ungaged Sites, 1993*. U.S.G.S., Reston, VA.

Klinesmith, D.E., and McCuen, R.H. (2004). "Discharge and Sediment Rating Curve Accuracy." *J. Floodplain Mgt.* 4(1): 1-10

Mendenhall, W. and Sincich, T. (1992). *Statistics for Engineering and the Sciences*. Dellen Publ. Co., San Francisco.

McCuen, R.H. (2003). *Modeling Hydrologic Change*. Lewis Publishers, CRC Press, Boca Raton, FL.

Roscoe, J.T. (1975). *Fundamental Research Statistics for the Behavioral Sciences* (2nd ed.) Holt, Rinehart and Winston, Inc., New York.

Stedinger, J.R. Vogel, R.M., and Foufoula-Georgiou, E. (1993). "Frequency Analysis of Extreme Events," Chap. 18 of *Handbook of Hydrology* (D.R. Maidment Ed.). McGraw-Hill Book Co., New York.

6

STORMWATER QUALITY

NOTATION

C	= concentration
C	= crop management factor
C_D	= drag coefficient
C_F	= runoff correction factor
D	= diameter
E	= soil loss
F_b	= buoyant force
F_D	= drag force
g	= gravitational acceleration
h	= depth of flow
K	= soil erodibility factor
L	= mass loading
L	= length
M	= mass
M_t	= cumulative mass
P	= annual precipitation
P	= conservation practice factor
Q	= flow rate
q	= water discharge per unit width
q_b	= bedload discharge per unit stream width
R	= erosivity index
R(t)	= ratio of runoff mass fraction and volume fraction
R_h	= hydraulic radius
R_v	= runoff coefficient
S	= field slope
S_c	= stream gradient
t	= time
T	= topographic factor
v	= volume
V	= velocity

V = total runoff volume
V_c = critical velocity
V_t = cumulative runoff volume
W = weight
X_t = runoff volume fraction
Y_t = pollutant mass fraction
γ = specific weight
ρ = density
μ = viscosity
τ = tractive force
ω = stream power per unit bed area

6.1 INTRODUCTION

Many of the physical and chemical parameters that are used to define water quality were discussed in Chapter 2. Recently, studies have been completed to evaluate water quality parameters in runoff from urban areas to quantify stormwater runoff as a source of pollution to water bodies. The goals are to characterize the variation of water quality constituents for a variety of land uses associated with urban and suburban development, to evaluate the impact of the runoff on the water quality of the receiving stream, to investigate possible treatment technologies to improve water quality, and understand the sources of the pollutants so that pollution prevention techniques may be implemented to improve the runoff water quality.

6.2 POLLUTANT LEVEL DETERMINATIONS

Because or the variability and unpredictability in the magnitude and frequency of runoff flows, sampling of runoff in order to investigate quality can be difficult. When investigating the input of pollutants to waters, a number of different measurements are possible. Knowledge of concentrations entering a receiving water are of importance in evaluating acute toxicity and immediate water quality concerns in a water during a precipitation event. Long-term pollutant inputs, however, are best quantified using mass loads. Load calculations require detailed knowledge of both pollutant concentrations and runoff flow rates as a function of each storm event, and a good data set should cover many events.

It is important to emphasize that the number of storms alone is not the only criterion essential to good decision-making. Storm characteristics are important, as stormwater treatment processes respond differently to storms with different characteristics. The magnitude and duration of runoff quantity will influence the water quality response of a treatment practice. Therefore, it is important to sample storms with different characteristics, including storms to which the treatment practice response is most sensitive.

6.2.1 Grab Sample Measurements

The simplest water quality measurements are grab samples taken at various times during a storm event. Grab samples are the easiest and least costly samples to obtain. The sample is collected in an appropriately prepared container and evaluated either directly in the field or after it is taken to a laboratory. By themselves, grab sample measurements do not provide adequate information about the pollutant load carried in the runoff. The concentration of any parameter in runoff is not constant over the entire storm event. The grab samples only represent an instantaneous measurement of what is happening at that point in time and space. The samples will not be representative of the entire event. They do, however, provide a simple indication of the pollutant concentration entering receiving water.

The temporal dynamics of pollutant concentrations during a storm event will depend on the storm type, duration, and intensity, as well as conditions of the drainage area. In a typical situation, the pollutant concentration is initially low, as light rainfall begins. The concentration increases to a high level early in the event as the rainfall detaches and transports pollutants that have accumulated over the watershed since the last significant storm event. Subsequently, the concentration decreases, since the majority of pollutants have already been mobilized. This rise and fall in concentration is schematically presented in Figure 6.1a and is often referred to as a *pollutograph*. The exact shape of this curve will depend on the characteristics of the storm event and the catchment area. Increases and decreases in rainfall intensity will alter the detachment and mobilization of pollutants from the surface and variations in runoff rates cause variations in the transport and additional scouring of pollutants. This is evident in the temporal rise and fall of pollutant concentrations that are typically noted. Coupled with the variability in runoff flow rate during an event, as discussed in Chapter 4, it is easy to see that a single grab sample provides little information about runoff water quality in either the time or space scale.

Collecting several grab samples over a single event can provide more reliable estimates of pollutant concentrations in runoff for that storm. Again, however, one must be aware of what such measurements mean. Collecting many samples over a single event can map out a pollutograph such as that shown in Figure 6.1a. Nonetheless, the characteristics of the pollutograph will be different for each storm event, depending on the rainfall, drainage area characteristics, and the amount of material available to be washed off. A high intensity burst of rainfall may detach pollutants that would not have been mobilized by a less intense storm pattern. The volume of runoff is an important factor in determining how much of the detached material is transported to the site where the pollutograph measurements are made. Consequently, while a pollutograph provides good information about a single event, large differences usually exist among different events, especially of events with different rainfall intensities and intensity patterns.

Similarly, selecting a single grab sample from many events provides a snapshot about water characteristics for each event, but will not tell the entire story about the entire curve for any one event. The value of single grab samples is sensitive to the point in time where the grab sample is made.

6.2.2 Composite Sample Measurements

To fully understand the pollutant levels transported in stormwater runoff, composite measurements must be completed. However, concentrations that are composited temporally, again, do not show the full impact of the runoff. A composited sample will give the mean concentration in the runoff. This mean, however, does not provide much information since the flow rate carrying the concentration is always changing.

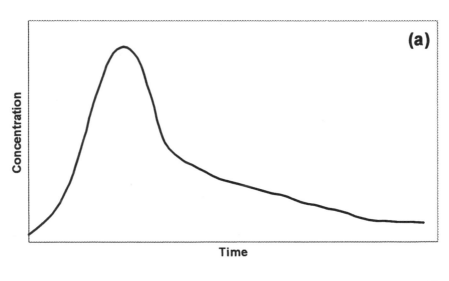

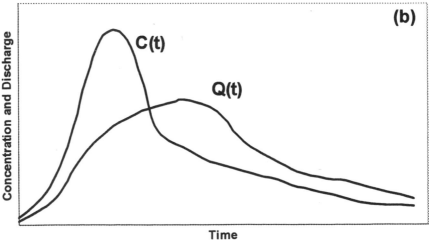

Figure 6.1. Schematic representation of (a) a pollutograph, and (b) a pollutograph and discharge hydrograph that illustrate a first flush.

Collecting many samples during an event, along with information on the flow rates can allow the calculation of a much more useful composite concentration known as the *event mean concentration,* or EMC. The EMC is a flow-weighted average of the concentration and is representative of the pollutant concentration over an entire event. The event mean concentration is computed by:

$$EMC = \frac{\int_0^{t_z} CQdt}{\int_0^{t_z} Qdt} \approx \frac{\sum_{all\,t} C_i Q_i \Delta t_i}{\sum_{all\,t} Q_i \Delta t_i} \tag{6.1}$$

in which C is the concentration distribution over time, Q is the time distribution of runoff discharge, C_i are individual measurements of the concentration, Q_i is the discharge at the time concentration C_i was measured, and Δt_i is the time interval associated with C_i. The numerator calculates the total mass of a pollutant in the runoff during the precipitation event. The denominator sums the total volume of runoff. If all runoff from a drainage area were collected in a large well-mixed tank during a storm, the pollutant concentration in this tank would correspond to the EMC.

From a long-term water quality basis, knowing the total mass of pollutant that enters the water body is important. Thus, with information on concentrations and flows, mass inputs, either from individual events or summed over a yearly period, provide useful information and can be compared with loads from other point or nonpoint pollutant sources to the water body.

Example 6.1. An automated sampler has been set up to monitor runoff from a large parking lot. The sampler measures runoff flow rates and take samples that are analyzed for various pollutant levels. The results for the concentration of oil from a storm event are presented in Table 6.1. The event mean concentration, EMC, of the oil and the total amount of oil carried in the runoff for this event using the summation of Eq. (6.1). With the data given, both integrals are solved numerically. The numerator of Eq. (6.1) will provide the total mass of oil. Table 6.1 is set up to solve both the numerator and denominator using the trapezoidal rule summed over the storm event. Zero discharges are shown for times of 10:30 am and 3:30 pm to complete the curves to zero at each end. $\overline{Q}$ and $\overline{C}$ are means of the current value and the value from the immediate prior measurement.

The total mass of oil transported is 333,000 mg, or 333 g, and the total volume of runoff is 88,000 L. Therefore, the EMC = 333,000 mg / 88,000 L = 3.78 mg/L, which is a reasonable value if it is compared to the individual oil concentrations over the sampling time (ranged from 0.8 to 5.8 mg/L). Note that it is not correct to find the mean of the oil concentrations in Table 6.1, which is 3.21 mg/L. The discharge-weighted mean is a more representative value experienced by the stream.

Table 6.1. Calculation of event mean concentration for a storm event (Example 6.1).

Time	Flow Rate (L/s) Q	Oil (mg/L) C	Δt (s)	$\overline{Q}$ (L/s)	$\overline{C}$ (mg/L)	$\overline{Q}\overline{C}\Delta t$ (mg)	$\overline{Q}\Delta t$ (L)
10:30 am	0						
				1.05	1.6	3020	1890
11:00 am	2.1	3.2	1800				
				3.4	3.85	23,600	6120
11:30 am	4.7	4.5	1800				
				6.4	5.15	59,300	11,520
noon	8.1	5.8	1800				
				8.55	5.1	78,500	15,390
12:30 pm	9.0	4.4	1800				
				8.9	4.1	65,700	16,020
1:00 pm	8.8	3.8	1800				
				7.95	3.95	56,500	14,310
1:30 pm	7.1	4.1	1800				
				6.1	3.1	34,000	10,980
2:00 pm	5.1	2.1	1800				
				4.0	1.45	10,400	7200
2:30 pm	2.9	0.8	1800				
				2.0	0.5	1800	3600
3:00 pm	1.1	0.2	1800				
				0.55	0.1	99	990
3:30 pm	0		1800				
						$\sum \overline{Q}\overline{C}\Delta t$ (mg) 333,000	$\sum \overline{Q}\Delta t$ (L) 88,000

6.3 STORMWATER RUNOFF QUALITY DATA

Table 6.2 has been compiled from a variety of sources to demonstrate typical water quality parameters found in urban runoff. In most cases, these values represent EMCs; in a few cases, grab samples are reported. The values in Table 6.2 should not be considered as comprehensive. More data continue to be collected on runoff quality. Most large jurisdictions in the U.S. have collected data on runoff quality in order to meet U.S. EPA reporting requirements. Water quality characteristics will depend on the location of the site, rainfall characteristics, the details of the land use, the presence of water quality improvement practices, and other parameters. Nonetheless, a few trends are apparent from Table 6.2 and will be briefly discussed. Some of the parameters in Table 6.2 can be compared to water quality limits presented and discussed in Chapter 2.

6.3.1   pH

Although rainfall pH can be very acidic in some areas of the U.S., as low as 4.1 or so, contact with the land surface tends to buffer the pH of the runoff. The pH of highway runoff does not vary much from neutral. From a rainfall pH of 3.8 to 4.5, the pH values of runoff from pavements were all greater than 5.6 (Sansalone and Buchberger 1997).

6.3.2 Suspended Solids and Oil and Grease

A wide range of suspended solids concentrations has been noted, from less than 1 to over 700 mg/L. Many of the results showed TSS values over 100 mg/L, which is quite high and can result in serious water quality concerns. Oil and grease measurements ranged over an order of magnitude, from about 0.4 to 4 mg/L. Like most pollutants, suspended solids and oil/grease are not found in the rainfall itself. These pollutants are picked up during the overland runoff and channel flow. Concentrations of these parameters will vary greatly with land use and storm characteristics.

6.3.3 Organic Carbon/Oxygen Demand

Organic carbon is usually a minor pollutant concern for urban runoff. Table 6.2 shows that TOC and TC values are usually less than 100 mg/L. The values for BOD are all less than 30 mg/L, which is usually the standard for discharge of treated municipal wastewater. Overall, urban runoff is not expected to contribute a major oxygen demand to water from biodegradable organic compounds.

6.3.4 Nutrients

Since high nutrient levels are contributing to eutrophication problems in many water bodies, nitrogen and phosphorus are major pollutant concerns in runoff. Although these compounds are typically considered as originating in runoff from agricultural fields and animal lots, high levels have been consistently measured in runoff from developed areas. Ammonia-nitrogen concentrations ranged from 0.2 to nearly 3 mg/L (Table 6.2). Oxidized nitrogen (nitrite + nitrate) ranged from 0.1 to over 2 mg/L. Measured values for TKN are generally about 1 mg/L. The data for both ortho-P and total P showed lower values on the order of 0.1 mg/L, but one ortho-P measurement was nearly 3 mg/L.

It is obvious that urbanized areas must be given consideration when evaluating nutrient loads to water bodies. Apparently, these compounds are resulting from overfertilization and other issues related to poor management of urban and suburban green space. Some nitrogen is also present in rainfall, as discussed in Section 2.4.3.

Table 6.2. Event mean concentrations of pollutants in urban and highway stormwater runoff.

Pollutant	Measured Values	Ref.
Traditional Pollutants		
pH	6.78, 6.98, 6.47	A
	4.4 - 12	C
	269, 236, 275	A
	0.5 - 750	C
TSS (mg/L)	135	D
	129, 91, 19, {0}	F
	283, 93, 30	G
TDS (mg/L)	120, 118, 137	A
	157, 88, 216	G
VSS (mg/L)	36, 18, 9, {0}	F
COD (mg/L)	130, 39, 37, {6}	F
	70, 39, 22	G
BOD (mg/L)	30, 18, 28	A
	12, 5, 4, {ND[b]}	F
TOC (mg/L)	14	H
TC (mg/L)	46, 21, 20, {ND}	F
DTC (mg/L)	25, 11, 18, {ND}	F
TKN (mg/L)	0.88	D
	1.42, 1.18, 1.00	G
NO2+3-N (mg/L)	0.78[a], 0.96[a], 0.61[a]	A
	1.07[a], 0.71[a], 0.37[a], {0.52[a]}	F
	2.25, 0.22, 0.14	G
NH3-N (mg/L)	2.39, 1.92, 2.75	A
	0.22	D
	0.83, 0.76, 0.52	G
TP (mg/L)	0.14	D
	0.33, 0.11, 0.10, {0.05}	F
	0.43, 0.52, 0.47	G
ortho-P (mg/L)	2.95, 1.60, 2.1	A
	0.10	D
	0.15, 0.30, 0.17	G
O&G (mg/L)	4.2, 1.4, 0.4, {ND}	F
	4.4, 2.4, 1.3	G
	6.7	H
	0.80 - 31.3	I

(continued)

Metals

Total Fe (mg/L)	6.11	D
	2.82, 1.40, 0.25, {0.079}	F
Soluble Fe (mg/L)	0.24	D
Total Al (µg/L)	<5-71300	C
Total Zn (µg/L)	733	B
	<1-13100	C
	66	D
	3880±5620	E
	222, 44, 24, {19}	F
Soluble Zn (µg/L)	20	D
Total Cu (µg/L)	<1-1830	C
	<30	D
	37, 7, 12, {3}	F
	24.2, 11.5, 4.6	G
Total Pb (µg/L)	142	B
	<1-330	C
	<100 (d); 53, 15, 3, {ND}	D
	21.0, 13.9, 6.5	F
		G
Soluble Cd (µg/L)	9.2±8.9	E
Total Cd (µg/L)	<0.1-220	C
	<5, <5, <5	G
Total Cr (µg/L)	15	B
	<0.1-710	C
	8.1, 3.5, <5	G
Total Ni (µg/L)	<1-170	C
	8.1, <5, 3.5	G

Organic Compounds

Benzene (µg/L)	0.022	H
Toluene (µg/L)	0.047	H
Acetone (µg/L)	24	H
Methyl Ethyl Ketone (MEK) (µg/L)	4.6	H
PCB's (ng/L)	130-633	B
Lindane (ng/L)	32-116	B
Chlordane (µg/L)	<0.3-2.9	C

(continued)

1,3-Dichlorobenzene (μg/L)	<0.5-120	C
Fluoranthrene (μg/L)	<0.5-130	C
Pyrene (μg/L)	<0.5-120	C
Benzo(b)fluoranthrene (μg/L)	<0.5-260	C
Benzo(k)fluoranthrene (μg/L)	<0.5-103	C
Benzo(a)pyrene (μg/L)	<0.5-300	C
Bis(2-chloroethyl) ether (μg/L)	<0.5-200	C
Bis(2-chloroisopropyl) ether (μg/L)	<0.5-400	C
Naphthalene (μg/L)	<0.5-300	C
Benzo(a)anthrcene (μg/L)	<0.5-73	C
Butylbenzyl phthalate (μg/L)	<0.5-130	C
1,2,3,4,6,7,8 dioxin[c] (pg/L)	4 -1100	J
1,2,3,4,6,7,8,9 dioxin[c] (pg/L)	35 -7700	J
1,2,3,4,6,7,8 furan[d] (pg/L)	<1 -260	J
1,2,3,4,6,7,8,9 furan[d] (pg/L)	<5 -570	J

Pathogens

Fecal Coliforms (#/100 mL)	41×10^4, 41×10^4, 6.5×10^4	A
Total Coliforms (#/100 mL)	5.6×10^6, 2.2×10^6, 2.0×10^6	A

[a]NO$_3$ only
[b]Not Detected
[c]chlorodibenzodioxin
[d]chlorodibenzofuran
A) Cordery 1977 (arithmetic mean) (3 sites)
B) Granier *et al.* (1990)
C) Pitt et al. (1995) (range of 87 samples)
D) Wu *et al.* (1996)
E) Sansalone and Buchberger (1997)
F) Barrett *et al.* (1998) (3 sites + rainfall {last value in list})
G) Wu *et al.* (1998), North Carolina
H) Lopes *et al.* (2000), Minnesota

6.3.5 Metals

A few measurements for iron and aluminum have been reported (Table 6.2). Values for these metals show wide variations and in some cases are very high as compared to typical fresh water values. Nonetheless, these metals are both major inorganic constituents of soils and are not considered toxic to humans. From a health perspective, these metals are not a major concern.

The other metals listed are more of a problem. Copper, lead, cadmium, nickel, chromium, and zinc are toxic to either humans or some plant and animal species. Concentration ranges for these metals are wide. In some cases they are low, but in others, reported levels are very high and exceed typical fresh water quality criteria, as presented in Table 6.3. Concentrations depend on the nature of the land use. Areas that experience high automobile use tend to have high metal concentrations in the runoff.

The fate, toxicity, and treatment efficiencies of heavy metals in runoff will depend on the chemical form of the metal. Most metals bind easily to natural soil and sediment materials, and a significant fraction of runoff metals is affiliated with the TSS. Thus, their fate is controlled by TSS transport. Even those metals that are not attached to the particulates (the dissolved metals) are likely part of a chemical complex, where the metal is bound to an inorganic species, such as carbonate or chloride, or a natural organic compound (Sedlak *et al.* 1997). This complexation can also affect the fate of the metal in the environment and the efficiency of treatment processes.

6.3.6 Toxic Organics

Only a few studies have documented concentrations of toxic organic compounds in urban runoff. Table 6.2 shows the wide ranges of values found. Lindane and chlordane are common pesticides; most of the other organic compounds are common components of fuels, oils, and unburned exhausts. Low levels of dioxins and furans were found in runoff. The presence and concentrations of these compounds will depend on their use in the catchment area, which in turn is a function of the land use. For example, a high fraction of lawns or other managed grass areas (such as golf courses) will generally show high pesticide levels. Areas with a high fraction of roadways and parking lots will contribute fuels, oils, and other hydrocarbons to the runoff.

6.3.7 Pathogens

Currently, little information is available on pathogen levels in urban runoff. However, the data that are available indicate high coliform levels. This is somewhat unexpected since fecal coliforms and other pathogens originate from animals wastes and animal densities are generally low on developed lands. However, apparently wastes from birds and other "urban animals," including pets, are being accumulated on impervious surfaces to be washed away with subsequent rainfall. Abandoned pet waste may be an important source of pathogens also.

When evaluating environmental measurements such as Table 6.2, the date of collection should be carefully considered. Changing environmental regulations over time have forced the reductions of several pollutant discharges to the environment and old measurements may not be representative of current conditions. A prime example is lead. The addition of lead to gasoline was phased out and eventually banned in the U.S. in the 1970's. Atmospheric deposition of lead from automobile combustion was significant prior to the phase out. Since the phase out, atmospheric lead deposition has declined significantly and accordingly, so have stormwater runoff levels, dropping by about an order of magnitude. Other pollutant levels may have declined as changes in societal use has shifted from one material to another.

Table 6.3. Fresh water acute and chronic surface water quality criteria for inorganic substances in Maryland.

Substance	Fresh Water Criteria (µg/L)		Substance	Fresh Water Criteria (µg/L)	
	Acute	Chronic		Acute	Chronic
Arsenic	340	150	Lead	65	2.5
Cadmium	2.0	0.25	Mercury	1.4	0.77
Chlorine	19	11	Nickel	470	52
Chromium III	570	74	Selenium	20	5
Chromium VI	16	11	Silver	3.2	-
Copper	13	9	Zinc	120	120
Cyanide	22	5.2			

COMAR (Code of Maryland Regulations) 26.08.02.03

6.4 POLLUTANT MASS LOADS

From a long-term planning point of view, instantaneous concentrations of pollutants, or even averages over individual storm events provide only a limited amount of information for a true water quality and ecosystem analysis. The total mass of a pollutant that enters a water body over an extended period of time can be used to evaluate long-term impacts on receiving waters and ecosystems. This value can also be normalized to the drainage area to produce a specific annual load, L, with units such as kg/ha-yr or lb/acre-yr:

$$L = \frac{\sum_{0}^{N} \int_{0}^{t_t} CQ \, dt}{A} \tag{6.2}$$

The integral calculates the total mass of a pollutant in the runoff during a single precipitation event. The summation adds the pollutant mass for each storm event over the one-year period; N is the number of events over the year. A is the drainage area. Appropriate conversion factors will need to be applied to Eq. (6.2).

As a simplification of Eq. (6.2), annual pollutant loads can be estimated using a mean runoff concentration and mean annual precipitation. This procedure, known at the *simple method,* estimates annual loads as:

$$L = \frac{P\, C_F\, R_V\, C}{100} \tag{6.3}$$

where L is the annual load (kg/ha-yr), P is the average annual precipitation (mm/yr), C_F is a correction factor that corrects for events that do not produce runoff (typically around 0.9 for impervious area), R_v is the runoff coefficient for the drainage area, and C is the pollutant event mean concentration (flow weighted average) in mg/L.

Example 6.2. Annual rainfall in the Washington, DC, area is about 42 inches. Using the simple method, the annual mass of oil from a 4.5-acre commercial area with a runoff coefficient of 0.8 can be estimated. The event mean concentration of oil, as given in Example 6.1, is 3.8 mg/L. The rainfall value of 42 inches is equal to 1067 mm/yr. Therefore:

$$L = \frac{(1067\ \text{mm/yr})(0.9)(0.8)(3.8\ \text{mg/L})}{100} = 29.2\ \frac{\text{kg}}{\text{ha - yr}} \tag{6.4}$$

Since 4.5 acres is equal to 1.82 ha, the computed total oil loading from this area is 53.1 kg/yr (117 lb/yr).

6.5 THE FIRST FLUSH

The concept of a first flush of pollutants is important to the management and improvement of stormwater quality. Nevertheless, the quantification of a first flush is complex, and the consistent occurrence of a first flush is debatable. The definition of what constitutes a first flush in stormwater runoff is not universally accepted. Additionally, while in a number of runoff events, a first flush of pollutants has been noted (using some definition), in others a clear first flush has not been apparent.

Conceptually, a first flush is an initial high level of pollutants that is carried in the first flow of the runoff. This concept is based on the idea that pollutants build up on land areas during the antecedent dry days through atmospheric deposition, dripping of automobile fluids, discharges from vehicle brakes, pesticide and fertilizer applications, and so forth. These pollutants are easily dislodged and transported by first contact with the runoff. Therefore, high pollutant concentrations are observed in the initial part of a pollutograph. The impact of rain and runoff later in the storm does not mobilize as much pollution because less pollution mass is available. Therefore, pollution levels drop off after the first flush period. This is illustrated in Figure 6.1b.

Understanding and quantifying the first flush is necessary for predicting environmental impacts to receiving waters and for the efficient design of treatment practices.   Organisms in streams must contend with the highest pollutant concentrations to which they are exposed. The first flush washoff usually has the highest concentrations of pollutants, so it is this flush that can prove detrimental to a healthy stream.

Stormwater treatment practices must also consider the first flush. When a first flush occurs, a significant fraction of the overall pollutant mass is carried in the initial small volume of water. Thus, designing and constructing a system to intercept and treat just this small volume can have a major impact on water quality improvement. For example, by assuming that the first half-inch of runoff contains most of the pollutants that are entrained in washoff from surfaces, treatment facilities can be sized to address just this first half-inch of runoff. The pollutant loads in runoff after this half-inch are assumed to be much smaller and should not have significant impacts on downstream ecology. The facility can be designed to allow all runoff in excess of one half-inch to bypass the storage or treatment unit, resulting in smaller treatment practice design. Also, by allowing the excess flow to bypass the facility, the washout of pollutants captured and accumulated from the first flush can be minimized. This approach has been taken by many state regulatory agencies, defining a first flush volume or a "water quality volume" that must be treated.

6.5.1 Defining the First Flush

Quantitatively defining a first flush is difficult and stating that a first flush occurs is, in many cases, based on an operative definition. Generally, the approach is to evaluate ratios of the total pollutant mass to those of the total runoff volume. As defined in Eq. (6.1), the total runoff volume for a rainfall event is given by:

$$V = \int_0^{t_s} Q dt \qquad (6.5)$$

The cumulative runoff volume from the beginning of the storm to any time, t_t, is found by:

$$V_t = \int_0^{t_t} Q dt \qquad (6.6)$$

Therefore, the fraction of the total runoff volume that has occurred by time t_t for an event, known as the volume fraction, X_t, is given by:

$$X_t = \frac{V_t}{V} \qquad (6.7)$$

The total pollutant mass (M) and the total cumulative mass (M_t) at any time t_t during the storm event are defined similarly:

$$M = \int_0^{t_s} CQdt \qquad (6.8)$$

$$M = \int_0^{t_t} CQdt \qquad (6.9)$$

Accordingly, the mass fraction (Y_t) is given as:

$$Y_t = \frac{M_t}{M} \qquad (6.10)$$

The ratio of the mass and volume fractions, R, can be used to define the first flush:

$$R(t) = \frac{Y_t}{X_t} \qquad (6.11)$$

In Eq. (6.11), the ratio is shown to be a function of time, and it is used to compute R(t) for all times and show its variation over the course of the storm event. Values of R that are greater than 1 indicate that a greater proportion of the pollutant mass has occurred than the same proportion of water. Therefore, a value of R greater than 1 at the earlier times during the storm indicate the occurrence of a first flush. When the value of R is 1, the ratio of pollutant mass to runoff volume is equal to that for the entire storm; therefore, the total mass-to-volume ratio at this point is equal to the EMC. When R is less than 1, the cumulative mass lags that of the corresponding volume, which corresponds to an event in which the concentration is initially less than the EMC and where a high concentrations are found late in the event. At the end of the event, Y_t, X_t, and R are all equal to 1.

First flush characteristics can be examined visually through a plot of Y_t as a function of X_t. If the pollutant washoff occurs at the same rate as the water discharge, then $Y_t = X_t$ which is shown in Figure 6.2 as storm event 1. Obviously, this event is not characterized by a first flush. If Y_t precedes corresponding values of X_t, then the pollutant mass is being exported faster than runoff flow. The initial slope is greater than 45°, and the initial pollutant concentration is greater than the EMC, thus indicating that some degree of first flush has occurred (see event 3 in Figure 6.2). Therefore, a plot of Y_t vs. X_y that shows a concave downward profile is indicative of a first-flush event. If little pollution has accumulated since the last storm event, then a first flush will not occur and the harder-to-detach pollutants will likely occur later in the storm event. Thus Y_t does not keep pace with X_t (see event 2 in Figure 6.2). The curve does not indicate a first flush and, in fact, suggests a "late flush."

Example 6.3.   Using the data collected in Example 6.1, the first-flush characteristics for this particular runoff event can be evaluated. The cumulative oil mass and runoff volume are calculated in Table 6.1. The total runoff volume is

88,000 L and the total oil mass is 333,000 mg. The mass ratio, Y_t, and runoff ratio, X_t, are calculated in Table 6.4. Both equal 1.0 at the end of the rain event. R(t) is calculated as the ratio of these two parameters, and the "R" plot is developed and shown in Figure 6.3, with the R=1 line for comparison. The R(t) curve is concave up in the beginning of the event and then shifts to concave down, suggesting that a first flush does not occur.

A value of R(t) greater than 1, or a concave down Y_t vs. X_t curve, indicates that the pollutant mass being exported is larger than average. But to have a significant first flush, a threshold magnitude for this export should be defined. A number of definitions have been used to define a significant first flush. One of the most common is when at least 60% of a pollutant mass is washed off by only 30% of the initial runoff volume. Others prefer 50% mass in the first 25% of volume (Wanielista and Yousef 1993) or 80% mass in the first 25% of volume (Bertrand-Krajewski et al. 1998).

Example 6.4. A quantitative examination of the oil data in Examples 6.1 and 6.3 can evaluate the fraction of oil mass transported in the first 30% of the runoff volume. By definition, 30% of the runoff occurs at $X_t = 0.3$. This occurred between noon and 12:30 pm. Linear extrapolation can be used to find the approximate time from Table 6.4:

$$12:00 + \frac{0.397 - 0.300}{0.397 - 0.222}(30 \text{ minutes}) = 12:17 \qquad (6.12)$$

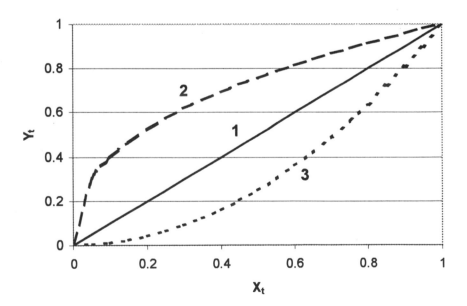

Figure 6.2 Cumulative pollutant mass fraction as a function of cumulative runoff volume fraction. Event 1 indicates a match between pollutant mass and runoff volume. Event 2 indicates a first flush. Higher pollutant mass output at the end of a storm is indicated by Event 3.

Table 6.4. Calculation of first-flush ratio R(t).

Time	$M_t = \sum \overline{Q}\,\overline{C}\,\Delta t$ (mg)	$Y_t = M_t / M$	$V_t = \sum \overline{Q}\,\Delta t$ (L)	$X_t = V_t / V$	$R(t) = Y_t / X_t$
10:30 am					
11:00 am	3020	0.0091	1890	0.021	0.43
11:30 am	26,620	0.080	8010	0.091	0.88
noon	85,920	0.258	19,530	0.222	1.16
12:30 pm	164,420	0.494	34,920	0.397	1.24
1:00 pm	230,120	0.691	50,940	0.579	1.19
1:30 pm	286,620	0.861	65,250	0.741	1.16
2:00 pm	320,620	0.963	76,230	0.866	1.11
2:30 pm	331,020	0.994	83,430	0.948	1.05
3:00 pm	332,820	0.999	87,030	0.989	1.01
3:30 pm	**333,000**	1.000	**88,000**	1.000	1.00

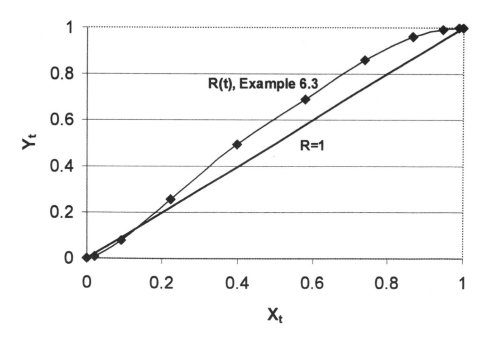

Figure 6.3. First-flush graphical analysis of ratio of runoff mass fraction and volume fraction for Example 6.3.

Thirty percent of the runoff had passed by 12:17 pm. Seventeen minutes is equal to 55.4% of the 30-minute time increment between noon and 12:30. The oil mass passed at this time, also calculated by liner extrapolation, is:

$$0.544 (0.494 - 0.258) + 0.258 = 0.389 \qquad (6.13)$$

Only 38.9% of the oil mass was carried in the first 30% of the runoff volume. This is not indicative of a strong first flush.

6.5.2 First Flush Measurements

Several published data sets have been collected to evaluate the occurrence of a first flush of pollutants from urban watersheds. Deletic (1998) did not find a strong indication of a first flush effect for suspended solids at two European sites. Some events showed a first flush effect; at Lund (Sweden), very strong first flushes were apparent for some events. However, other events showed the opposite effect with the majority of the suspended solids load being transported in the latter portion of the event. The mean curves for the 69 events at Lund show a slight first flush. The mean indicates the lack of a first flush for the 23 events recorded at Miljakovac (Yugoslavia). These curves show the storm-to-storm variation of the temporal distribution of suspended solids washoff. Many of the curves that demonstrated the large first flush characteristic resulted from high intensity events. Overall, these two cases suggest that the storm intensity plays a major role in suspended solids concentrations and loadings, with higher intensity rainfall dislodging large quantities of accumulated particulate matter in a short time.

Little evidence of first flush conditions for suspended solids was found in monitoring 75 runoff events for three highway sites in Texas (Barrett et al. 1998). A first flush was most visible (in terms of decreasing TSS concentration over the event) in events of short duration and constant rainfall intensity. During long durations, continued traffic flow provided a continuous source of contaminants to the drainage area. When the rainfall intensity varied, it is expected that variations in rainfall and flow energies modified pollutant pickup during the event. After the first 3-5 mm of runoff, TSS concentrations were mostly constant.

In measuring TSS directly from roadway pavements in Cincinnati, OH, very weak first flush characteristics (defined at $R > 1$) were noted (Sansalone et al. 1998). The most apparent first flush occurred in a very long duration event.

Less information is available for other pollutants. Measured concentrations of PCBs, Pb, and Zn from a residential/commercial area in France qualitatively showed a first flush, presented as falling pollutant concentrations as the storm duration continued (Granier et al. 1990). The first flush dependence is attributed to the affiliation of these pollutants to particles. Comparing two PCB congeners, (2,2',5,5' tetrachloro and 2,2',3,4,4',5,5' heptachloro) showed that concentrations of the lower chlorinated congener increased with event duration while levels of the highly chlorinated species decreased. Chromium and Lindane exhibited fairly constant concentrations throughout the storm events. Neither conductivity nor water temperature demonstrated first flush characteristics from two asphalt catchments in Europe (Deletic 1998).

Some recent studies of 32 storm events in Maryland in an ultra-urban area evaluated the occurrence of first flush, which was defined as more than 50% of the mass discharging with the first 25% of the runoff volume. First flush was found in eight events for nitrite (NO_2^-), six events for total phosphorus, five events for total Cu and TKN, four events for TSS and nitrate (NO_3^-), and three events for total Pb and Zn. Importantly, a "second flush," which is defined as 50% of the pollutant mass occurring in some 25% of runoff volume beyond the first 25%, was noted in several cases. Three second flush events were found for TSS, two for nitrate, and one each for nitrite, Pb, and Zn.

The strength of the first flush signal will be influenced by the location in the runoff flow path where the water quality measurements are made. When measurements are made close to the pollutant source, fewer factors are able to "smear" a first flush washoff. Factors that depress a first-flush signal include the turbulence in the runoff, dilution from clear water inflow at intermediate points, and the pollutant interaction mechanisms that can occur during overland travel.

The size of the drainage area is important in the appearance and predominance of a first flush. Small drainage areas tend to have a more defined first flush. In small areas, the washoff occurs and is carried rapidly to the discharge point. An example of a pronounced (qualitative) first flush for many compounds is given in the collection of runoff samples combined directly from three types of commercial roofs in Europe (Mason *et al.* 1999). Concentrations were significantly higher in the first few minutes of rainfall, but decreased significantly after that for every compound measured. Example values for one event are presented in Table 6.5.

In larger catchment areas pollutant deposition will not be homogeneous over the entire area. Times of concentration will differ and event intensities may vary over the catchment area. As a result, deposited pollutants are spread over a larger volume and a first flush does not appear at the outfall.

Some evidence indicates that suspended solids may behave differently than other pollutants. The mechanisms involved in the mobilization of suspended solids are different than those for pollutants that are transported in dissolved form. Rainfall and runoff energy will affect these pollutant categories in different ways. Add in the fact that many pollutants are transported as both dissolved and affiliated with particulates and the pollutant transport process can be very complex.

6.5.3 The Antecedent Dry Weather Period

The antecedent dry weather period (ADWP) is considered as an important parameter in stormwater runoff quality and in modeling. The ADWP is defined as the time between the end of a precipitation event and the beginning of another. The standard assumption for water quality modeling is that pollutants build up on surfaces throughout the watershed between the rainfall events at a constant rate. When an event comes, the pollutants that accumulated since the last storm event washoff over the duration of the event. Therefore, the longer the ADWP, the greater the quantity of pollutants deposition and the higher are the concentrations and loadings of pollutants. Many stormwater quality models use the ADWP in predicting available pollutant loads for an event.

Table 6.5. Rainfall component concentrations from roof runoff (combined runoff from three commercial roofs) in Switzerland comparing first flush and steady-state concentrations (Mason *et al.* 1999).

Parameter	Concentration	
	First flush (first few minutes after beginning of rain)	**Steady state**
Ca (mM)	0.4-1.3	0.1-0.5
Mg (mM)	0.06-0.44	0.01-0.04
Na (mM)	0.22-0.56	0.02-0.20
K (mM)	0.10-0.35	0.02-0.07
NH4 (mM)	0.14-0.29	0.02-0.07
Alkalinity (mM)	0.7-2.8	0.3-0.6
Cl (mM)	0.06-0.22	0.02-0.06
SO4 (mM)	0.1-0.2	0.01-0.1
ortho-P (μM)	1-2	0.1-1.0
NO3 (mM)	0.29-0.64	0.02-0.20
Total N (mM)	0.50-1.2	0.07-0.30
DOC (mM)	1-3	0.5-1.0
Acetate (μM)	40-135	8-20
Formate (μM)	35-62	0-18
Oxalate (μM)	10-17	3-10

Metals	*Dissolved*	*Particulate*	*Dissolved*	*Particulate*
Cd (μg/L)	<0.01-0.48	<0.01-0.11	<0.01-0.47	<0.01-0.07
Cu (μg/L)	2.0-56	0.2-13.8	1.0-11.6	0.1-14.7
Cr (μg/L)	0.2-2.6	0.2-9.6	0.2-2.2	0.1-4.0
Pb (μg/L)	0.06-2.73	0.33-29.9	<0.01-0.47	0.04-16.7
Zn (μg/L)	8.4-468	3.1-98	<0.05-87	<0.05-54

A few sampling investigations, however, have provided little support for pollutant load dependencies on ADWP. A slight dependence on the ADWP was found for runoff conductivity measured from two different sites in Europe, but dependence was not noted for suspended solids concentration (Deletic 1998). Suspended solids loading rate was found to be a function of rainfall intensity. Data of Granier *et al.* (1990) also do not support the concept of a dependence of loads on ADWP. Comparing two events of nearly equal flow volume, but one with 3 antecedent dry days and the other with 22, showed equal transported loads of PCBs (1.6 g) and Lindane (0.2 and 0.3 g). As with first flush, however, water quality data measurements taken closer to the pollutant source may exhibit a stronger relationship between loadings and ADWP. Total Cu and Pb (dissolved +

particulate) loads direct from roof runoff in Switzerland were found to increase with the duration of the event and with the length of the ADWP (Mason *et al.* 1999).

Rainfall intensity will play a role in pollutant mobilization and can confound simple ADWP correlations. A high intensity event produces rainfall and runoff flow with greater kinetic energy, able to detach, mobilize, and dissolve more particulate matter from the land and impervious surfaces. Consequently, more intense storms carry higher solids loadings than less intense storms of the same volume. Even though the total pollutant levels accumulated on the various watershed surfaces is high during long ADWPs, some of these pollutants can be mobilized only by a reasonably high intensity event. Pollutants that are not mobilized during a weak event will remain on the surface and contribute to the accumulation during the next ADWP. With this in mind, understanding pollutant buildup and discharge during a precipitation event may depend on antecedent conditions over a period of several previous events.

6.6 PARTICULATES IN STORMWATER RUNOFF

6.6.1 Physical Characteristics

Particle size will control the particle settling velocity, and accordingly, the trap efficiency in a treatment practice. Therefore, knowledge of the particle sizes present in runoff is necessary to evaluate the fate of these particles. Unfortunately, few careful studies that characterize particle sizes have been completed. Several recent investigations have evaluated particle size distributions in urban stormwater runoff. A summary of these data is presented in Table 6.6. The sizes are presented in terms of d_{10}, d_{30}, etc. Ten percent of the particles (by mass) are less than d_{10} and 30% are less than d_{30}, and so forth.

Table 6.6. Particle size distributions in urban stormwater runoff

Sizes (μm)	Street runoff (Wu et al. 1996)	Motorway runoff (Roger et al. 1998)	Roadway runoff (Sansalone et al. 1998)		Street swept sediments (Viklander 1998)
			Mean	Range	
d_{10}	1	-	110	70-190	120
d_{30}	3	-	300	210-410	-
d_{50}	7	-	600	350-800	1500
d_{70}	15	-	1000	600-1800	-
d_{90}	26	54	3300	2000-6000	6000

The size distributions are obviously very different among the four studies described in Table 6.6. The first three studies characterized particles captured in runoff, but using different methods to capture the flow and some may not have captured particles that are easily settled. The last study removed particulates directly from streets, but many of the particles on the streets may not be mobilized during a runoff event.

Particle accumulation on roadways or in runoff can be strongly dependent on the time that the measurements are made. In most areas that receive frozen precipitation, sand, cinders, and other materials are commonly spread on roadways to improve traction, along with deicing agents. These particles can be washed off with subsequent high intensity precipitation (Viklander 1998).

6.6.2 Chemical Characteristics

Two issues are important with respect to evaluating the chemistry of stormwater particulates. First, the materials that make up the particulates needs to be characterized, including the density of the particles, their toxicity, and the degree to which other materials may adsorb to them. The suspended solids in runoff are generally comprised of local soils that are deposited and mobilized. Soils consist primarily of clays, which are aluminosilicate minerals, along with oxides of silicon, iron, and aluminum. Organic matter from decaying vegetation is usually present as a small fraction. Other particulate matter that has been deposited on the drainage area or is mobilized can contribute to the suspended solids load. Many of these other materials have anthropogenic origin and include particulates that wear from pavements, building materials, and auto brakes and tires.

Grout *et al.* (1999) evaluated the chemical characteristics of colloidal materials in stormwater runoff from Houston, TX. The colloids were defined as being those less than 45 μm in size, but could be separated by ultrafiltration. The elements found to be mostly in the colloidal phase were C, Si, and Fe. Al, Ca, K, Mg, Ba, and Cu were found to be primarily in the dissolved state based on this fractionation. The make-up of the colloidal phase varied over the duration of the storm event. Baseflow colloids were primarily silica. At the beginning of runoff, the primary elements included C, Fe, and Al, but the latter two decreased with time. During the period of peak flow, the makeup shifted again, being dominated by Si, also with high levels of Al and Fe.

The second important chemical characteristic is the adsorbed load carried by particulates. In water, heavy metals, hydrophobic organic compounds, and many other pollutants affiliate very strongly with particulate matter. With metals, this affiliation results from sorption reactions between the metals and various inorganic and organic components of the solids. Usually these sorption reactions are strongly pH dependent, with the affiliation becoming stronger at higher pH for common cationic metals (Stumm and Morgan 1996). Many organic pollutants partition into the particulate organic matter; the greater the organic content of the particle, the higher is the expected organic pollutant load.

Based on limited information available from the U.S. and Europe, the partitioning of metals between the dissolved and particulate phases in urban runoff vary significantly from metal to metal. Copper, cadmium, chromium, and zinc have been found predominantly in the dissolved form. Lead, nickel, aluminum, and iron, however, were primarily affiliated with the particulate matter (Sansalone and Buchberger 1997, Sedlak *et al.* 1997, Mason *et al.* 1999). This partitioning can have a significant effect the on treatment efficiency for metals. Metals affiliated with particles will be removed by stormwater treatment practices that target sediment removal. Dissolved metals, however, will not be affected by physical treatment processes.

The affiliation of heavy metals with particulates is not homogeneous across all types of particles and sizes. Viklander (1998) determined metal contents in street sediments in Sweden. For two areas with a moderate traffic count, the highest fraction for each of the four metals examined, lead, copper, cadmium, and zinc, were found in the smallest particles analyzed, 75 μm and smaller. The levels of metals generally decreased with each larger size fraction. However, for a city center street with a high traffic count (20,000 vehicles/day), the highest levels of Cd and Pb were not found in the 75 μm size, but in the 250 μm fraction. The concentrations of metals in the sediments were also higher from this high traffic area. Therefore, even though metals are affiliated with particles, removal of a large amount of particles does not guarantee similar removal of metals.

6.7 POLLUTANT SOURCES

To address pollutant reduction in stormwater runoff, some understanding about the sources of the pollutants is necessary. It is important to know which land use areas contribute the most pollutants and the form of these pollutants. In a multiple land use watershed, several different pollutant sources are likely. Even quiet residential areas can contribute significant levels of pollutants to runoff.

6.7.1   Contributions from Different Land Uses

A detailed study was completed in Wisconsin in the early 1990's to evaluate runoff concentrations and loadings from different source areas within residential, commercial, and industrial land uses (Bannerman *et al.* 1993). Fecal coliform levels were found to be much higher from residential areas, supporting the hypothesis that pets and wildlife contribute to pathogen loadings. The residential area was divided into six source areas. Lawns constituted the largest fraction, 66.7% of the total residential area. Lawns were followed by roofs at 12.8%, feeder streets at 8.8%, driveways at 5.2%, and collector streets at 5.1%. The smallest component was sidewalks at 1.3%. Geometric means of contaminant concentrations from these areas are presented in Table 6.7.

Suspended solids and phosphorus levels from lawns were relatively high. High levels of zinc were discharged from roofs, which suggests the presence of galvanized downspouts and flashing. The highest concentrations of nearly all pollutants were found to originate from streets.

Weighting the concentration for each component by the drainage area allowed the calculation of the contribution of each source to the total pollutant load. The results of these calculations are presented in Table 6.8. Feeder and collector streets dominate every category. The sum of these two sources ranged from 58% (for copper) to 80% (TSS and zinc) of the pollutant load. Driveways were the second most important source in all categories. Other than phosphorus, pollutant loadings from lawns were less than 10%. These results demonstrate the significance of driveway/roadway/street impervious areas in contributing pollutants to runoff and reinforce the importance of automobiles as a non-point water pollutant source. They also indicate that grassy areas can help control overall pollution levels. Because of the different pollutant loads contributed from the different watershed components, optimal results can be obtained by integrating water quality control methods with land uses.

Similar information for suspended solids levels in runoff from various land used are presented in Table 6.9 (Clark *et al.* 1998). These values range over three orders of magnitude, from 10 to 10,000 mg/L. Not surprisingly, disturbed land on a construction site had the highest TSS level. Anytime bare soil is exposed, the potential for erosion is high. States and municipalities have established specific guidelines to minimize runoff from disturbed soils during construction operations.

Table 6.7. Geometric mean contaminant concentrations in a residential area in Wisconsin (Bannerman *et al.* 1993).

Contaminant	Lawns	Roofs	Feeder Streets	Driveways	Collector Streets
Suspended Solids (mg/L)	397	27	662	173	326
Total P (mg/L)	2.67	0.15	1.31	1.16	1.07
Total Cd (μg/L)	-	-	0.8	0.5	1.4
Total Cr (μg/L)	-	-	5	2	12
Total Cu (μg/L)	13	15	24	17	56
Total Pb (μg/L)	-	21	33	17	55
Total Zn (μg/L)	59	149	220	107	339
Fecal Coliform (cfu/100 mL)	42,100	294	92,100	34,300	56,600

Table 6.8. Contributions of watershed components to total pollutant loads based on data from a residential area in Wisconsin (Bannerman *et al.* 1993).

Contaminant	Lawns	Roofs	Feeder Streets	Driveways	Collector Streets	Sidewalks
Total Solids	7%	<1%	56%	12%	20%	5%
Suspended Solids	7%	<1%	62%	9%	18%	4%
Total P	14%	<1%	39%	20%	19%	8%
Total Cu	3%	1%	33%	13%	45%	5%
Total Zn	2%	2%	42%	11%	38%	5%
Fecal Coliform	5%	<1%	57%	12%	21%	5%

Table 6.9. Suspended solids concentrations expected in runoff from land uses and treatment BMPs (Clark *et al.* 1998).

Source area	TSS concentration (mg/L)
Roof Runoff	10
Paved Parking, Storage, Driveway, Streets, Walk Areas	50
Unpaved Parking and Storage Areas	250
Landscaped Areas	500
Construction Site Runoff	10,000

6.7.2 Specific Pollutant Sources

A more focused characterization of the contributions of pollutants to stormwater runoff requires identification of the specific source of the pollutants. When land is developed, buildings are added and automobile use becomes more widespread. Human intervention adds various materials that contribute pollutants to developed lands. To evaluate pollutant input from a fundamental mass balance perspective and to investigate pollution prevention issues for runoff, these direct sources need to be identified and quantified.

As alluded to in the previous section, a significant fraction of runoff pollution, especially toxics, can be attributed to various processes related to vehicles and their use. Wear from tires is one pollutant source. As tires wear during travel, the tire tread material is abraded by the pavement surface into very fine particles. Tire tread contains a high concentration of zinc, which accrues at the various accumulation points of these particles on and near roadways. Although it is originally bound in the tire matrix, the zinc is released in a more available form as the matrix is slowly biodegraded over time.

Automobile brakes are a second source of metals from vehicles. Brake pads are purposely designed to wear during the frictional braking process, emitting small particulates to the local wheel environment. The brake wear material contains metals such as copper, iron, lead, and zinc, which deposit on the vehicle and roadways and ultimately are washed into receiving water bodies.

Fluid leakage from automobiles can be responsible for major pollutant loadings. Leaks of motor oil, fuel, gear fluids, and coolants can be washed from impervious parking surfaces and impact the aquatic life in the receiving water bodies. Oil sheens are visible evidence of this source of pollution. In many cases, specific toxic organic compounds (and sometimes metals), which may persist in the environment, are present in these fluids.

Vehicle combustion emissions provide yet an additional source of pollutants to an urban environment and are especially detrimental to receiving aquatic ecosystems. Although tailpipe emissions are associated with air pollution problems, the fate of many of these pollutants leads back to the earth surface, through dry deposition, or wet deposition in the rainfall itself. In either case, toxic organic compounds from incomplete combustion, nitrate from NO_x emissions, and other pollutants are scavenged by rainfall and runoff. Some pollutants condense rapidly in the atmosphere after being discharged from the tailpipe and accumulate only short distances from the point where they were emitted, while others will travel much further. As a historical example, lead was used in gasoline for many years to help improve engine performance. The use of lead in gasoline was phased out in the U.S. in the 1970s and 1980s, but many soils near major roadways still have elevated lead levels due to these emissions and condensation and deposition processes. All in all, runoff from roadways and parking areas, from high automobile use, will contain high levels of many pollutants. These areas may have the greatest water quality impact in the drainage area and should be targeted in stormwater runoff treatment scenarios.

Buildings and other structures contribute low levels of pollutants in developed areas. Heavy metals can leach from paints or other coatings and from various materials used in construction. In older areas, lead paint still can be found on exterior building surfaces. Galvanized metals are employed in various types of construction, especially on roofs, which will gradually release zinc. Copper sheeting is commonly employed as flashing and as decorative covering. Other pollutants can be released by weathering of various asphaltic roofing materials. As these materials weather and corrode, particulate and dissolved pollutants are mobilized by impinging rainfall.

Gravel roofs can contribute high levels of Ca, Mg, Na, and K to the runoff flow (Mason *et al.* 1999). Results suggested that metals from atmospheric deposition were attenuated by the roof materials and that nitrification was taking place on the gravel roofs during long-duration storm events.

The release of the herbicide *R-mecoprop* and a similar compound, *S-mecoprop*, were detected from roof runoff from a number of buildings in Switzerland (Bucheli *et al.* 1998). These compounds were formed from the chemical and biological hydrolysis of a polymer compound used in bituminous roof sealants. Concentrations of the two *mecoprop* compounds ranged from 1 to 500 μg/L, with an average of 4 μg/L from a number of roofs. The Swiss and European Union drinking water standard for S-mecoprop is 0.1 μg/L.

Through sampling of buildings and automobile components, estimation of fluxes from literature information, and simple assumptions on rainfall and runoff hydrology, the flux of heavy metals into stormwater from residential areas was estimated, as shown in Figure 6.4 (Davis *et al.* 2001). Washoff from buildings was found to be a major contributor of lead. Brake emissions from automobiles and buildings were the major sources of copper. Most of the cadmium was attributed to atmospheric emissions. Several sources, including buildings and auto tire wear, were important for zinc, which is found at much higher concentrations because of the many sources of zinc in developed areas.

In a more focused study on lead, concentrations of lead in a controlled wash from painted buildings in the Washington, DC, area ranged from less than 2 μg/L to greater than 20,000 μg/L (Davis and Burns 1999). The sampling results were approximated by a log-normal distribution. Older, peeling paints produced higher lead levels than did newer paints. It was found that 10% of the samples produced 87% of the lead, suggesting that a few buildings could be contributing large amounts of lead to stormwater runoff in areas where exterior lead paint is still present.

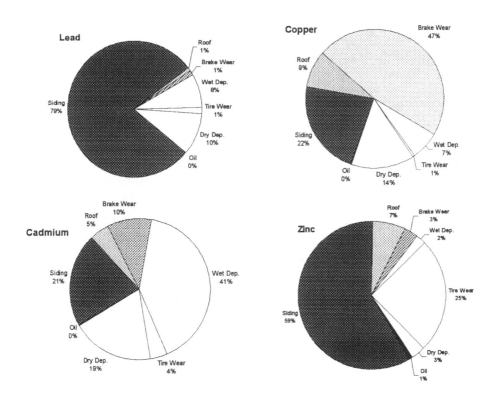

Figure 6.4. Estimated contributions of in urban runoff from various sources (Davis et al. 2001).

Intensive landscaping care can contribute several different pollutants into runoff from urban areas. Fertilizers are applied to managed lawns, possibly several times per year. If not administered properly, nitrogen and phosphorus compounds can be washed from these lawns. Fertilizers carelessly spread onto impervious surfaces are 100% runoff contributions. Various herbicides and insecticides are frequently employed to control weeds and damage from insects. Again, without careful control of application, excess pesticides will wash from the applied areas.

In high-population areas, wastes from pets, which if not cleaned up, can produce significant organic and pathogen loadings that may wash into waterways. A study that evaluated bacterial loadings in streams after precipitation events noted a strong dependency on the degree of development in the drainage basin (Young and Thackston 1999). In the four sites investigated, *E. Coli*, fecal coliform, and fecal strepticocci levels were much higher in the more developed watersheds following storm events. High bacterial counts were correlated with population, percent impervious area, and apparent domestic animal density. Ratios of fecal coliform to fecal strepticocci suggested that the bacteria were of animal origin. Evidence indicates that waste from pest, birds, squirrels, and other animals in residential areas can be a major contributor of pathogen pollution to waterways during precipitation events.

Other pollutants also originate from developed areas. Any type of disturbance that creates uncovered ground provides the potential for sediments to be detached by rain and mobilized by runoff, during or after construction. Soaps, detergents, and other cleaning products that are used exteriorly on cars, buildings, or other objects can be washed to storm drains and into streams.

Atmospheric deposition can be a significant source of a number of pollutants in runoff. Combustion processes, both stationary, such as incinerators and industrial furnaces, and mobile, such as automobiles, discharge trace pollutants to the atmosphere. Atmospheric pollutants are transported back to the land surface by both dry and wet deposition processes. Pollutants partition onto dust particulates and aerosols and settle out over land and water areas as dry deposition. Dry deposited materials accumulate until they are washed off by the next precipitation event. Pollutants present in atmospheric gases are scavenged by falling raindrops, transporting these pollutants along with the rain. When these depositions occur on impervious surfaces, the pollutants are directly available to mobilization by runoff.

Studies have indicated that less than 10% of the Fe; 10-30% of the TSS, TDS, TP, NO_{2+3}-N, Zn, and PCB's; 30-55% of the Cu, Cr, Pb, COD, and OP; 70-90% of TKN and NH_3-N; and nearly all of the Lindane in urban runoff may be attributed to rainfall contributions (Granier et al. 1990, Barrett et al. 1998, Wu et al. 1998). One estimate for nitrate was 72% and those for Cu ranged from 17 to 30-55%.

As part of the Chesapeake Bay Atmospheric Deposition Study in the early 1990's, both dry and wet deposition of trace metals and hydrophobic organic chemicals were measured (Leister and Baker 1994; Scudlark et al. 1994; Wu et al. 1994). Both sources contribute approximately equally to the metal deposition. Dry and wet atmospheric deposition of several species were also monitored for several months in Switzerland (Mason et al. 1999). The values for the Switzerland study, which was completed at an industrial city area, were found to be about an order of magnitude greater than those of the Chesapeake Bay study, which were based on studies of rural areas of the Bay, and wet deposition was found to be higher than dry in most cases.

With the introduction of catalytic converters to automobiles in the 1970's to reduce combustion pollutant emissions, along with the phase out of leaded gasoline, the characteristics of automobile tailpipe emissions have been significantly altered over the past few decades. The primary catalysts are made of rare metals: platinum, palladium, and rhodium. While these catalysts have been developed to address a major pollution problem, they are not without some degree of environmental price. During the operation of the automobile, high temperatures in the catalytic converter release minute amounts of the catalyst metals, which ultimately accumulate in runoff and roadside sediments. Concentrations of platinum, palladium, and rhodium in roadside sediments (< 63 μm) in Sweden were found to be approximately 190, 100, and 80 ng/g, respectively (Rauch et al. 2000). Although these levels are low, this buildup of rare metals bears watching.

Collectively, all of these pollutants from a large developed area can have a major impact on water bodies and various ecosystems during wet weather flows, and as a later consequence of wet weather loadings. Some of these pollutant sources are unavoidable with development as currently practiced. Heavy metals are commonly used in automobiles and structural components. Addressing these pollutants requires treatment of runoff before it enters the natural water environment, or elimination of toxic materials and other pollutants from our daily use.

6.8 EMPIRICAL HIGHWAY RUNOFF MODELS

As discussed throughout this chapter, highways, roads, and streets are major contributors of pollutants to urban runoff. Several expressions have been developed to estimate pollutant levels based on roadway use and climatic conditions. From studies in North Carolina, the TSS concentration in highway runoff could be predicted using an equation based on total traffic count (Wu *et al.* 1998):

$$TC_{TSS} = 0.86 \text{ (VDS)} - 650 \quad \text{(for VDS} < 756) \tag{6.14}$$

where TC_{TSS} is the specific loading of suspended solids, given in mg/m^2 of road surface and VDS is the total traffic count during the duration of a rain event. Dividing TC_{TSS} by the depth of runoff provides an estimate of the TSS EMC for the event.

Similar, more complex predictive regression expressions have been developed to predict loadings (g/m^2) of TSS, VSS, COD, BOD_5, oil and grease, phosphorus, nitrate, iron, zinc, lead, and copper in highway stormwater pollutant loads from Austin, TX (Irish *et al.* 1998). Parameters used in the regressions include storm duration, flow, intensity, vehicle travel parameters, and previous storm event parameters. These expressions are presented in Table 6.10.

Example 6.5. A 3-hr storm event with 16 mm of rainfall along a stretch of highway results in 14 mm of runoff. The traffic count during this time is 1800 vehicles. The EMC for suspended solids in the runoff from this roadway can be estimated. From Eq. (6.14), TC_{TSS} is determined from the traffic count.

$$TC_{TSS} = 0.86 \text{ (1800)} - 650 = 898 \text{ mg/m}^2 \tag{6.15}$$

$$\begin{aligned} EMC_{TSS} &= (898 \text{ mg/m}^2)/(0.014 \text{ m runoff}) = 64100 \text{ mg/m}^3 \\ &= 64.1 \text{ mg/L TSS in runoff.} \end{aligned} \tag{6.16}$$

6.9 STOKES LAW

The situation of a solid particle free falling through a liquid arises frequently in engineering applications. This is the case for a sediment particle suspended in water. Three forces act on the particle: gravity, buoyancy, and drag. The first force acts downward and the latter two forces act upward. Since the drag force is a function of settling velocity (as to be seen below), at some point the settling velocity becomes high enough so that the vector sum of the three forces is equal to zero:

Table 6.10. Regression equations used for the prediction of pollutant concentrations in highway runoff (Irish *et al.* 1998).

Number	Pollutant (g/m^2)	Equation
1	TSS	TSS = 0.2556 + 0.3068(*Flow*) + 2.018(*Intensity*) + 0.0037(*ADP*) - 2.986(*PINT*)
2	VSS	VSS = -0.0186 + 0.0348(*Flow*) + 0.1649(*Intensity*) + 0.0005(*ADP*) - 0.0069(*PFLOW*) - 0.6721(*PINT*)
3	COD	COD = -0.0613 + 0.0007(*Duration*) + 0.07773(*Flow*) + 0.7785(*Intensity*) - 0.0041(*ADP*) - 6x10^{-6}(*ATC*)
4	BOD	BOD = -0.0081 + 0.0035(*Flow*) + 0.0619(*Intensity*) + 1.1x10^{-5} (*VDS*) + 1.5x10^{-7}(*ATC*)
5	Oil & Grease	O&G = -0.0004 + 0.003(*Flow*) + 1.0x10^{-5} (*VDS*)
6	Phosphorus	P = -0.0005 + 3.3x10^{-6}(*Duration*) + 0.0002(*Flow*) + 0.0032(*Intensity*) - 5.1x10^{-9}(*ATC*)
7	Nitrate	NO3 = -0.0015 + 0.0006(*Flow*) + 0.0086(*Intensity*) + 1.2x10^{-8}(*ATC*)
8	Iron	Fe = -0.0028 + 0.0042(*Flow*) + 0.0282(*Intensity*) + 2x10^{-5}(*ADP*)
9	Zinc	Zn = 0.0002 + 2.5x10^{-6}(*Duration*) + 0.0001(*Flow*) + 4.9x10^{-9}(*ATC*) - 3.2x10^{-6} (*PDUR*) + 0.0003(*PFLOW*) - 0.0241(*PINT*)
10	Lead	Pb = 0.0008 + 6.5x10^{-5}(*Flow*) - 0.002(*Intensity*) + 8.0x10^{-8}(*VDS*) - 0.0023(*PINT*)
11	Copper	Cu = 1.9x10^{-5} + 3.8x10^{-6}(*Duration*) + 2.4x10^{-5}(*Flow*) - 2.4x10^{-7}(*VDS*)
Flow	L/m^2	total volume of flow per unit area of watershed during the storm event
Duration	min	total duration of storm event
Intensity	L/m^2-min	*Flow divided by Duration*
ADP	hours	total duration of the antecedent dry period
ATC	vehicles/lane	average number of vehicles using the highway during the *ADP* in a single lane
PDUR	min	total duration of preceding storm
PFLOW	L/m^2	total volume of flow per unit area of watershed during the preceding event
PINT	L/m^2-min	preceding event intensity, = *PFLOW/PDUR*
VDS	vehicles/lane	average number of vehicles using the highway during the storm duration in a single lane

$$\sum F = 0 = F_b + F_D - W \qquad (6.17)$$

in which F_b is the buoyant force, F_D is the drag force, and W is the weight (gravitational force). For a sphere, these forces are computed by the following:

$$F_b = \rho g v = \gamma \left(\pi D^3 / 6 \right) \qquad (6.18)$$

$$F_D = C_D \frac{1}{2} \rho A V^2 = \rho C_D \pi V^2 D^2 / 8 \qquad (6.19)$$

$$W = \rho_s g v = \gamma_s \left(\pi D^3 / 6 \right) \qquad (6.20)$$

in which v is the particle volume, D is the diameter of the sphere, γ is the specific weight of the fluid, V is the free-fall terminal velocity, γ_s is the specific weight of the soil particle, and C_D is the drag coefficient. For low Reynolds Number, less than about 5, the drag coefficient is approximately:

$$C_D = \frac{24}{\mathbf{R}} = \frac{24\mu}{\rho V D} \qquad (6.21)$$

in which $\mathbf{R}$ is the Reynolds number, ρ is the water density, and μ is the water viscosity. Substituting Eqs. (6.18), (6.19), (6.20), and (6.21) into Eq. (6.17) yields:

$$0 = \gamma \left(\frac{\pi D^3}{6} \right) + 3\pi \mu V D - \gamma_s \left(\pi \frac{D^3}{6} \right) \qquad (6.22)$$

$$= \gamma D^2 + 18 \mu V - \gamma_s D^2 = 18 \mu V - D^2 (\gamma_s - \gamma)$$

Solving for the terminal velocity gives Stokes Law:

$$V = \frac{D^2 (\gamma_s - \gamma)}{18 \mu} \qquad (6.23)$$

where γ_s and γ are the specific weights of the soil and the liquid medium, respectively. Stokes Law has several inherent assumptions: (1) the particle is not influenced by other particles or nearby objects and surfaces; (2) the particles are spherical; and (3) the specific gravity of the soil particle and the viscosity of the fluid are known exactly. These assumptions, especially the first two, do not usually hold but Stokes Law is commonly used in spite of these problems. For materials that are falling in water and that have a specific weight near that of soil grains (typically around 165 lb/ft^3, Stokes Law applies for particles that have diameters between 0.0002 mm and 0.2 mm.

The driving force for the settling is the difference in specific weight (or density) between the particle and the water. The greater this difference, the faster the particle will settle. Also, the importance of the particle size is apparent in Eq 6.23.

The settling velocity is proportional to the square of the diameter of the particle. Thus, larger particles, sand and coarse silt, readily settle from water and are relatively easily removed. Smaller particles, clays and fines, settle very slowly and may be transported long distances before settling to the channel bottom. They are also very difficult to remove in treatment practices since the treatment times are usually short.

Sedimentation is a very important mechanism in the removal of sediments from stormwater runoff. Detention ponds and wetlands are specifically designed to promote sedimentation of particulates. In areas where runoff flows are slowed, such as filter strips or swales, increased time for sedimentation is available which enables this water quality improvement mechanism to be exploited.

Stokes law is also valid for evaluating a floating object. Therefore, it can be applied to the design and analysis of stormwater management devices for the removal of oil droplets, grease, and other small flotable materials.

Example 6.6. Runoff into a small stream transports sediment with a mean diameter of 0.000164 ft (0.05 mm) at 55°F. The stream enters a wetland with an average depth of 2.5 ft. Stokes Law is used to find the settling velocity. The water specific weight and viscosity are found from the water properties table given in Appendix B:

$$V = \frac{D^2(\gamma_s - \gamma)}{18\mu} = \frac{(0.000164 \text{ ft})^2(165 - 62.4)\text{lb/ft}^3}{18(2.5 \times 10^{-5}\text{lb} - \text{sec/ft}^2)} \tag{6.24}$$
$$= 0.00613 \text{ ft/sec}$$

Thus, the time required for this particle to settle the 2.5 ft wetland depth is:

$$t = \frac{L}{V} = \frac{2.5 \text{ ft}}{0.00613 \text{ ft/sec}} = 408 \text{ sec} = 6.8 \text{ min} \tag{6.25}$$

If the water moves within the wetland at a horizontal velocity of 1 ft/sec, then the particle will move horizontally for a distance of 408 ft during the time that the particle moves 2.5 ft vertically.

Example 6.7. The wash load, which consists of silt and clay particles, settles very slowly in a river. Consider a silt particle with a diameter of 0.3×10^{-4} ft. At 55°F it has a settling velocity of:

$$V_s = \frac{D^2(\gamma_s - \gamma)}{18\mu} = \frac{(0.3 \times 10^{-4})(165 - 62.4)}{18(2.5 \times 10^{-5})} = 0.0002 \text{ ft/s} \tag{6.26}$$

For it to settle 1 ft requires 5,000 sec. In this time the particle could travel 20,000 ft downstream in flow with a velocity of 4 ft/sec. Taking resuspension into consideration, silt and clay particles are not likely to settle to the river bed.

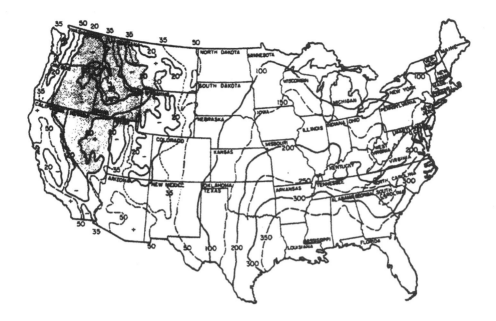

Figure 6.5. Average annual values of the rainfall-erosivity factor, R.

6.10 UNIVERSAL SOIL LOSS EQUATION

Sheet erosion is defined as erosion that results from a combination of splash action forces and forces associated with thin-layered surface runoff. A number of methods have been proposed to estimate sheet erosion rates. The Universal Soil Loss Equation is the most widely used for agricultural for vegetated landscapes.

In the Universal Soil Loss Equation (USLE), the soil loss (E) in tons/acre/year is a function of the erosivity index (R), a soil erodibility factor (K), the field slope (S) and length (L), a crop management factor (C), and a conservation practice factor (P):

$$E = RKTCP \qquad (6.27)$$

in which T is the topographic factor, which depends on L and S.

The erosivity index R is a summation of the product of the kinetic energy of rainfall (hundreds of ft-tons/acre) and the maximum 30-min rainfall intensity (in./hr) for all significant storms in a year. The value of R has been computed for locations within the United States and is given in Figure 6.5.

The soil erodibility factor (K) is the average soil loss (tons per acre per unit of the rainfall factor R) from a particular soil in cultivated continuous fallow with standard values of both the plot length and the slope, which have been selected somewhat arbitrarily as 73 ft and 9%, respectively. The value of K is a function of

(1) the percent silt plus very fine sand, (2) the percent sand, (3) the percentage of organic matter, (4) a soil structure index, which is presented on an ordinal scale, and (5) a permeability index, which is also measured on an ordinal scale. For U.S. mainland soils, the value of K is obtained from Figure 6.6.

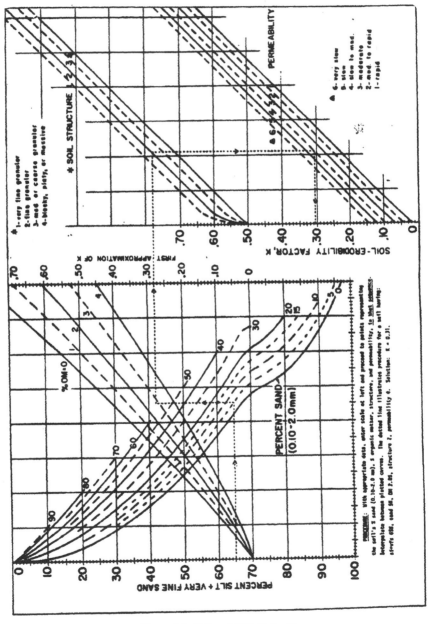

Figure 6.6. Soil erodability factor, K

Table 6.11. Values of the USLE's Topographic Factor, T, for specified combinations of slope length and steepness

Slope (%)	Slope Length (ft)											
	25	50	75	100	150	200	300	400	500	600	800	1000
0.5	0.07	0.08	0.09	0.10	0.11	0.12	0.14	0.15	0.16	0.17	0.19	0.20
1	0.09	0.10	0.12	0.13	0.15	0.16	0.18	0.20	0.21	0.22	0.24	0.26
2	0.13	0.16	0.19	0.20	0.23	0.25	0.28	0.31	0.33	0.34	0.38	0.40
3	0.19	0.23	0.26	0.29	0.33	0.35	0.40	0.44	0.47	0.49	0.54	0.57
4	0.23	0.30	0.36	0.40	0.47	0.53	0.62	0.70	0.76	0.82	0.92	1.0
5	0.27	0.38	0.46	0.54	0.66	0.76	0.93	1.1	1.2	1.3	1.5	1.7
6	0.34	0.48	0.58	0.67	0.82	0.95	1.2	1.4	1.5	1.7	1.9	2.1
8	0.50	0.70	0.86	0.99	1.2	1.4	1.7	2.0	2.2	2.4	2.8	3.1
10	0.69	0.97	1.2	1.4	1.7	1.9	2.4	2.7	3.1	3.4	3.9	4.3
12	0.90	1.3	1.6	1.8	2.2	2.6	3.1	3.6	4.0	4.4	5.1	5.7
14	1.2	1.6	2.0	2.3	2.8	3.3	4.0	4.6	5.1	5.6	6.5	7.3
16	1.4	2.0	2.5	2.8	3.5	4.0	4.9	5.7	6.4	7.0	8.0	9.0
18	1.7	2.4	3.0	3.4	4.2	4.9	6.0	6.9	7.7	8.4	9.7	11.0
20	2.0	2.9	3.5	4.1	5.0	5.8	7.1	8.2	9.1	10.0	120	13.0
25	3.0	4.2	5.1	5.9	7.2	8.3	10.0	12.0	13.0	14.0	17.0	19.0
30	4.0	5.6	6.9	8.0	9.7	11.0	14.0	16.0	18.0	20.0	23.0	25.0
40	6.3	9.0	11.0	13.0	16.0	18.0	22.0	25.0	28.0	31.0	-	-
50	8.9	13.0	15.0	18.0	22.0	25.0	31.0	-	-	-	-	-
60	12.0	16.0	20.0	23.0	28.0	-	-	-	-	-	-	-

The topographic factor (T) is a function of the slope (%) and the overland length (ft), which is usually less than 400 ft. The slope is the average land gradient. The length is the average distance from the point of overland flow to whichever of the following limiting conditions occurs first: (1) the point where the slope decreases to the extent that deposition begins or (2) the point where runoff enters well-defined flow areas, such as rills. Values of the topographic factor are given in Table 6.11. For the index slope of 9% and the index length of 73 ft, T equals 1.

The crop management factor (C) is the ratio of the soil quantities eroded from land that is cropped under specific conditions to that which is eroded from clean-tilled fallow under identical slope and rainfall conditions. Values for C are given in Table 6.12 for the 37 states east of the Rocky Mountains. The baseline conditions for which C equals 1 are continuous fallow, tilled up and down slope. For other cover and management conditions, C is less than 1. The conservation practice factor (P) is a function of the support practice and the land slope. Values for P are given in Table 6.13 and represent a comparison of the land use practice to straight-row farming.

Table 6.12. Generalized values of the Cover and Management Factor, C, in the 37 states east of the Rocky Mountains.

		Productivity level	
		High	Med.
	Crop, Rotation, and Management	C value	
	Base value: continuous flow, tilled up and down slope	1.00	1.00
CORN	C, RdR, fall TP, conv (1)	0.54	0.62
	C, RdR, spring TP, conv (1)	0.50	0.59
	C, RdL, fall TP, conv (1)	0.42	0.52
	C, RdR,we seeding, spring TP, conv (1)	0.40	0.49
	C, fall shred stalks, spring TP, conv (1)	0.35	0.44
	C(silage), W(RdL, fall TP)(2)	0.31	0.35
	C, RdL, fall chisel, spring disk, 40-30% (1)	0.24	0.30
	C(silage), W we seeding, no-til pl c-k W(1)	0.20	0.24
	C, fall shred stalks, chisel pl, 40-30% re (1)	0.19	0.26
	C-C-C-W-M, RdL, TP for C, disk for W (5)	0.17	0.23
	C, RdL, strip till row zones, 55-40% re (1)	0.16	0.24
	C-C-W-M, RdL, TP for C, disk for W (4)	0.12	0.17
COTTON			
	Cot, conv (Western Plains)(1)	0.42	0.49
	Cot, conv (South)(1)	0.34	0.40
MEADOW	Grass and legume mix	0.004	0.01
	Alfasfa, lespedeza, pr Sericia	0.020	
	Sweet clover	0.025	
SORGHUM, GRAIN (Western Plains)			
	RdL, spring TP, conv (1)	0.43	0.53
	No-till pl is shredded 70-50%re	0.11	0.18
SOYBEANS			
	B, RdL, spring TP, conv (1)	0.48	0.54
	C-B, TP annually, conv (2)	0.43	0.51
	B, no-til pl	0.22	0.28
	C-B, no till pl, fall shred C stalks (2)	0.18	0.22
WHEAT			
	W-F, fall TP after W (2)	0.38	
	W-F, scubble mulch, 500 lb rc(2)	0.32	
	Spring W, RdL, Sept TP, conv (N. Dak and S. Dak(1)	0.23	
	Winter W, RdL, Aug TP, conv (Kans) (1)	0.19	
	Spring W, stubble mulch, 750 lb rc(1)	0.15	
	Spring W, stubble mulch, 1250 lb rc(1)	0.12	
	Winter W, stubble mulch, 750 lb rc(1)	0.11	
	Winter W, stubble mulch, 1250 lb rc(1)	0.10	

(continued)

Table 6.12. (continued) Generalized values of the Cover and Management Factor, C, in the 37 states east of the Rocky Mountains.

- This table is for illustrative purposes only and is not a complete list of cropping systems or potential practices. Values of C differ with rainfall pattern and planting dates. These generalized values show approximately the relative erosion-reducing effectiveness of various crop systems, but locationally derived C values should be used for conservation planning at the field level. Table of local values are available from the Natural Resources Conservation Service.
- High level is exemplified by long-term yield averages greater than 75 bu of corn or 3 tons of grass and legume hay: or cotton management that regularly provides good stands and growth.
- Numbers in parentheses indicate number of years in the rotation cycle. (1) designates a continuous one-crop system.
- Abbreviations

B: soybeans	F: fallow	C: corn
M: grass and legume hay	c-k: chemically killed	pl: plant
conv: conventional	W: wheat	cot: cotton

we: winter cover

lb rc: pounds of crop residue per acre remaining on surface after new crop seeding
% rc: percentage of soil surfaces covered by residue mulch after new crop seeding
70-50% rc: cover for C values in first column: 50% for second column
RdR: residues (from clover, straw, etc.) removed or burned
RdL: all residues left on the field (on surface or incorporated)
TP: turn plowed (upper 5 or more inches of soil inverted, covering residues)

Table 6.13. Values of Support-Practice Factor, P

	Land Slope (%)				
	1.1-2	2.1-7	7.1-12	12.1-18	18.1-24
Practice	Factor P				
Contouring, P_e	0.60	0.50	0.60	0.80	0.90
Contour strip cropping,[a] P_{sc}					
R-R-M-M	0.30	0.25	0.30	0.40	0.45
R-W-M-M	0.30	0.25	0.30	0.40	0.45
R-R-W-M	0.45	0.38	0.45	0.60	0.68
R-W	0.52	0.44	0.52	0.70	0.90
R-O	0.60	0.50	0.60	0.80	0.90
Contour listing or ridge planting, P_{cl}	0.30	0.25	0.30	0.40	0.45
Contour terracing,[b,c] P_t	$0.6n^{-1/2}$	$0.5n^{-1/2}$	$0.6n^{-1/2}$	$0.8n^{-1/2}$	$0.9n^{-1/2}$
No support practice	1.0	1.0	1.0	1.0	1.0

[a]R, rowcrop; W, fall-seeded grain; 0, spring-seeded grain; M, meadow. The crops are grown in rotation and so arranged on the field that rowcrop strips are always separated by a meadow or winter-grain strip.
[b]These P_t values estimate the amount of soil eroded to the terrace channels and are used for conservation planning. For prediction of off-field sediment, the P_t values are multiplied by 0.2.
[c]n, number of approximately equal-length intervals into which the field slope is divided by the terraces. Tillage operations must be parallel to the terraces.

Predicted values of E represent average, time-invariant estimates. Given that R is based on an average number and distribution of storms per year, actual values of erosive soil loss would vary from year to year depending on the number, size, and timing of erosive rainstorms and other weather conditions. Although any one predicted value of E may not be highly accurate, the USLE should be more reliable when it is used to measure either relative effects or long-term sheet and rill erosion rates.

Example 6.8. The general data requirements for making soil loss estimates with the USLE equation are (1) site location (to get R); (2) soil properties (to get K); (3) flow length and slope (to get T); (4) crop, rotation, and management practices (to get C); and (5) slope and support practice (to get P). The estimation process will be illustrated using a hypothetical example. The site, which has a drainage area of 2 acres, is located in central Illinois. The site of interest has an average slope of 2.0% and a flow length of 300 ft. A soil analysis indicates 25% sand, 2% organic matter, 35% silt and very fine sand, a medium granular structure, and moderate permeability. The plot is used for corn, with a crop management value of 0.31, which was determined from information published locally. A support practice is not provided.

For central Illinois, Figure 6.5 suggests a value for R of 200. A value for the topographic factor can be obtained from Table 6.11; for a slope of 2.0% and a length of 300 ft, T equals 0.28. Using the soil characteristics given, a value for the soil erodibility factor of 0.17 is obtained from Figure 6.6. Since there is no support practice, P equals 1.0. Thus the average annual soil loss is:

$$E = RKTCP = 200(0.17)(0.28)(0.31)(1.0) = 2.95 \ tons/acre/year \qquad (6.28)$$

For the 2-acre plot, the gross erosion would be 5.9 tons/yr.

Example 6.9. The USLE can be used to assess the effect of the support practice on erosion amounts. Using the information and results from Example 6.8, the effect of contouring can be examined. For a slope of 2.0%, Table 6.13 indicates a support-practice factor of 0.6 when the field is contoured. This would reduce the unit erosion to 1.8 tons/acre/yr, or 3.5 tons/yr for the 2-acre plot. Thus contouring reduced gross erosion by 40%.

6.11 PROBLEMS

6.1   Using the runoff flow and pollutant concentration data in the table on the next page, calculate the total mass of TSS in the runoff event.

6.2   Using the runoff flow and pollutant concentration data in the table on the next page, calculate the total mass of phosphorus in the runoff event.

6.3   Using the runoff flow and pollutant concentration data in the table on the next page, calculate the total mass of TKN in the runoff event.

6.4   Using the runoff flow and pollutant concentration data in the table on the next page, calculate the total mass of Pb in the runoff event.

Sample time	Flow rate (L/s)	Concentrations			
		TSS (mg/L)	Phosphorus (mg/L)	TKN (mg/L)	Pb (μg/L)
11:50	2.8	700	1.8	20	870
12:10	3.3	250	0.55	7.4	430
12:30	4.9	110	0.24	3.5	440
12:50	3.6	110	0.32	2.4	620
13:10	5.7	150	0.24	1.7	1000
13:30	5.6	160	0.28	3.8	900
13:50	20	760	0.29	7.0	790
14:10	15	450	0.30	5.6	790
14:30	5.4	270	0.24	4.2	770

6.5   Using the runoff flow and pollutant concentration data in the table above, calculate the EMC for TSS in this runoff event.

6.6   Using the runoff flow and pollutant concentration data in the table above, calculate the EMC for phosphorus in this runoff event.

6.7   Using the runoff flow and pollutant concentration data in the table above, calculate the EMC for TKN in this runoff event.

6.8   Using the runoff flow and pollutant concentration data in the table above, calculate the EMC for Pb in this runoff event.

6.9   Using the runoff flow and pollutant concentration data in the table above, make a plot similar to Figure 6.2 and discuss the first flush ramifications for TSS.

6.10 Using the runoff flow and pollutant concentration data in the table above, make a plot similar to Figure 6.2 and discuss the first flush ramifications for phosphorus.

6.11 Using the runoff flow and pollutant concentration data in the table above, make a plot similar to Figure 6.2 and discuss the first flush ramifications for TKN.

6.12 Using the runoff flow and pollutant concentration data in the table above, make a plot similar to Figure 6.2 and discuss the first flush ramifications for Pb.

6.13   Using the runoff flow and pollutant concentration data in the table above, calculate the percent of total TSS mass that is exported in the first 25% of runoff volume. Discuss this value with respect to first flush for TSS.

6.14   Using the runoff flow and pollutant concentration data in the table above, calculate the percent of total phosphorus mass that is exported in the first 25% of runoff volume. Discuss this value with respect to first flush for phosphorus.

6.15   Using the runoff flow and pollutant concentration data in the table above, calculate the percent of total TKN mass that is exported in the first 25% of runoff volume. Discuss this value with respect to first flush for TKN.

6.16   Using the runoff flow and pollutant concentration data in the table above, calculate the percent of total Pb mass that is exported in the first 25% of runoff volume. Discuss this value with respect to first flush for Pb.

6.17   Estimate the annual mass of TSS from a 3-acre drainage area using the Simple Method. The TSS EMC is 250 mg/L, the runoff coefficient is 0.88 and the average annual precipitation is 38 in.

6.18   Using the Simple Method, estimate the total annual phosphorus load from a 20-ha residential subdivision. The phosphorus EMC is 1.6 mg/L, the runoff coefficient is 0.68 and the average annual precipitation is 34 in.

Parameter	Units	Value
Flow	L/m^2	20
Intensity	L/m^2-min	0.05
Duration	hr	6
Average Dry Period	days	7
Average Traffic Count During Dry Period	vehicles/lane	63000
Vehicles During Storm	vehicles/lane	1200
Previous Storm Intensity	L/m^2-min	0.02

6.19   Using the information in the table above and the regression equations of Table 6.10, estimate the deposited TSS load washed from the highway.

6.20   Using the information in the table above and the regression equations of Table 6.10, estimate the deposited phosphorus load washed from the highway.

6.21   Using the information in the table above and the regression equations of Table 6.10, estimate the deposited nitrate load washed from the highway.

6.22   Using the information in the table above and the regression equations of Table 6.10, estimate the deposited COD load washed from the highway.

6.23   Using the information in the table above and the regression equations of
       Table 6.10, estimate the deposited oil & grease load washed from the
       highway.

6.24   Using the information in the table above and the regression equations of
       Table 6.10, estimate the deposited BOD_5 load washed from the highway.

6.25   Using the information in the table above and the regression equations of
       Table 6.10, estimate the deposited copper load washed from the highway.

6.26   Using the information in the table above, but assuming that the previous
       storm intensity was 0.05 and the current intensity is 0.02 L/m^2-min, along
       with the regression equations of Table 6.10, estimate the deposited TSS load
       washed from the highway.

6.27   Assume that the storm event shown in Figure 6.1 has a duration of runoff of
       6 hours. (a) Where in time would you need to take measurements in order to
       characterize the pollutograph?   (b) How many measurements would be
       adequate? (c) Would your response be the same if the duration of runoff was
       15 minutes?

6.28   Table 6.2 includes 11 measurements of TSS (sources A, D, F, and G). If 80
       mg/L or less is considered "good" water quality, is it reasonable to conclude
       that the mean of the 11 values reflect "good" water quality? Use a 5% level
       of significance.

6.29   Table 6.2 includes 9 measurements of total phosphorus (TP). If 0.25 mg/L is
       considered "good" water quality, is it reasonable to conclude that the mean of
       the 8 values reflect "good" water quality? Use a 10% level of significance.

6.30   Provide examples of sheet erosions, rill erosion, and gully erosion in your
       local community.

6.31   Using a summation of the buoyant, drag, and weight forces, derive Stokes
       Law (Eq. 6.23).

6.32   Compute the ratios of the fall velocity of coarse sand to fine sand and
       medium to fine sand.

6.33   If a piece of fine sand is placed on the surface of a stream that is flowing at a
       velocity of 1.2 ft/sec at a depth of 1 ft, what longitudinal distance will the
       particle travel before it touches the river bed? Assume that the fall velocity
       follows Stokes Law.

6.34   Determine the relative importance of the buoyant, drag, and weight forces on
       a particle of medium sand.

6.35   A particle of fine sand enters a wetland forebay at an elevation of 1.5 ft above the bottom. What length of time must the particle remain in the forebay for it to settle to the bottom?

6.36   Estimate the soil loss using the USLE for a square 2-acre plot at a 3% slope for the northwest corner of Ohio. Assume the soil is 35% silt plus very fine sand, 10% sand (0.1 < d < 2 mm), 1% organic matter, fine granular structure, and moderate permeability. Assume a sweet clover meadow with no support practice.

6.37   For the conditions of Problem 6.36 show the variation of the soil loss amount as the percentage of sand varies from 0 to 30%

6.38   For the basic conditions of Problem 6.36, determine the variation in each of the inputs that will cause a 10% increase in the computed soil loss.

6.12 REFERENCES

Bannerman, R.T., Owens, D.W., Dodds, R.B., and Hornewer, N.J. (1993) "Sources of Pollutants in Wisconsin Stormwater," *Water Sci. Technol.*, **28**(3-5), 241-259.
Barrett, M.E., Irish, Jr., L.B., Malinia, Jr., J.F., and Charbenuea, R.J. (1998) "Characterization of Highway Runoff in Austin, Texas, Area," *J. Environ. Engg., ASCE,* **124,** 131-137.
Bertrand-Krajewski, J.-L, Chebbo, G. and Saget, A. (1998) "Distribution of Pollutant Mass vs Volume in Stormwater Discharges and the First Flush Phenomenon." *Water Research, 32*(8), 2341-2356
Bucheli, T.D., Müller, S.R., Voegelin, A., and Schwarzenbach, R.P. (1998) "Bituminous Roof Sealing Membranes as Major Sources of the Herbacide (r,S)-Mecoprop in Roof Runoff Waters: Potential Contamination of Groundwater and Surface Waters," *Environ. Sci. Technol., 32,* 3465-3471.
Clark, S.E., Pitt, R., Brown, P., and Field, R. (1998) "Treatment by Filtration of Stormwater Runoff Prior to Groundwater Recharge," *WEFTEC 98*, Orlando FA.
COMAR (Code of Maryland Regulations) 26.08.02.03.
Cordery, I. (1977) "Quality Characteristics of Urban Storm Water in Sydney, Austrailia," *Water Resour. Res., 13,* 197-202.
Davis, A.P. and Burns, M. (1999) "Evaluation of Lead Concentration in Runoff from Painted Structures." *Water Research, 33*(13), 2949-2958
Davis, A.P., Shokouhian, M., and Ni, S. (2001) "Loadings of Lead, Copper, Cadmium, and Zinc in Urban Runoff from Specific Sources," *Chemosphere,* **44**(5), 997-1109.
Deletic, A. (1998) "The First Flush Load of Urban Surface Runoff." *Water Research, 32*(8), 2462-2470.
DuBoys, P. (1879) "LeRhone et Les Rivieres a Lit Affouillable," *Annales des Ponts et Chaussees.*
Egiazarov, J.V. (1965) "Calculation of Nonuniform Sediment Concentrations," *J. Hydraulics Division, ASCE,* **91**(HY2).
Fisher, T.S., Hayward, D.G., Stephens, R.D., and Stenstrom, M.K. (1999) "Dioxins and Furans Urban Runoff," *J. Environ. Engg., ASCE,* **125,** 185-191.
Goncharov, V.N. (1964) *Dynamics of Channel Flow*, Israel Program for Scientific Translations.
Granier, L., Chevreuil, M., Carru, A.-M., and Létolle, R. (1990) "Urban Runoff Pollution by Organochlorines (Polychlorinated Biphenyls and Lindane) and Heavy Metals (Lead, Zinc and Chromium)" *Chemosphere,* **21,** 1101-1107.
Grout, H., Wiesner, M.R., and Bottero, J.-Y. (1999) "Analysis of Colloidal Phases in Urban Stormwater Runoff," *Environ. Sci. Technol., 33,* 831-839.
Irish, Jr., L.B., Barrett, M.E., Malina, Jr., J.F., and Charbeneau, R.J. (1998) "Use of Regression Models for Analyzing Highway Storm-Water Loads," *J. Environ. Engg., ASCE,* **124,** 987-993.
Leister, D.L. and Baker, J.E. (1994) "Atmospheric Deposition of Organic Contaminants to the Chesapeake Bay," *Atmos. Environ., 28,* 1499-1520.
Lopes, T.J., Fallon, J.D., Rutherford, D.W., and Hiatt, M.H. (2000) "Volatile Organic Compounds in Storm Water from a Parking Lot," *J. Environ. Engg., ASCE,* **126,** 1137-1143.

Mason, Y., Ammann, A.A., Ulrich, A. and Sigg, L. (1999) "Behavior of Heavy Metals, Nutrients, and Major Components during Roof Runoff Infiltration," *Environ. Sci. Technol.*, 33, 1588-1597.

Meyer-Peter, E. and Muller, R. (1948) "Formulas for Bed Load Transport," *Proceedings*, IAHR, 2nd Congress, Stockholm.

Pitt, R., Field, R., Lalor, M., and Brown, M. (1995) "Urban Stormwater Toxic Pollutants: Assessment, Sources, and Treatability," *Water Environ. Res.*, 67, 260-275.

Rauch, S., Morrision, G.M., Motelica-Heino, M., Donard, O.F.X., and Muris, M. (2000) "Elemental Association and Fingerprinting of Traffic-Related Metals in Road Sediments," *Environ. Sci. Technol.*, 34, 3119-3123.

Roger, S., Montrejaud-Vignoles, M., Andral, M.C., Herremans, L., and Fortune, J.P. (1998) "Mineral, Physical, and Chemical Analysis of the Solid Matter Carried by Motorway Runoff Water," *Water Res.*, 32, 1119-1125.

Sansalone, J.J., and Buchberger, S.G. (1997) "Partitioning and First Flush of Metals in Urban Roadway Storm Water," *J. Environ. Engg., ASCE*, 123, 134-143.

Sansalone, J.J., Koran, J.M., Smithson, J.A., and Buchberger, S.G. (1998) "Physical Characteristics of Urban Roadway Solids Transported during Rain Events," *J. Environ. Engg., ASCE*, 124, 427-440.

Schoklitsch, A. (1914) "Uber Schleppkraft and Geschiebebewegung," Engelmann, Leipzig.

Schoklitsch, A. (1934) "Geschiebetrieb and Geschiebefracht," *Wasserkraft and Wasserwirtschaft*, Jahrgand 39, Heft 4..

Scudlark, J.R., Conko, K.M., and Church, T.M. (1994) "Atmospheric Wet Deposition of Trace Elements to Chesapeake Bay: CBAD Study Year 1 Results," *Atmos. Environ.*, 28, 1487-1498.

Sedlak, D.L., Phinney, J.T., and Bedsworth, W.W. (1997) "Strongly Complexed Cu and Ni in Wastewater Effluents and Surface Runoff," *Environ. Sci. Technol.*, 31, 3010-3016.

Shamov, G.I. (1959) *River Sediments*, Gidrometeiozdat, Leningrad.

Shields, A. (1936) *Anwendung der Ahnlich Keitmechanik und der Turbulanzforschung auf die Geschiebebewegung*, (Application of Similitude and Turbulence Research to Bed Load Movement), Mitteilungen der Prussischen Versuchanstalt fur Wasserbau and Schiffbau (Berlin), No. 26.

Stenstrom, M.K., Silverman, G.S., Bursztynsky, T.A. (1984) "Oil and Grease in Urban Stormwaters," *J. Environ. Engg., ASCE*, 110, 58-72.

Stumm, W., and Morgan, J.J. (1996) *Aquatic Chemistry*, Wiley, NY.

Viklander, M. (1998) "Particle Size Distribution and Metal Content in Street Sediments," *J. Environ. Engg., ASCE*, 124, 761-766.

Wanielista, M.P. and Yousef, Y.A. (1993) *Stormwater Management*, Wiley, NY.

Wischmeier, W.H. and Smith, D.D. (1965) *Predicting Rainfall-Erosion Losses from Cropland East of the Rocky Mountains*, Agriculture Handbook #282, U.S. Department of Agriculture, Washington, D.C.

Wu, J.S., Allan, C.J., Saunders, W.L., and Evett, J.B. (1998) "Characterization and Pollutant Loading Estimation for Highway Runoff," *J. Environ. Engg., ASCE*, 124, 584-592.

Wu, J.S., Holman, R.E., and Dorney, J.R. (1996) "Systematic Evaluation of Pollutant Removal by Urban Wet Detention Ponds," *J. Environ. Engg., ASCE*, 122, 983-988.

Wu, Z.Y., Han, M., Lin, Z.C., and Ondov, J.M. (1994) "Chesapeake Bay Atmospheric Deposition Study, Year 1: Sources and Dry Deposition of Selected Elements in Aerosol Particles," *Atmos. Environ.*, 28, 1471-1486.

Young, K.D.and Thackston, E.L. (1999) "Housing Density and Bacterial Loading in Urban Streams," *J. Environ. Engg., ASCE*, 125, 1177-1180.

7

IMPROVEMENT OF STORMWATER QUALITY

7.1 INTRODUCTION

The improvement of urban runoff water quality presents a formidable challenge because of the many constraints and criteria that must be met. Engineering issues must be addressed, but other factors such as economics and politics can also play deciding roles. Traditionally, stormwater quality improvement has been managed in the same way as waste water discharges, specifically, to design and construct some type of treatment or management structure at the point of discharge into the receiving waters. Several different technologies have been developed as end-of-pipe treatment practices, such as retention ponds and sand filters, as to be discussed in upcoming chapters. Recently, however, just as industries are investigating ways to reduce the production of wastes, rather than just treat them immediately prior to discharge, those responsible for stormwater management are beginning to develop methods to minimize the discharge of runoff from developed lands, to minimize the mobilization of pollutants into runoff, and to address ecological issues related to development and stormwater management.

Effective solutions to these problems require more than just traditional structural measures. Concepts such as *Smart Growth* and *Low Impact Development* take a more holistic and sustainable approach to water quality improvement. Where and how land development occurs is important in determining the quality of the runoff. This includes both the regional siting of clusters of homes and commercial areas, as well as the location of structures and driveways on individual lots. Decisions made by the developer in laying out the lot and the homeowner in maintaining the lot will impact runoff quantity and quality, as well as the stream that receives runoff from the site. Integrating on-site land use management practices with more traditional stormwater management devices and facilities will be necessary to protect the environment, while continuing to develop land to meet this ever increasing demand.

When dealing with stormwater, a number of constraints must be recognized in the implementation of any type of treatment/management practice. First, any practice or facility must consider the physical processes that control hydrologic runoff. For example, during rainfall events, infiltration rates are limited by ambient soil and landscape characteristics. New management methods cannot expect unreasonably high rates. Second, public safety and the minimization of property damage must be considered. For example, any runoff flow must be transported away from the design area without causing flooding. Third, the traditional concerns of the public must be considered in the design. For example, the public generally does not find the ponding of runoff near their homes to be desirable. Therefore, any local storage must have standing water for only a short time. Fourth, space is very limited in urban areas. Thus any practice that is implemented must have a small footprint. This has economic implications as well as those for hydrology and water quality.

Treatment requirements for managing urban stormwater can be complex. The public wants both the control of large floods and the reduction of pollutants often associated with small storms. Therefore, treatment should be effective at very low flow rates, such as that which occurs with a light shower, as well as at very high flow rates produced from a high-intensity storm. Pollutants accumulate on land surfaces during periods between storm events. Thus management methods should be designed to minimize the washoff of accumulated pollutants during subsequent intense rainfall and high runoff flows. As discussed in the previous chapter, many different pollutants are found in runoff, all at different concentrations. The concentrations vary from event to event and even throughout a single event. Therefore, treatment processes must be sufficiently sophisticated to provide significant pollutant removal within the spatial and temporal variations noted in the concentrations of pollutants.

For reasons of practicality, the treatment must be passive; that is, there should not be moving parts or chemical additions. Such active treatment methods would not be practical from economic and management perspectives. The processes need to be able to withstand temperature ranges expected in winter and in summer and to maintain effectiveness during long stretches of wet weather and drought. The selected technology must be easy and inexpensive to maintain. Costs for maintenance of stormwater management technologies are probably the primary concern of local governments and developers. The facility itself should be aesthetically pleasing and should not present any unreasonable safety hazards. Aesthetical factors can significantly influence, positively and negatively, the sale price of residential properties and commercial areas. Certainly, overall, the cost must be low.

7.2 BEST MANAGEMENT PRACTICES

A number of treatment techniques are commonly employed for stormwater management. Collectively, these are known as "Best Management Practices" or "BMPs." Some BMPs are physical devices or systems that are specifically designed for the treatment of runoff. For example, detention/retention basins have traditionally been used to manage stormwater discharge rates. Some newer BMPs include land management practices that are not physical structures, but are designed to minimize the runoff that leaves a developed site.

Physical devices that are designed to be put on-line to improve runoff quality are primarily based on the treatment principle of gravity separation. Therefore, they would not be effective in the removal of dissolved pollutants. As discussed previously, many of the pollutants found in stormwater runoff are affiliated with suspended solids, but a significant fraction remains dissolved. Thus physical devices may be appropriate in cases where appropriate pollutant removal can be achieved through gravity separation, but are minimally effective for sites where dissolved pollutant concentrations are significant. A summary of pollutant removal efficiencies of several common stormwater management devices is presented in Table 7.1.

The removal efficiencies of the management processes listed in Table 7.1 are often labeled trap efficiencies. Trap efficiency is discussed in detail in subsequent chapters on management practices. Mathematically the trap efficiency is the ratio of the amount of a pollutant retained by a treatment device to the amount in the inflow. Since it is easier to measure the amount of a pollutant in the outflow than it is to measure the amount retained, employing mass balance principles, the trap efficiency, TE, is given as:

$$TE = 1 - \frac{C_{out}}{C_{in}} \qquad (7.1)$$

A mass-based trap efficiency can be similarly defined:

Table 7.1. Pollutant removal efficiencies of different processes (Schueler 1987, Schueler *et al.* 1992, U.S. EPA 1990). L = Low, M = Moderate, H = High, VH = Very High.

| Pollutant | Process | | | | | |
	Dry Detention Basins	Wet Retention Ponds	Infiltration Devices	Sand Filters	Vegetative Practices	Constructed Wetlands
Nutrients	L	M to H	H	M	L	M to H
Sediment	H	H	VH	VH	M	VH
Metals	M to H	M to H	VH	VH	M	H
Organic Matter	M	M	VH	M	L	M
Oil and Grease	M	M	H	H	M	VH
Bacteria	H	H	VH	M	L	H

$$TE_m = 1 - \frac{L_{out}}{L_{in}} \qquad (7.2)$$

where L represents the pollutant mass loads.

When evaluating pollutant treatment efficacies, however, a focus on just the percentage or fractional removal may not give an adequate environmental picture of a time-varying process. Because of the dynamic nature of precipitation events, a steady-state condition cannot be obtained. Input and output rates of both runoff flow and pollutant concentrations for a treatment practice will vary over the time of an event and with every event. The accumulation of pollutants and storage of water in the practice will affect trap efficiencies on an event-to-event basis. Thus removals should also be presented and discussed on a mass basis or as a change in EMC. For small events, it is possible that all of the runoff is held in detention. In this case, 100% removal of pollutants can be experienced, since water is not discharged from the practice.

Pollutant removal from runoff will result in the accumulation of pollutants in the treatment device. Thus the possibility exists for the washout of these pollutants during a subsequent storm event, which can cause output levels to be higher than those in the input. The result is a "negative removal" of pollutants in the subsequent event. This potential discharge must be considered in the design of a treatment device. Proper maintenance and cleaning of the practice also is necessary to prevent or minimize the release of captured pollutants. Chemical and biological processes can also convert one form of a pollutant to another, causing washout or negative removal. This is a fairly common occurrence with nitrogen compounds.

For sampling and monitoring, flow rates and pollutant concentrations are usually measured on a discreet sample-point basis rather than continuously. The time interval may be constant, such as every 15 minutes, or somehow be weighted to flow rate. In this latter case, more samples can be taken on the rising limb and near the peak of the hydrograph, with fewer points taken on the recession of the hydrograph where the system acts more linear. For each time interval, using the trapezoidal rule, average flow rates and concentrations can be computed and summed to compute loads. In this manner, the load, L, during a period Δt for average flow rate and average concentration is:

$$L = \overline{CQ}\,\Delta t \qquad (7.3)$$

The total load is given by the summation over all points for the flow duration T_d:

$$L_T = \sum_1^{T_d} L = \sum_1^{T_d} \overline{CQ}\,\Delta t \qquad (7.4)$$

Example 7.1. Stormwater runoff flow and suspended solids concentration for the influent and effluent of a stormwater detention pond are given in the following table. The mass of suspended solids accumulated in the pond from this event can be found.

Time	Flow into pond		Flow leaving pond	
	Flow rate (L/s)	TSS (mg/L)	Flow rate (L/s)	TSS (mg/L)
2:30 pm	0	0	0	0
3:00 pm	88	88	0	0
3:30 pm	122	234	34	68
4:00 pm	154	178	88	98
4:30 pm	102	122	76	132
5:00 pm	34	76	54	150
5:30 pm	0	0	26	64
6:00 pm	0	0	0	0

The total mass into and out of the pond are found using the integral of the flow rates and TSS concentrations:

Time	Δt (s)	Flow rate, Q (L/s)	$\overline{Q}$ (L/s)	TSS, C (mg/L)	$\overline{C}$ (mg/L)	$L = \overline{C}\,\overline{Q}\Delta t$ (mg) x10^6	$L_T = \overline{C}\,\overline{Q}\Delta t$ (mg) x10^6
Input							
2:30 pm		0		0			
3:00 pm	1800	88	44	88	44	3.48	3.48
3:30 pm	1800	122	105	234	161	30.4	33.9
4:00 pm	1800	154	138	178	206	51.2	85.1
4:30 pm	1800	102	128	122	150	34.6	119.6
5:00 pm	1800	34	68	76	99	12.12	131.8
5:30 pm	1800	0	17	0	38	1.16	132.9
Output							
2:30 pm		0		0			
3:00 pm	1800	0	0	0	0	0	0
3:30 pm	1800	34	17	68	34	1.04	1.04
4:00 pm	1800	88	61	98	83	9.11	10.15
4:30 pm	1800	76	82	132	115	16.97	27.1
5:00 pm	1800	54	65	150	141	16.50	43.6
5:30 pm	1800	26	40	64	107	7.70	51.3
6:00 pm	1800	0	13	0	32	0.75	52.1

The total suspended solids mass into the pond is 1.329×10^8 mg (=133 kg) and in the outflow is 5.21×10^7 (=52.1 kg). Therefore, the mass of solids captured by the pond is 133 - 52.1 = 80.9 kg, which yields a trap efficiency of 80.9/133 = 0.61 = 61%.

The total water inflow into and discharged from the facility can also be computed with average discharges using the trapezoidal rule:

$$V_T = \sum_{1}^{T_d} \overline{Q} \Delta t \tag{7.5}$$

In this case, 9×10^5 L of water flowed into the basin, while 5×10^5 L was discharged. Thus, the difference of 4×10^5 L was stored in the detention facility during the storm. If this storage volume had not existed in the basin, then the trap efficiency would have been lower than the 61% computed for the storm event.

Example 7.1 may suggest that the trap efficiency is the primary criterion to be used in the assessment of water quality. While the removal efficiency is important, the load discharged is also important to the health of the down-gradient receiving stream. A focus just on fractional pollutant removal may not provide for the best water quality. Important to the receiving waters is the mass or concentration of pollutant being discharged. Water quality standards are generally based on the concentration of a pollutant in a receiving water body, not the removal efficiency of the treatment facilities. The amount discharged into the water body depends not only on the removal efficiency of the treatment facility, but also on the input mass and concentrations of the pollutants. If the input water quality is relatively good, a treatment device that provides only marginal removal efficiency may be adequate. For a very poor quality inflow, however, even an excellent fractional removal (e.g., 99%) may still discharge water of poor quality to the receiving stream, which can be detrimental to the quality and ecology of the stream.

Example 7.2. The input TSS EMC to a stormwater treatment facility is 115 mg/L. The facility is 80% effective (based on EMC) in the removal of TSS. Since the facility captures 80% of the solids, 20% is discharged. The effluent EMC is, therefore, 115 mg/L x (0.20) = 23 mg/L. The removal to provide 23 mg/L in the effluent at an influent TSS EMC of 276 mg/L is:

$$\frac{276 - 23}{276} = 0.917 = 91.7\% \tag{7.6}$$

Example 7.3. Over a period of two years, measurements of discharge (ft^3/s) and total phosphorus (mg/L) were made at the inflow and outflow ports of a detention facility for nine storm events (see Table 7.2). These values are assumed to be representative of 80 runoff-producing events that occur each year at the site. Compute (a) the mass-based trap efficiencies for the total phosphorus for each storm event; (b) the long-term expected trap efficiency; (c) the total phosphorus load expected per year in the receiving stream.

Table 7.2. Calculation of mass trap efficiency (TE) for individual storm events in Example 7.3

Storm event	Inflow rate (ft^3/s)	Outflow rate (ft^3/s)	Inflow conc. (mg/L)	Outflow conc. (mg/L)	Load in (lb/day)	Load out (lb/day)	TE_m (%)
1	32.2	14.0	0.62	0.21	107	15.8	85.3
2	66.3	37.9	1.24	0.52	443	106	76.0
3	47.0	24.3	0.88	0.36	223	47.2	78.8
4	81.9	52.7	1.37	0.61	605	173	71.3
5	26.5	10.6	0.55	0.14	78.6	8.23	89.5
6	57.1	30.8	0.93	0.45	286	74.7	73.9
7	60.6	34.1	0.90	0.50	294	91.9	68.7
8	29.4	15.6	0.52	0.26	82.4	21.9	73.5
9	51.8	25.5	0.79	0.38	221	52.2	76.3
Total					2340	591	77.0 Mean

(a) The product of the phosphorus concentration and the flow rate give the load for both the input and output. A factor of 5.39 is used to convert from (ft^3/s)(mg/L) to (lb/day). The trap efficiencies can then be computed using Eq (7.2), with the values shown in the table. The storm event phosphorus trap efficiencies vary from 68.7% to 89.5%

(b) The average trap efficiency is found by summing the loads in the inflow and outflow and using the sum to compute $TE_m=(1-(\Sigma L_o/\Sigma L_i))$. Thus, for the data shown in Table 7.2:

$$TE_m = \left(1 - \frac{591}{2340}\right) = 0747 = 74.7\% \qquad (7.7)$$

(c) The average storm event load discharged to the receiving stream is 591/9=65.7 lbs/day. For 80 events per year, the total phosphorus load would be 5253 lbs if the nine storm events are representative of the other storms.

Generally, stormwater management facilities cannot be designed to handle major storm events without consuming large land areas. Therefore, a runoff volume to be treated associated with the smaller, more frequent storms is generally specified by local regulatory agencies. This volume is usually the first inch of runoff, or some similar value, from the contributing drainage area. Using this simple indicator of the first flush, a significant fraction of the total pollutant loading should be contained in this "water quality volume." Provisions to hold and capture pollutants from this inflow are important and a treatment practice should be designed so that larger flows do not wash out any pollutants captured from the initial treatment of the water quality volume. For example, an emergency spillway can be used to pass all runoff above the first-flush volume around a facility.

Some general studies of runoff from urban areas have demonstrated that reductions in runoff toxicity are possible through the use of some simple physicochemical treatments. Several laboratory treatability tests were completed on runoff samples from Birmingham, AL (Pitt *et al.* 1995). The settling of particulate matter for 24 hours greatly decreased toxicity for many samples (as measured by Microtox®, a simple microbial toxicity measurement), which demonstrates the importance of the capture of suspended solids by a treatment practice to improve water quality. Toxicity reductions were also found by screening (< 40 μm) and aeration or aeration combined with a photodegradation process.

Important to the success of any stormwater treatment and management practice is a strict maintenance program. The use of treatment BMPs fundamentally change the flow of nonpoint source pollutants in the environment. Without treatment, pollutants are washed from roadways, parking lots, and lawns, ultimately finding their way to streams, lakes, and bays. Making use of a BMP, the captured pollutants are retained in the uplands. At this point, they must be properly managed so as to not cause new problems at the site of their capture. If pollutants like suspended solids accumulate, provisions must be made so that these solids are periodically removed. If not, they can build up, which may decrease the efficiency of the practice, or these captured materials may be later washed out of the BMP during a major storm event. This would negate the long-term usefulness of the facility, at least with respect to the objective of water quality enhancement.

Other maintenance activities may include cutting and maintenance of vegetation in vegetated facilities, cleaning filters to maintain infiltration capabilities, and ensuring proper flow pathways. In some practices, chemical and biological processes may degrade captured pollutants, which would provide some degree of sustainability. However, in most systems, the accumulation of pollutants must be addressed by their planned removal. Also, as discussed earlier, some chemical and biological transformations can convert pollutants from one form to another, which can alter the efficacy at which the pollutant can be removed from the capture facility.

7.3 PROBLEMS

7.1   List some of the criteria that must be used for the selection of a stormwater BMP to address water quality improvement.

7.2   A stormwater BMP is being monitored for its efficiency in the capture of lead. Four samples of runoff input and output from the BMP are given in the table that follows.
    (a) Estimate the input and output EMC of lead.
    (b) Determine the % removal of lead by the BMP facility based on EMCs.
    (c) Determine the % removal of lead by the BMP facility based on the mass of lead.

	Input			Output	
Sample Time	Flow Rate (L/s)	Concentration of Lead (µg/L)	Sample Time	Flow Rate (L/s)	Concentration of Lead (µg/L)
1:00 pm	25	40	1:30 pm	8	10
1:30 pm	30	80	2:00 pm	18	15
2:00 pm	20	50	2:30 pm	22	20
2:30 pm	10	20	3:00 pm	20	15

7.3 The EMCs for TSS in the runoff from a commercial parking lot are presented in the table below. A stormwater BMP that is being installed to treat this runoff will provide 80% TSS removal.
   (a) Determine the effluent TSS EMC for each event.
   (b) Determine the input and output TSS total mass for all eight events.

	Input	
Event	TSS EMC (mg/L)	Event Total Volume (ft^3 x10^3)
1	66	11
2	123	2.2
3	43	13
4	168	4.4
5	88	5.5
6	110	7.8
7	76	1.0
8	92	2.6

7.4 A different BMP is being evaluated to treat the runoff from the commercial parking lot of Problem 7.3. This alternative BMP will provide an average removal of 40 mg/L TSS for each event.
   (a) Determine the effluent TSS EMC for each event.
   (b) Determine the input and output TSS total mass for all eight events.
   (c) Compare the values with those from Problem 7.3 and discuss.

7.5 Inflow to (Q_i) and outflow from (Q_o, ft^3/s) a detention basin are recorded along with TSS loads (lbs/s) and used to calibrate the following equations:

$$L_i = 7.32x10^{-3}Q_i^{1.43}$$
$$L_o = 0.447x10^{-3}Q_o^{1.82}$$

Discharge measurements from six storm events are:

	Discharge (ft^3/s)		Duration (min)	
Event	Q$_i$	Q$_o$	In	Out
1	112	51	46	62
2	72	28	31	53
3	143	96	83	110
4	58	20	22	41
5	96	39	37	66
6	83	33	51	80

Compute: (a) the trap efficiency for each storm and (b) the average trap efficiency. (c) Discuss the effect of the largest storm event on the overall TE.

7.6 For a particular detention basin, the trap efficiency (T$_e$) and discharge (Q, ft^3/s) are related by: $T_e = 11.6 \, Q^{-0.76}$. (a) What is the trap efficiency for an inflow of 100 cfs? (b) If the local government wants to increase the trap efficiency from 52% to 60%, what reduction in discharge will be needed?

7.4 REFERENCES

Pitt, R., Field, R., Lalor, M., and Brown, M. (1995) "Urban Stormwater Toxic Pollutants: Assessment, Sources, and Treatability," *Water Environ. Res.*, 67, 260-275.

Schueler, T.R. (1987) "Controlling Urban Runoff: A Practical Manual for Planning and Design of Urban BMPs." Publication no. 87703, Metropolitan Washington Council of Governments, Washington, D.C.

Schueler, T.R., Kumble, P.A., and Heraty, M.A. (1992) *A Current Assessment of Urban Best Management Practices: Techniques for Reducing Non-Point Source Pollution in the Coastal Zone.* Publication no. 92705, Metropolitan Washington Council of Governments, Washington, D.C.

U.S. Environmental Protection Agency (1990) "Urban Targeting and BMP Selection. Information and Guidance Manual for State Nonpoint Source Program Staff Engineers and Managers." The Terrene Institute. EPA 68-C8-0034.

8

STORAGE AND FLOW CONTROL

NOTATION

A   = drainage area
A   = cross-sectional area
A_o = area of orifice
$C_{\bullet}$ = orifice coefficient
C_d = discharge coefficient
C_w = weir coefficient
D   = particle diameter
D   = conduit diameter
D_e = capture depth of first-flush
d_c = critical depth
E   = specific energy
E_c = elevation of pipe centerline
E_d = dead storage elevation
E_i = invert elevation
E_0 = invert elevation for low stage
E_1 = low-stage flood elevation
E_2 = high-stage flood elevation
F   = Froude number
f   = fraction of imperviousness
g   = acceleration of gravity
H   = height of settling basin
H_o = height of orifice
h   = height
I   = average imperviousness
K_p = friction head loss coefficient
L   = length of settling basin
L_w = weir length
n   = number of outlets
p   = pressure

Q = flow through volume rate
Q_a = post-development depth of direct runoff
q = discharge rate
q_e = design discharge rate
q_p = peak discharge
q_{pa} = post-development peak discharge rate
q_{pb} = pre-development peak discharge rate
q_{o2} = discharge from low-stage outlet during high-stage event
q_1 = discharge rate through one orifice
q_u = unit width discharge rate(cfs/ft)
q_u = unit discharge (cfs/mi^2/in.)
R_i = removal fraction
R_q = ratio of pre- to post-development peak discharge rate
R_s = ratio of storage depth to runoff depth
T_e = first-flush detention duration
t = time
$\bar{t}$ = mean travel time
V = velocity
V_c = critical velocity
V_d = volume of dead storage
V_e = first-flush storage volume
V_h = horizontal velocity
V_o = velocity through orifice
V_s = volume of storage
V_{st} = volume of storage
V_v = vertical (fall) velocity
W_0 = width of orifice
W_1 = bottom width of storage facility
W_2 = top width of storage facility
w = settling basin width
z = elevation head
Δt = time increment
α = ratio of outflow to inflow discharge rates
γ = specific weight
μ = viscosity

8.1 INTRODUCTION

In the 1950s and 1960s, hydrologists began to note the hydrologic effects of urbanization. Flood peaks and channel erosion increased, while times of concentration and the time between out-of-bank flows decreased. Just as large dams had been the initial answer to flooding on large rivers, detention basins became a response to the flooding on small urban watersheds. Just as the upstream-downstream controversy arose with respect to the large dams, a similar controversy arose with respect to detention basins. Some wanted larger regional basins located downstream because this would be a solution that involved lower maintenance costs. Others wanted small upland detention basins because initially it was thought that this would prevent the small upland streams from being scoured. But the general belief was that storage facilities represented the best solution to controlling increases in the magnitude and frequency of urban flooding.

At that time, water quality issues were not a central concern. Only the control of channel erosion and the resulting loss of stream clarity were viewed as quality issues. During the 1970s, the interest in nonpoint source pollution related to the urban environment represented the first broad concern for water quality issues other then channel erosion. It was the belief of some that the urban detention basin could lead to water quality improvements if changes in design practices were incorporated. The initial attempts were concentrated on small islands that were located between the principal inlets and the outlets. The inclusion of island baffles was thought to increase the detention time and, therefore, allow greater time for suspended soil to settle. Slow release risers were also thought to be a mechanism for improving downstream water quality. Small orifices in the riser allowed the active storage area to drain over a period of many days, thus reducing storage requirements and simultaneously increasing detention time.

Storage volume control is still a primary option for controlling the water quantity and quality effects of land development. Detention basins are generally inexpensive to construct, and maintenance, while costly, seems to be acceptable. Small ponds, if properly designed and maintained, can contribute to increases in property values. They also have the advantage that trap efficiencies are easy to document.

Water quality improvements have been assumed for SWM basins due to the quiescent detention of the water. It is assumed that these ponds act as sedimentation basins, removing suspended solids from the water. Also, any pollutants that are attached or affiliated with suspended solids are concurrently removed. The analysis of TSS removal in detention ponds is done using classical sedimentation theory, as discussed later in this chapter. Stokes Law for settling particles is used in conjunction with flow assumptions to relate particle removals for various particle sizes to the pond overflow rate (flow rate/surface area). With stormwater runoff, however, the flow rate depends on the intensity of the storm. For low intensity events, the flow rate is low and very good particle removal can occur. However, high intensity storms have short detention times and provide little opportunity for the settling of particles. Very high intensity events can scour the bottom of detention ponds and carry out sediments that have been captured from previous events. During these extreme events, particle loadings leaving a pond can be higher than the input loading.

A few recent studies have examined other water quality parameters in the outflow of ponds. The results have been quite variable because of the many different factors involved, including different detention times for different storm events, the concentrations of pollutants entering the basins from different land uses, and environmental factors, such as temperature. Many chemical and biological processes can take place in the ponds, which can impact water quality, either beneficially or deleteriously.

8.1.1 Effects of Urban Development

It is widely recognized that land development, especially in urban areas, is responsible for significant changes in runoff characteristics. Within the context of the hydrologic cycle, land development decreases the natural storage of a watershed. The removal of trees and vegetation reduces the volume of interception storage. Grading of the site reduces the volume of depression storage and often decreases the permeability of the surface soil layer, which reduces infiltration rates and the potential for storage of rainfall in the soil matrix. In urban areas, increased impervious cover also reduces the potential for infiltration and soil storage of rainwater. This reduction of natural storage (that is, interception, depression, and soil storage) causes changes in runoff characteristics. Specifically, both the total volume and the peak of the surface (or direct) storm runoff increase. The loss of natural storage also causes changes in the timing of runoff, specifically a decrease in both the time to peak and the time of concentration. Runoff velocities are increased, which can increase surface rill and gully erosion rates.

In addition to change in the physical processes associated with overland runoff, land development significantly alters the processes associated with flow in receiving streams. Land development is often accompanied by changes to drainage patterns and channel characteristics. For example, channels may be cleared of vegetation and straightened, with some also being lined with concrete or riprap or replaced by a pipe. Higher stream velocities may also increase rates of bed-load movement. Modifications to the channel may result in decreases in channel storage and roughness, both of which can increase flow velocities and the potential for flooding at locations downstream from the developing area.

Recognizing the potential effects of these changes in runoff characteristics on the inhabitants of local communities, various measures have been proposed to offset these reductions in natural storage. The intent of stormwater management (SWM) has been to mitigate the hydrologic impacts of this lost natural storage, usually using manmade storage. Although a variety of SWM alternatives have been proposed, the stormwater management basin remains the most popular. The SWM basin is frequently referred to as a detention or retention basin, depending on its effects on the inflow hydrograph. For our purpose, the terms will be used interchangeably because the fundamental concepts for estimating volumes to control discharge rates are the same.

8.1.2 SWM Policy Considerations

To mitigate the detrimental effects of land development, SWM policies have been adopted with the intent of limiting peak flow rates from developed areas to that which occurred prior to development. In addition to specifying the conditions under which SWM methods must be used, these policies indicate the intent of SWM. The intent can be interpreted to mean that the flood frequency curve for the post-development conditions coincides with the curve for the pre-development conditions for each exceedance frequency. Such policies often use one or two return periods such as the 2-, 10-, or 100-year discharges as the target points on the frequency curve. Where channel erosion is of primary concern, a smaller return period such as the 6-month event may serve as the target event.

Policies should also identify a specific design method for use in the design of a SWM control method. Although data do not exist to show that any one method is best, the specification of a specific method as part of a SWM policy will ensure design consistency. The Rational formula and Soil Conservation Service, SCS (now National Resources Conservation Service, NRCS) methods are commonly used.

Recently, data have indicated that policies designed with a focus on limiting peak flows have not been adequate to provide improvements in water quality. In some cases, the flow control has been inadequate and channel erosion has been severe, along with high sediment transport during storm events. Additionally, flow control measures tend to focus on larger events. On a mass loading basis, however, smaller events--the common storms that occur every few days–contribute significant pollutant mass to receiving waters. These smaller events are not adequately controlled by traditional SWM practices.

8.1.3 Elements of SWM Structures

Figure 8.1 shows a schematic of the cross section of a detention basin with a single-stage riser. A pool is formed behind the detention structure. The flood runoff enters the pool at the upper end of the detention basin. The inflow to the detention basin is generally considered to be the hydrograph for the post-development conditions. Water can be discharged from the pool through a pipe that passes through or around the detention structure. The invert elevation of the riser can serve to limit the outflow rate, thus forming a permanent pool, with the permanent pool elevation changing only through evaporation and infiltration losses. The use of a permanent pool has a number of advantages, including water quality control, aesthetic considerations, and wildlife habitat improvement. Of course, a permanent pool also increases the total storage volume, which requires both a larger retaining structure and a larger commitment of land, both of which increase the cost of the project. Storage below the riser invert elevation is sometimes referred to as *dead storage* or the permanent pool. Storage allocated for flood runoff is referred to as *active storage*.

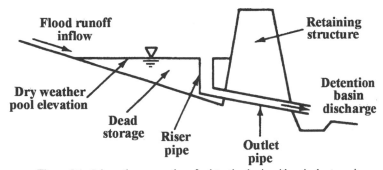

Figure 8.1. Schematic cross section of a detention basin with a single-stage riser.

Figure 8.1 does not show several other elements of detention basin design. The riser inlet should be fitted with both an antivortex device and a trash rack. The anti-vortex device prevents the formation of a vortex, thus maintaining the hydraulic efficiency of the outlet structure. The trash rack prevents trash (and people) from being sucked into the riser by high-velocity flows. At least one antiseep collar is fitted to the outside of the discharge pipe to prevent erosion about the pipe within the retaining structure. All detention basins should have an emergency spillway to pass runoff from very large flood events, so that the detention structure is not overtopped and washed out. The elevation of the bottom of the emergency spillway, which will pass high flows around the retaining structure, is above the elevation of the riser outlet but below the top of the retaining structure.

8.1.4 Analysis Versus Synthesis

The problem of analysis versus synthesis is best evaluated in terms of systems theory. The problem is viewed in terms of the input (inflow runoff hydrograph), output (outflow hydrograph), and the transfer function (stage-storage-discharge relationship). In the analysis phase, the two hydrographs would be measured at an existing stormwater management facility, and it would be necessary to calibrate the stage-storage-discharge relationship. While the stage-storage relationship is determined from topography, the stage-discharge relationship would have to be analyzed (calibrated). For a given storage facility, the physical characteristics of the outlet facility would be known. Therefore, the analysis would involve determining the best values of the weir and/or orifice coefficients for the outlet. Given the cost involved in data collection, analyses are rarely undertaken; therefore, only the synthesis case will be discussed in this chapter. This is the design case.

In the synthesis case, the objective is to make estimates of either the outflow hydrograph or the necessary characteristics of the proposed riser. For watershed studies where a detention basin exists, it would be of interest to synthesize the outflow flood hydrograph; in this case, the inflow hydrograph is estimated from a design storm. The standard procedure is to assume a design storm and a unit hydrograph and convolve the rainfall excess with the unit hydrograph. The resulting runoff hydrograph is used as the input (inflow runoff hydrograph). Weir coefficients are assumed along with the linear storage equation to compute the outflow hydrograph.

The second case of synthesis, which will be referred to as the problem of design, has the objective of estimating the characteristics of the riser/outlet facility in order to meet some design objective. Very often, the design objective is for the post-development peak discharge to be constrained by the storage facility to equal the pre-development discharge rate. In this case, the output of the design problem is either the area of the orifice or the weir length, the riser and conduit diameters, and the outlet facility elevation characteristics. Unlike the analysis case, the weir and/or orifice coefficients are assumed, as is the design criteria, unlike the watershed evaluation case outlined in the previous paragraph.

8.1.5 Planning versus Design

A number of detention basin planning methods for estimating the required volume of detention storage are introduced in this chapter. Methods will be provided for estimating the size of the outlet structure for both single-stage and two-stage risers. A single-stage facility controls flooding for one return period. A two-stage facility controls flooding for two return periods. The methods of this chapter will be classed as planning methods, although they are frequently used for design. Design techniques differ in two ways from the SWM planning methods. First, the planning methods only require peak discharge estimates, as opposed to requiring entire flood hydrographs. Thus, routing hydrographs through the detention basin is not necessary when using these planning methods. Second, since routing is not required, a stage-storage-discharge relationship is not required; instead, the storage-discharge relationship is inherent in the planning methods. A design method should use flood hydrographs, storage-indication routing, and a site-specific stage-storage-discharge relationship. For this reason the design method is more accurate than the planning methods. The terms planning and design are used to distinguish between approaches to SWM problem solving that reflect differences in expected accuracy, as well as the cost and effort involved. The following summarizes the differences between methods: The problem of planning the detention facility is separated into two parts: estimating the volume of storage and separately sizing the characteristics of the outlet facility. In planning, they are treated separately, whereas in design their determination is made simultaneously.

The dynamic nature of a storm event and the routing of runoff through a SWM basin prevents simple analysis of suspended solids removal. The basins will be efficient during small events when flows are low. In this case, the holding time in the basin will be long and reasonable time for solids to settle will be available. With large events, however, removal fractions will be small as the holding time will be short.

The storage of water between storm events can also play a role in water quality improvement (or degradation). Physical, chemical, and biological processes that are too slow to occur during the retention time for a storm event may be important during the time between events. Thus, the water quality may improve in the dead storage between events. This cleaner water can be washed out during the subsequent event–overall a lower mass of pollutants may enter the receiving water body.

8.2 WEIR AND ORIFICE EQUATIONS

Weirs and orifices are engineered devices that can be used to control and measure flow rates. While these devices can occur naturally, for the context of engineering design, the discussion will center on the equations used in the design of hydrologic/hydraulic facilities, such as wetlands and detention basins.

8.2.1 Orifice Equation

When an inlet is submerged, as a grate inlet would be during a heavy rainfall, it acts as an orifice, which is an opening to an area with a relatively large volume, such as a tank or manhole. Figure 8.2 shows a schematic of a tank with a hole of area A_2 in its side. If we assume all losses can be neglected, Bernoulli's equation can be written between a point on the surface of the pool (point 1) and a point in the cross section of the orifice (point 2):

$$\frac{p_1}{\gamma} + \frac{V_1^2}{2g} + z_1 = \frac{p_2}{\gamma} + \frac{V_2^2}{2g} + z_2 \qquad (8.1)$$

This can be simplified by making the following assumptions: (1) The pressure at both points is atmospheric, therefore $p_1 = p_2$; (2) The surface area of the pool A_1 is very large relative to the area of the orifice A_2, so from the continuity equation, V_1 is essentially 0; and (3) $z_1 - z_2 = h$. Thus, Eq. (8.1) becomes:

$$h = V_2^2 / 2g \qquad (8.2)$$

Solving for V_2 and substituting it into the continuity equation yields:

$$q = AV = A\sqrt{2gh} \qquad (8.3)$$

Equation 8.3 depends on two assumptions that are not always true: the losses in the system are negligible and the pressure is atmospheric across the opening of the orifice; it is actually atmospheric at a point below the orifice, called the *vena contracta*, where the cross-sectional area is a minimum. Because of these violations, the discharge will be less than that given by Eq. (8.3). The actual discharge through the orifice is estimated by applying a discharge coefficient C_d to Eq. (8.3):

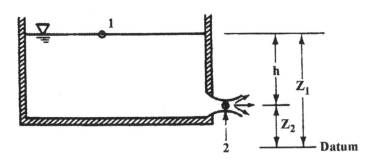

Figure 8.2 Schematic diagram of flow through an orifice.

$$q = C_d A \sqrt{2gh} \qquad (8.4)$$

in which C_d is called the discharge coefficient. Equation 8-4 is referred to as the orifice equation. Values of C_d range from 0.5 to 1.0, with a value of 0.6 often used. C_d is dimensionless.

Example 8.1. A forklift is accidentally driven into the side of an above-ground storage tank and it punctures an 11-in.^{2} hole 6 in. above the ground. In order to complete the environmental report, the engineer needs to estimate the amount of petroleum that leaked out of the tank during the 25 minutes between the time of the puncture and the time that the hole was plugged. The surface elevation of the petroleum above the ground was 8.46 ft after the leak was stopped. The tank has a diameter of 40 ft. The engineer estimates the discharge coefficient is 0.55, which is less than the 0.6 value because of the additional losses created by the jagged edges of the hole. Thus, the estimated discharge is:

$$q = C_d A \sqrt{2gh} = 0.55 \left(\frac{11}{144} \right) \left[2(32.2)(8.46 - 0.5) \right]^{0.5} = 0.951 \, \text{ft}^3/\text{s} \qquad (8.5)$$

Thus, the volume (V_0) discharged was approximately:

$$V_0 = q \Delta t = 0.951 \frac{ft^3}{\text{sec}} \times 25 \, \text{min} \times \frac{60 \, \text{sec}}{\text{min}} = 1427 \, \text{ft}^3 \qquad (8.6)$$

The volume used the final fluid height of 8.46 ft to estimate the discharge rate. At the time of puncture, the height and the discharge rate were higher. To evaluate the possible effect on the accuracy of the estimated volume, the change in height (Δh) from the time of puncture to the time of plugging is computed from geometry:

$$\Delta h = \frac{V_0}{A_0} = \frac{1427 \, \text{ft}^3}{\pi (40)^2 / 4} = 1.136 \, \text{ft} \qquad (8.7)$$

If the initial height was $8.46 + 1.14 = 9.60$ ft, then the initial discharge would have been:

$$q = 0.55 \left(\frac{11}{144} \right) \left[2(32.2)(9.60 - 0.5) \right]^{0.5} = 1.017 \, \text{ft}^3/\text{sec} \qquad (8.8)$$

This discharge yields an estimate of the volume of:

$$V_0 = q \Delta t = 1.017 \frac{ft^3}{\text{sec}} \times 25 \, \text{min} \times \frac{60 \, \text{sec}}{\text{min}} = 1526 \, \text{ft}^3 \qquad (8.9)$$

Thus, the actual volume discharged from the tank was less than 1526 ft^3 but more than 1427 ft^3.

Example 8.2. Because of corrosion in the sheet piling of a cofferdam, a 5-in.[2] hole is opened, which allows water from the river that surrounds the cofferdam to enter the construction site. The river stage is 11.6 ft above the level of the hole. Assuming a discharge coefficient of 0.6, the discharge rate is:

$$q = C_d A \sqrt{2gh} = 0.6 \left(\frac{5}{144} \right) [2(32.2)(11.6)]^{0.5} = 0.57 \ ft^3/sec \qquad (8.10)$$

8.2.2 Weir Equation

Consider the flow cross section shown in Figure 8.3. Point 1 is located at a point upstream of the obstruction at a distance where the obstruction cannot influence the flow characteristics at the obstruction. Point 2 is at the obstruction. The following analysis assumes (1) ideal flow, (2) frictionless flow, (3) critical flow conditions at the obstruction, and (4) the obstruction has a unit width perpendicular to the direction of flow. For the critical flow conditions, the following equations describe hydraulic conditions at the obstruction:

$$F = 1 = \frac{V_c}{(gd_c)^{0.5}} \qquad (8.11)$$

$$d_c = \left(\frac{q_u^2}{g} \right)^{1/3} \qquad (8.12)$$

$$E = d_c - \frac{V_c^2}{2g} = \frac{3}{2} d_c \qquad (8.13)$$

where F is the Froude number, V_c is the critical velocity, d_c is the critical depth, q_u is the discharge rate per unit width, and E is the specific energy. If hydrostatic pressure is assumed at sections 1 and 2, then $p_i/\gamma = h_i$. Thus, Bernoulli's equation is:

$$h_1 + \frac{V_1^2}{2g} + z_1 = h_2 + \frac{V_2^2}{2g} + z_2 \qquad (8.14)$$

Letting $Z = z_2 - z_1$ and assuming that the velocity head at section 1 is much smaller than the velocity head at section 2, Eq. (8.14) reduces to:

$$h_1 = \frac{V_c^2}{2g} + d_c + Z \qquad (8.15)$$

Using Eqs. (8.11) and (8.13), the velocity head is $V_c^2/2g = d_c/2$. Letting $h = h_1 - z$, then $h = 0.5d_c + d_c = 1.5d_c$ or $d_c = 2h/3$. Solving Eq. (8.12) for q_u, it then follows that:

$$q_u = \left(gd_c^3\right)^{0.5} = \left[g\left(\frac{2}{3}h\right)^3\right]^{0.5} = \left(\frac{8g}{27}\right)^{0.5} h^{3/2} \qquad (8.16)$$

For q_u [=] ft^3/sec, h [=] ft, and g [=] ft/sec^2, Eq. (8.16) yields:

$$q_u = 3.088h^{1.5} \qquad (8.17)$$

Letting $q = q_u L$ and replacing the constant 3.088 with the weir coefficient C_w yields the general weir equation:

$$q = C_w L h^{1.5} \qquad (8.18)$$

in which L is the length (ft) of the weir. Values of C_w can range from 2.3 to 3.3, depending on the losses that occur at the weir, but values from 2.6 to 3.1 are common for English units. An efficiency factor f can also be applied to reduce the discharge q for other losses.

Example 8.3. An engineering research unit is installing weirs on small experimental watersheds. Prior to installation they conduct laboratory tests of the weirs to determine the weir coefficients. For one test, a volume of 273 ft^3 passes over the rectangular weir in 30 sec. This yields a discharge of 273 ft^3/30 sec = 9.1 ft^3/sec. The water surface is 0.5 ft above the 8-ft weir. Thus, Eq. (8.18) can be used to estimate the weir coefficient:

$$C_w = \frac{q}{Lh^{1.5}} = \frac{9.1}{8(0.5)^{1.5}} = 3.22 \qquad (8.19)$$

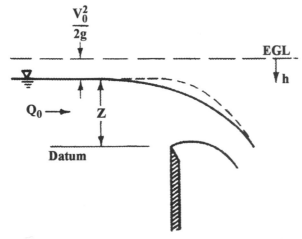

Figure 8.3 Schematic diagram of flow over a sharp-crested weir.

Example 8.4. In making a water balance of a wetland constructed adjacent to a development project, the engineer needs to estimate the discharge from the wetland during the design storm. The outlet from the wetland is a 400-ft grassed berm. The engineer estimates that during the design storm the water would be 0.6 ft above the top of the berm. The loss coefficient for the berm is estimated to be 2.9. Thus, the expected discharge is:

$$q = C_w L h^{1.5} = 2.9(400)(0.6)^{1.5} = 64.1 \, \text{ft}^3/\text{sec} \qquad (8.20)$$

8.3 DETENTION PONDS

The two primary design elements of a detention basin are the volume of storage and the characteristics of the riser. Simple design methods perform separate calculations in their design. Methods that size the two simultaneously require greater effort but are believed to be more accurate. Only the former is considered herein.

8.3.1 Storage Volume Estimation

Chapter 6 of Technical Release 55, or TR-55 (SCS, 1986), provides a method for quickly analyzing the effects of a storage reservoir on peak discharges. It is based on average storage and routing effects for many structures that were evaluated using the computerized TR-20 method (SCS, 1984). The ratio of the volume-of-storage to the volume-of-runoff (V_s/Q_a) is given as a function of the ratio of the peak rate of outflow to the peak rate of inflow, α. The relationship between V_s/Q_a and α is shown in Figure 8.4. Mathematically, the relationship is given by:

$$R_s = \frac{V_s}{Q_a} = C_0 + C_1\alpha + C_2\alpha^2 + C_3\alpha^3 \qquad (8.21)$$

in which C_0, C_1, C_2, and C_3 are coefficients (see Table 8.1) that are a function of the SCS rainfall distribution. The computations can be made on Worksheet 8.1. The equivalent depth of storage (in.) is computed by:

$$V_s = R_s Q_a \qquad (8.22)$$

For a drainage area of A acres, the volume in acre-ft is:

$$V_{st} = V_s A / 12 \qquad (8.23)$$

Example 8.5. A community is planning a 4.5-acre recreation center that will include basketball courts, tennis courts, a small building, a softball field, and a paved parking lot. The pre-development and post-development curve numbers are 79 and 87, respectively. The design 24-hour rainfall depth in 4.2 in. This yields runoff depths of 2.13 and 2.82 inches for the pre-development and post-development conditions, respectively. The pre-and post-development times of concentration are 12 and 6 min., respectively. The I_a/P ratios are 0.13 and 0.1, respectively. The unit peak discharges computed using the SCS Graphical method are 784 and 1010 $ft^3/sec/mi^2/in.$, respectively. Thus, the peak discharges are:

$$q_{pb} = q_{ub}AQ_b = 784((4.5/640)(2.13\,in.)) = 11.7\,cfs \qquad (8.24)$$

$$q_{pa} = q_{ua}AQ_a = 1010((4.5/640)(2.82\,in.)) = 20.0\,cfs \qquad (8.25)$$

Thus, the ratio $q_{pb}/q_{pa} = 0.585$, which is α in Eq. (8.21). The storage volume ratio R_s is:

$$R_s = 0.682 - 1.43(0.585) + 1.64(0.585)^2 - 0.804(0.585)^3 = 0.246 \qquad (8.26)$$

Using the post-development runoff depth of 2.82 in., the required volume of storage is

$$V_{st} = R_sQA/12 = 0.246(2.82)(4.5)/12 = 0.260\,ac\text{-}ft \qquad (8.27)$$

If topography permits an average depth of 4 ft, then the pond will have a surface area of 0.065 acres, which represents 1.4% of the total drainage area.

Table 8.1. Coefficients for the SCS Detention Volume Method

Rainfall Distribution	C_0	C_1	C_2	C_3
I or IA	0.660	-1.76	1.96	-0.730
II or III	0.682	-1.43	1.64	-0.804

8.3.2 Sizing Riser Structures

The estimation of the volume of detention storage is the first step. The second necessary step in sizing a detention basin is the determination of the physical characteristics of the outlet structure. The outlet may be based on a weir or an orifice, or both. Figure 8.1 shows a schematic of a basin with a pipe outlet. In addition to determining the diameter of the pipe barrel for a pipe outlet facility, it is also necessary to establish elevations of the pipe inlet and outlet. For those policies that require a permanent pool (wet pond), both the volume of dead storage and the corresponding elevation of the permanent pool must be set. Storage volumes computed with the method of Section 8.3.1 are active storage estimates, which is added to the dead storage to estimate the total storage. Both the size and effectiveness of a detention basin are largely dependent on the exceedance frequency (return period). Studies have shown that a basin designed to control the frequent events (that is, 2- or 5-yr events) will tend to overcontrol the less frequent events (that is, 50- or 100-yr events). Conversely, a basin designed to control the less frequent events will tend to undercontrol the more frequent events. An outlet facility sized to pass the 2-yr event will not allow the 100-yr event to pass with the same speed that a pipe outlet sized for a 100-yr event will pass through; thus, overcontrol results. In the past, most SWM policies have required a single-stage riser. More enlightened SWM policies have been developed that require two-stage control because of the problems of undercontrol and overcontrol associated with single-stage risers. The sizing of both single-stage and multi-stage risers will be discussed here.

In the sizing of risers, it is necessary to determine both the required volume of storage and the physical characteristics of the riser. The physical characteristics include the outlet pipe diameter, the riser diameter, the length of the weir or the area of the orifice, and the elevation characteristics of the riser. Single-stage risers with weir flow and orifice flow are shown in Figures 8.5a and b, respectively. For weir flow control, the weir equation, (Eq. 8.18), defines the relationship between the discharge (q) and (1) the depth in feet (h) above the weir, (2) the discharge or weir coefficient C_w, and (3) the length of the weir L_w. Values of C_w depend on the characteristics of both the weir and the discharge rate, with values ranging from 2.6 to 3.3; however, a constant value of about 3.0 is often used. The general formula for flow through an orifice is given by Eq. (8.4).

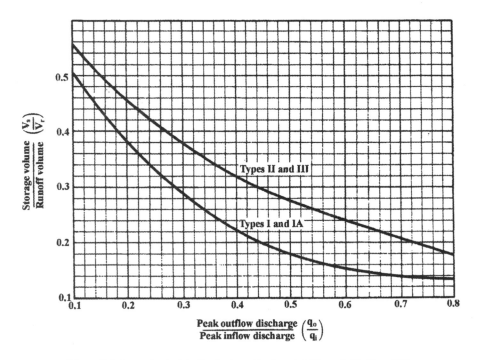

Figure 8.4 Approximate detention basin routing for rainfall types I, IA, II, and III.

Worksheet 8.1 Detention Volume Estimation: SCS TR-55 Method

Variable	Pre-development	Post-development
Drainage are, A (ac)		
Drainage area, A_m (mi^2)		
Time of Conc., t_c (hr)		
Curve number, CN		

Step	Variable	Low stage	High stage
1	Rainfall, P (in.)		
2	I_a/p (pre-development)		
	I_a/p (post-development)		
3	Runoff depth, Q_b (pre-dev.) (in.)		
	Runoff depth, Q_a (post-dev.) (in.)		
4	Unit peak, q_{ub} (pre-dev.) (ft^3/s/mi^2/in.)		
	q_{ua} (post-dev.) (ft^3/s/mi^2/in.)		
5	Peak discharge, q_{pb} (pre-dev.) (ft^3/s)		
	q_{pa} (post-dev.) (ft^3/s)		
6	q_p ratio $= q_{pb}/q_{pa} = R_q$		
7	Storage ratio, R_v		
8	Depth of storage, d_s (in.)		
9	Volume of storage, V_s (ac-ft)		
10	Elevation (ft)		

(2) $I_a = 0.2 * S$ where $S = (1000 / CN) - 10$
(3) $Q = (P - 0.2 * S)^2 / (P + 0.8 * S)$
(4) Obtain unit peak discharge rates from Eqs. (4.53)
(5) $q_p = q_u A_m Q$
(7) $R_v = C_0 + C_1 R_q + C_2 R_q{}^2 + C_3 R_q{}^3$

Rainfall Type	C_0	C_1	C_2	C_3
I, IA	0.660	-1.76	1.96	-0.730
II, III	0.862	-1.43	1.64	-0.804

(8) $d_S = R_v Q_q$
(9) $V_S = d_S A/12$
(10) Use stage-storage relationship

8.3.2.1 Sizing Single-Stage Risers

To estimate the characteristics of a riser require the following inputs:
1. Watershed characteristics, including area, pre- and post-development times of concentration, and pre- and post-development curve numbers.
2. Rainfall depth for the design storm.
3. Characteristics of the riser structure, including pipe roughness (n), length, and an initial estimate of the diameter.
4. Elevation information, including the stage-versus-storage relationship, the wet pond elevation, and the elevation of the centerline of the pipe.

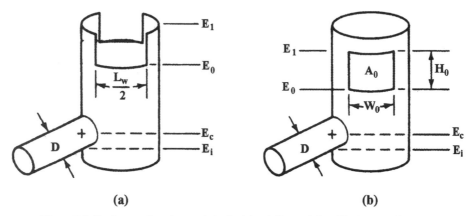

Figure 8.5 Single-stage riser characteristics for (a) weir flow and (b) orifice (or port) flow

5. Hydrologic and hydraulic models, including a model for estimating peak discharges and runoff depths, a model for estimating the volume of storage as a function of pre- and post-development peak discharges, and a model for estimating weir and orifice coefficients, as necessary.

The output from the analysis yields the following: (1) the length of weir or area of the orifice; (2) the depth and volume of storage; (3) elevations of riser characteristics; and (4) the diameter of the outlet pipe.

The following steps can be used to size a single-stage riser using the SCS method, with Worksheet 8.2:

1. a. Using the 24-hour rainfall depth and the pre-development CN, find the runoff depth, Q_b.
   b. Using the 24-hour rainfall depth and the post-development CN, find the runoff depth, Q_a.
2. a. Determine the pre-development peak discharge q_{pb}.
   b. Determine the post-development peak discharge q_{pa}.
3. Compute the discharge ratio:

$$R_q = \frac{q_{pb}}{q_{pa}} \qquad (8.28)$$

4. Enter the R_s-versus-R_q (Figure 8.4) curve with R_q to find the storage volume ratio R_s.
5. a. Compute the equivalent depth of storage in inches:

$$V_s = Q_a R_s \qquad (8.29)$$

   b. Convert V_s to ac-ft by multiplying by $A/12$, where A = area (ac).

Worksheet 8.2. Single-Stage or Two-Stage Riser Computations

Variable	Pre-development	Post-development
Drainage area, A (ac)		
Drainage area, A_m (mi^2)		
Time of Conc., t_c (hr)		
Curve number, CN		

OUTLET CHARACTERISTICS			
Var	Units	Value	Comment
n	—		
L	ft		
D	ft		Initial estimate
K_p	—		Table 8.2
$Kp*L$	ft		
C_*			Eq. (8.31)
E_0	ft		Wet Pond
E_c	ft		Centerline
E_t	ft		Tailwater elev.
C_w	—		Weir coeff.

Step	Variable	Low stage	High stage
1	Rainfall, P (in.)		
2	I_a/P (pre-dev.) I_a/P (post-dev.)		
3	Runoff depth, Q_b (pre-dev.) (in.) Runoff depth, Q_a (post-dev.) (in.)		
4	Unit peak, q_{ub} (pre-dev.) (ft^3/s/mi^2/in.) q_{ua} (post-dev.) (ft^3/s/mi^2/in.)		
5	Peak discharge, q_{pb} (pre-dev.) (ft^3/s) q_{pa} (post-dev.) (ft^3/s)		
6	q_p ratio = $q_{pb}/q_{pa} = R_q$		
7	Storage ratio, R_v		
8	Depth of storage, d_s (in.)		
9	Volume of storage, V_s (ac-ft)		
10	Dead storage, V_d (ac-ft)		
11	Total storage, Vt (ac-ft)		
12	Elevation, E_j (ft)		
13	Diameter, D (ft)		
14	Orifice width, W_o Orifice area, A_o (ft^2) Orifice height, H_o (ft) Orifice discharge, q_{o2} (ft^3/s)		
15	Low-stage weir length, L_{w1} (ft) Weir discharge, q_{o2} (ft^3/s)		
16	High-stage weir length, L_{w2} (ft)		
17	Invert elevation, E_i (ft)		

(10) from stage-storage, with E_o

(11) $V_t = V_s + V_d$

(12) from stage-storage, with V_t

(13) $D = C_* q_{pb2}^{0.5}/(E_1 - E_c)^{0.25}$

(14) $W_o \sim 0.75 * D$
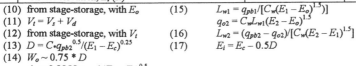
$A_o = 0.2283 q_{pb1} / (E_1 - E_o)^{0.5}$
$H_o = A_o/W_o$
$q_{o2} = 4.82 A_o (E_2 - E_1)^{0.5}$

(15) $L_{w1} = q_{pb1}/[C_w(E_1 - E_o)^{1.5})]$
$q_{o2} = C_w L_{w1}(E_2 - E_o)^{1.5}$

(16) $L_{w2} = (q_{pb2} - q_{o2})/[C_w(E_2 - E_1)^{1.5}]$

(17) $E_i = E_c - 0.5D$

Table 8.2. K_p Values for reinforced concrete ($n = 0.013$) and corrugated metal pipe ($n = 0.024$)

Pipe Diameter (in.)	RCP	CMP
24	0.01240	0.0423
27	0.01061	0.0362
30	0.00922	0.0314
36	0.00723	0.0246
42	0.00589	0.0201
48	0.00493	0.0168
54	0.00421	0.0144
60	0.00366	0.0125

For other values of n and d, K_p can be computed by $K_p = 5087\, n^2\, d^{-4/3}$, where d [=] inches.

6. Using the elevation E_0 of either the weir or the bottom of the orifice, obtain the volume of dead storage V_d from the stage-storage curve.
7. Compute the total storage in acre-feet:

$$V_t = V_d + V_s \qquad (8.30)$$

8. Enter the stage-storage curve with V_t to obtain the maximum water surface elevation, E_1.
9. a.   Obtain the friction head-loss coefficient K_p from Table 8.2.
   b.   Using the product LK_p, which is denoted as X, compute C_*:

$$C_* = 0.456 + 0.047X - 0.0024X^2 + 0.00006X^3 \qquad (8.31)$$

   c.   Compute the conduit diameter:

$$D = C_* q_{pb}^{0.5} h^{-0.25} \quad \text{where} \quad h = E_1 - E_c \qquad (8.32)$$

   (assuming that E_c is greater than the tailwater elevation).
   d.   Adjust D to the nearest larger commercial pipe size.
10. If the outlet is an orifice, determine characteristics of the orifice:
   a.   Set the orifice width, W_0; as a rule of thumb, try $0.75D$.
   b.   Compute the area of the orifice:

$$A_0 = \frac{0.2283 q_{pb}}{(E_1 - E_0)^{0.5}} \qquad (8.33a)$$

   c.   Compute the height of the orifice:

$$H_0 = \frac{A_0}{W_0}$$ (8.33b)

11. If the outlet is a weir, determine the weir length:

$$L_w = \frac{q_{pb}}{C_w (E_1 - E_0)^{1.5}}$$ (8.34)

where a value of C_w must be assumed.

12. Compute the conduit invert elevation (ft) at the face of the riser:

$$E_i = E_c - \frac{D}{2}$$ (8.35)

Example 8.6. The 23-acre watershed of Figure 8.6 can be used to illustrate the sizing of a single-stage riser. Given the small drainage area, a corrugated metal pipe will be used. A rectangular orifice will be cut into the riser.

The post-development peak discharge is 21 ft³/s. For the pre-development conditions, the CN is 61, and a time of concentration of 1.25 hr was computed using a flow velocity of 0.2 ft/sec over a length of 900 ft. For a CN of 61 and rainfall of 4.8 in., the runoff depth is 1.25 in. and the I_a/P equals 0.266. The unit peak discharge from Eqs. (4.53) is 290 cfsm/in., so the pre-development peak discharge is:

$$q_{pb} = 290 \left(\frac{23}{640} \right) (1.25) = 13 \text{ cfs}$$ (8.36)

Thus, the pre- to post-development peak discharge ratio is 0.619. Using this as input to Figure 8.4 yields a storage volume ratio of 0.235. Thus, the volume of active flood storage is:

$$V_s = R_s Q_a = 0.235(1.59) = 0.374 \text{ in.}$$ (8.37)

$$V_{st} = V_s A / 12 = 0.374(23)/12 = 0.716 \text{ ac-ft}$$ (8.38)

The stage-storage relationship at the site of the detention structure is

$$V_s = 0.0444 h^{2.17}$$ (8.39)

in which h is the stage (ft) measured above the datum and V_s is the storage (ac-ft).

Based on the invert elevation of the orifice, which is $E_0 = 2$ ft, the dead storage can be computed from Eq. (8.39) as 0.2 ac-ft. The total storage, active plus dead, is

0.916 ac-ft (that is, $V_t = V_{st} + V_d = 0.716 + 0.2$). Using V_t with Eq. (8.39), the depth at flood stage can be computed by solving for h:

$$h = \left(\frac{V_t}{0.0444}\right)^{1/2.17} = \left(\frac{0.916}{0.0444}\right)^{0.461} = 4.03 \text{ ft} \qquad (8.40)$$

The diameter of the outlet pipe is computed with Eq. (8.32):

$$D = 0.63(13)^{0.5}(4.03 - 1.00)^{-0.25} = 1.7 \text{ ft} \qquad (8.41)$$

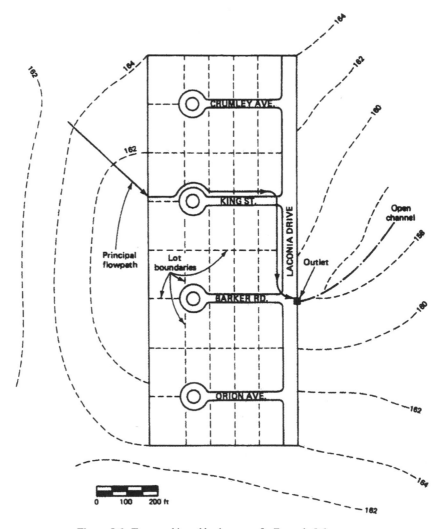

Figure 8.6. Topographic and land use map for Example 8.6.

Worksheet 8.3 Sizing of single-stage riser for Example 8.6

Variable	Pre-development	Post-development
Drainage area, A_m (mi^2)	0.03594	0.03594
Time of Conc., t_c (hr)	1.25	0.82
Curve number, CN	61	66

OUTLET CHARACTERISTICS			
Var	Units	Value	Comment
n	—	0.024	
L	ft	90	
D	ft	2	Initial estimate
K_p	—	0.05	Table 8.2
$Kp*L$	ft	4.5	
C_*		0.63	Eq. (8.31)
E_0	ft	2	Wet Pond
E_c	ft	1	Centerline
E_t	ft	—	Tailwater elev.
C_w	—	3.1	Weir coeff.

Step	Variable	Low stage	High stage
1	Rainfall, P (in.)	4.8	
2	I_a/P (pre-development)	0.266	
	I_a/P (post-development)	0.210	
3	Runoff depth, Q_b (pre-dev.) (in.)	1.25	
	Runoff depth, Q_a (post-dev.) (in.)	1.59	
4	Unit peak, q_{ub} (pre-dev.) (ft^3/s/mi^2/in.)	290	
	q_{ua} (post-dev.) (ft^3/s/mi^2/in.)	370	
5	Peak discharge, q_{pb} (pre-dev.) (ft^3/s)	13	
	q_{pa} (post-dev.) (ft^3/s)	21	
6	q_p ratio = $q_{pb}/q_{pa} = R_q$	0.619	
7	Storage ratio, R_v	0.235	
8	Depth of storage, d_s (in.)	0.374	
9	Volume of storage, V_s (ac-ft)	0.716	
10	Dead storage, V_d (ac-ft)	0.20	
11	Total storage, Vt (ac-ft)	0.916	
12	Elevation, E_j (ft)	4.03	
13	Diameter, D (ft)	1.72 (use 1.75)	
14	Orifice width, W_o	1.25	
	Orifice area, A_o (ft^2)	2.083	
	Orifice height, H_o (ft)	1.666	
	Orifice discharge, q_{o2} (ft^3/s)	—	
15	Low-stage weir length, L_{w1} (ft)		
	Weir discharge, q_{o2} (ft^3/s)		
16	High-stage weir length, L_{w2} (ft)		
17	Invert elevation, E_i (ft)	0.14	

(10) from stage-storage, with E_o
(11) $V_t = V_s + V_d$
(12) from stage-storage, with V_t
(13) $D = C_* q_{pb2}^{0.5}/(E_1 - E_c)^{0.25}$
(14) $W_o \sim 0.75 * D$
$\quad A_o = 0.2283 q_{pb1} / (E_1 - E_o)^{0.5}$
$\quad H_o = A_o/W_o$
$\quad q_{o2} = 4.82 A_o (E_2 - E_1)^{0.5}$

(15) $L_{w1} = q_{pb1}/[C_w(E_1 - E_o)^{1.5}]$
$\quad q_{o2} = C_w L_{w1}(E_2 - E_o)^{1.5}$
(16) $L_{w2} = (q_{pb2} - q_{o2})/[C_w(E_2 - E_1)^{1.5}]$
(17) $E_i = E_c - 0.5D$

As a rule of thumb, the width (W_0) of the orifice is taken as 75% of the conduit diameter; therefore, W_0 equals 1.25 ft. The area of the orifice is:

$$A_0 = 0.2283(13)/(4.03 - 2.0)^{0.5} = 2.083 \text{ ft} \qquad (8.42)$$

Thus, the height of the rectangular orifice is 2.08/1.25 = 1.666 ft. The invert of the outlet conduit (E_i) is:

$$E_i = E_c - 0.5D = 1.0 - 0.5(1.75) = 0.125 \qquad (8.43)$$

The diameter of the riser barrel is usually 2 to 3 times the diameter of the outlet conduit, so a 4.5-ft (54-in.) corrugated metal pipe can be used for the riser. The computations are summarized on Worksheet 8.3.

8.3.2.2 Sizing of Two-Stage Risers

Where stormwater or drainage policies require control of flow rates of two exceedance frequencies, the two-stage riser is an alternative for control. The structure of a two-stage riser is similar to the single-stage riser except that it includes either two weirs or a weir and an orifice (see Figure 8.7). For the weir/orifice structure, the orifice is used to control the more frequent event, and the larger event is controlled using the weir. The runoff from the smaller and larger events are also referred to as the low-stage and high-stage events, respectively. Values for variables at low and high stages may be followed by a subscript 1 or 2, respectively; for example q_{pb2} will indicate the pre-development peak discharge for the high-stage event. Recognizing that the two events will not occur simultaneously, both the low-stage weir or orifice and the high-stage weir are used to control the high-stage event.

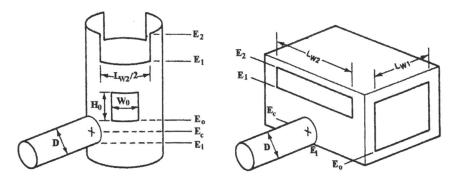

Figure 8.7. Two-stage outlet facilities.

The sizing of a two-stage riser is only slightly more complicated than the sizing of a single-stage riser. The procedure follows the same general format as for the single-stage riser, but both the high-stage weir and low-stage outlet characteristics must be determined. The input for sizing a two-stage riser is same as that for a single-stage riser, but many of the values must be computed for both low-stage and high-stage events. The input consists of watershed characteristics, rainfall depths, site characteristics, outlet characteristics, and the stage-storage relationship for the location. The input requirements are summarized in Worksheet 8-2. The following steps can be used to size a two-stage riser for the cases where the low-stage outlet is either a weir or an orifice:

1.   a. Using the 24-hr rainfalls and the pre-development CN, find the runoff depth for both the low- and high-stage events, Q_{b1} and Q_{b2}.

      b. Using the 24-hr rainfalls and the post-development CN, find the runoff depth for both the low- and high-stage events, Q_{a1} and Q_{a2}.

2.   a. Determine the pre-development peak discharges for both the low- and high-stage events, q_{pb1} and q_{pb2}.

      b. Determine the post-development peak discharges for both the low- and high-stage events, q_{pa1} and q_{pa2}.

3.   Compute the discharge ratios for both the low- and high-stage events:

$$R_{q1} = \frac{q_{pb1}}{q_{pa1}} \qquad\qquad (8.44a)$$

$$R_{q2} = \frac{q_{pb2}}{q_{pa2}} \qquad\qquad (8.44b)$$

4.   Enter the R_s-versus-R_q curve (Figure 8.4) with R_{q1} and R_{q2} to find the storage volume ratios R_{s1} and R_{s2}.

5.   a. Compute the equivalent depth of active storage in inches for both the low- and high-stage events:

$$V_{s1} = Q_{a1}R_{s1} \qquad\qquad (8.45a)$$

$$V_{s2} = Q_{a2}R_{s2} \qquad\qquad (8.45b)$$

      b. Convert V_{s1} and V_{s2} to acre-feet by multiplying by $A/12$, where A is the drainage area in acres.

6.   Using the elevation E_0 obtain the volume of dead storage V_d from the elevation-storage curve.

7.   Compute the total storage (acre-feet) for both the low- and high-stage events:

$$V_{t1} = V_d + V_{s1} \qquad\qquad (8.46a)$$

$$V_{t2} = V_d + V_{s2} \qquad\qquad (8.46b)$$

8. Enter the stage-storage curve with V_{f1} and V_{f2} to obtain the low- and high-stage water surface elevations, E_1 and E_2.
9. a. Obtain the friction head-loss coefficient K_p from Table 8.2.
    b. Using the product LK_p obtain C_* from Eq. (8.31).
    c. Compute the conduit diameter:

$$D = C_* q_{pb2}^{0.5} h^{-0.25} \quad \text{where } h = E_1 - E_c \qquad (8.47)$$

(assuming that E_c is greater than the tailwater elevation).
    d. Adjust D to the nearest larger commercial pipe size.
10. If the low-stage control is an orifice, determine characteristics of the orifice:
    a. Set the orifice width, W_0; as a rule of thumb, try $0.75*D$.
    b. Compute the area of the orifice:

$$A_0 = \frac{0.2283 q_{pb1}}{\left(E_1 - E_0\right)^{0.5}} \qquad (8.48)$$

    c. Compute the height of the orifice:

$$H_0 = \frac{A_0}{W_0} \qquad (8.49)$$

    d. Estimate the flow through the low-stage orifice during the high-stage event:

$$q_{02} = 4.82 A_0 \left(E_2 - E_1\right)^{0.5} \qquad (8.50)$$

11. If the low-stage control is a weir, determine the characteristics of the weir:
    a. Compute the weir length:

$$L_{w1} = \frac{q_{pb1}}{C_w \left(E_1 - E_0\right)^{1.5}} \qquad (8.51)$$

    b. Compute the flow over the low-stage weir during the high-stage event:

$$q_{02} = C_w L_{w1} \left(E_2 - E_0\right)^{1.5} \qquad (8.52)$$

12. Compute the high-stage weir length:

$$L_{w2} = \frac{q_{pb2} - q_{02}}{C_w(E_2 - E_1)^{1.5}} \qquad (8.53)$$

13. Compute the conduit invert elevation (ft) at the face of the riser:

$$E_i = E_c - \frac{D}{2} \qquad (8.54)$$

Example 8.7. Consider the forested (fair condition) watershed shown in Figure 8.8. A 52.8-acre tract within the watershed (dashed lines) is to be developed as a commercial/business center, with the total watershed having an area of 128.5 acres. To control the runoff rates from the developed site, an off-site detention structure is planned for the total watershed outlet. The local drainage policy requires control of both the 2- and 10-yr peak discharges. The SCS two-stage riser method can be used to develop a planning estimate of the storage volume and outlet facility characteristics. The input and calculations are given in Worksheet 8.4.

For the watershed of Figure 8.8, the pre-development CN, assuming a C soil group, is 73 (see Table 4.7). Assume that the 2-yr, 24-hr and 10-yr, 24-hr rainfall depths are 3.0 and 5.0 in., respectively; therefore, the runoff depths are 0.86 and 2.28 in., respectively. From Eq. (4.25), the I_a/P ratios are 0.25 and 0.15, respectively, for the 2- and 10-yr events. The watershed has a total length of 3900 ft, with 50 ft of sheet flow in the forest, 250 ft of overland flow in the forest, 500 ft of upland gully flow, and 3100 ft of channel flow. Equation (4.11) can be used to compute the travel time for sheet flow. The flow velocities for the overland and gully flows can be obtained from Figure 4.7, with the velocity method used to compute the travel time. Channel flow velocities should be computed using Manning's equation. The following tabular summary shows the calculation of the pre-development time of concentration:

Flow Path	n	L (ft)	S (ft/ft)	i (in./hr)	R_h (ft)	V (fps)	T_t (sec)
Sheet flow	0.2	50	0.050	5	—	—	289
Forest (heavy litter)	—	250	0.050	—	—	0.56	447
Concentrated (gully)	—	500	0.050	—	—	5.25	95
Channel	0.065	1200	0.044	—	1.2	5.43	221
Channel	0.060	1900	0.040	—	1.3	5.92	321
		3900					1373

This yields a time of concentration of 0.381 hr. Based on a t_c of 0.381 hr and the previously given values of I_a/P, unit peak discharge estimates of 540 and 580 ft³/sec/mi²/in. are obtained from Eq. (4.23) for the 2- and 10-yr events, respectively. Thus the 2- and 10-yr peak discharges are:

Worksheet 8.4 Two-stage riser computations for Example 8.7

Variable	Pre-development	Post-development
Drainage area, A (ac)	128.5	128.5
Drainage area, A_m (mi^2)	0.2008	0.2008
Time of Conc., t_c (hr)	0.381	0.28
Curve number, CN	73	82

OUTLET CHARACTERISTICS			
Var	Units	Value	Comment
n	—	0.013	
L	Ft	225	
D	Ft	5	Initial estimate
K_p	—	0.00366	Table 8.2
$Kp*L$	Ft	0.824	
$C.$		0.493	Eq. (8.31)
E_0	Ft	2	Wet Pond
E_c	Ft	-6	Centerline
E_t	Ft		Tailwater elev.
C_w	—	3.1	Weir coeff.

Step	Variable	Low stage	High stage
1	Rainfall, P (in.)	3.0	5.0
2	I_a/P (pre-development)	0.25	0.15
	I_a/P (post-development)	0.15	0.10
3	Runoff depth, Q_b (pre-dev.) (in.)	0.86	2.28
	Runoff depth, Q_a (post-dev.) (in.)	1.38	3.08
4	Unit peak, q_{ub} (pre-dev.) (ft^3/s/mi^2/in.)	540	580
	q_{ua} (post-dev.) (ft^3/s/mi^2/in.)	660	685
5	Peak discharge, q_{pb} (pre-dev.) (ft^3/s)	93.2	265.5
	q_{pa} (post-dev.) (ft^3/s)	182.9	423.6
6	q_p ratio = $q_{pb}/q_{pa} = R_q$	0.510	0.627
7	Storage ratio, R_v	0.272	0.230
8	Depth of storage, d_s (in.)	0.375	0.708
9	Volume of storage, V_s (ac-ft)	4.02	7.59
10	Dead storage, V_d (ac-ft)	1.41	—
11	Total storage, Vt (ac-ft)	5.43	9.00
12	Elevation, E_j (ft)	4.90	6.87
13	Diameter, D (ft)	4.42 (use 4.5)	—
14	Orifice width, W_o	8.33	
	Orifice area, A_o (ft^2)	12.5	
	Orifice height, H_o (ft)	1.5	
	Orifice discharge, q_{o2} (ft^3/s)	84.6	
15	Low-stage weir length, L_{w1} (ft)		
	Weir discharge, q_{o2} (ft^3/s)		
16	High-stage weir length, L_{w2} (ft)		21.1
17	Invert elevation, E_i (ft)	-8.25	

(10) from stage-storage, with E_o

(11) $V_t = V_s + V_d$

(12) from stage-storage, with V_t

(13) $D = C \cdot q_{pb2}^{0.5}/(E_1 - E_c)^{0.25}$

(14) $W_o \sim 0.75 * D$

   $A_o = 0.2283 q_{pb1} / (E_1 - E_o)^{0.5}$

   $H_o = A_o/W_o$

   $q_{o2} = 4.82A_o (E_2 - E_1)^{0.5}$

(15) $L_{w1} = q_{pb1}/[C_w(E_1 - E_o)^{1.5})]$

   $q_{o2} = C_w L_{w1}(E_2 - E_o)^{1.5}$

(16) $L_{w2} = (q_{pb2} - q_{o2})/[C_w(E_2 - E_1)^{1.5}]$

(17) $E_i = E_c - 0.5D$

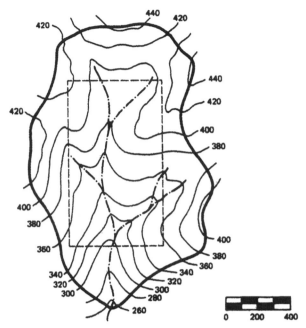

Figure 8.8. Watershed layout for Example 8.7

$$q_{pb1} = (540 \text{ ft}^3/\text{sec/mi}^2/\text{in.}) \left(\frac{128.5}{640} \text{ mi}^2 \right)(0.86 \text{ in.}) = 93.2 \text{ ft}^3/\text{sec} \quad (8.55)$$

$$q_{pb2} = (580 \text{ ft}^3/\text{sec/mi}^2/\text{in.}) \left(\frac{128.5}{640} \text{ mi}^2 \right)(2.28 \text{ in.}) = 266 \text{ ft}^3/\text{sec} \quad (8.56)$$

For the post-development condition, 52.8 acres will be developed for commercial/business. Initial site plans indicate 88% of the developed area will be impervious cover, with the remainder in lawn cover (good condition). Thus the weighted CN for the 128.5 acre watershed is:

$$CN = \frac{52.8}{128.5}[0.88(98) + 0.12(74)] + \frac{75.7}{128.5}(73) = 82 \quad (8.57)$$

For rainfall depths of 3.0 and 5.0 in., the runoff depths are 1.38 and 3.08 in., respectively. The post-development I_a/P ratios are 0.15 and 0.1, respectively. Using the assumption that the t_c decreased to 0.28 hr, the unit peak discharges are 660 and 685 ft^3/sec/mi^2/in., and the peak discharges are:

$$q_{pa1} = \left(660\,\text{ft}^3/\text{sec/mi}^2/\text{in.}\right)\!\left(\frac{128.5}{640}\,\text{mi}^2\right)\!(1.38\,\text{in.}) = 183\,\text{ft}^3/\text{sec} \qquad (8.58)$$

$$q_{pa2} = \left(685\,\text{ft}^3/\text{sec/mi}^2/\text{in.}\right)\!\left(\frac{128.5}{640}\,\text{mi}^2\right)\!(3.08\,\text{in.}) = 424\,\text{ft}^3/\text{sec} \qquad (8.59)$$

Worksheet 8.4 provides a summary of the sizing of the detention structure. Using the ratios of the pre-development to post-development peak discharges with the SCS detention relationship (Figure 8.4), the depth of active storage for low-stage and high-stage control are 0.375 and 0.708 in., respectively, which translate to storage volumes of 4.02 and 7.59 acre-ft.

At the site of the detention structure, the stage-storage relationship is $V = 0.5\,h^{1.5}$ where V [=] ac-ft and h [=] ft. For a permanent pond depth of 2 ft, the dead storage is 1.41 acre-ft. Thus the total volume of storage at the low-stage and high-stage flood conditions would be 5.43 and 9.00 acre-ft, respectively; from the stage-storage curve these volumes correspond to depths of 4.90 and 6.87 ft.

In addition to the storage volumes, the riser must also be sized. Using Equation (8.47), the diameter of the pipe outlet is:

$$D = 0.493\!\left[\frac{266}{\left(4.9-(-6)\right)^{0.5}}\right]^{0.5} = 4.42\,\text{ft} \quad (\text{use 54 in.}) \qquad (8.60)$$

Since the difference $E_2 - E_1$ is only 2 ft, the height of the orifice opening H_0 will be set at 1.5 ft. The area of the orifice that would be required to limit the discharge through the orifice to the pre-development peak discharge of 93.2 ft^3/sec can be computed with Eq. (8.48):

$$A_0 = \frac{93.2}{4.38(4.9-2)^{0.5}} = 12.5\,\text{ft}^2 \qquad (8.61)$$

Therefore, the width of the rectangular orifice is 8.33 ft, which is computed using Eq. (8.49). To compute the length of the weir with Eq. (8.53), the discharge (q_r) through the orifice when the high-stage event occurs must be estimated with Eq. (8.50):

$$q_r = 4.82(12.5)(6.87-4.90)^{0.5} = 84.6\,\text{ft}^3/\text{sec} \qquad (8.62)$$

Thus, with Eq. (8.53), the weir length is:

$$L_w = \frac{265.5-84.6}{3.1(6.87-4.9)^{1.5}} = 21.1\,\text{ft} \qquad (8.63)$$

8.4 EXTENDED DETENTION

Stormwater management basins have traditionally been designed for flood control. Wet ponds have been used with the hope that they would provide some water quality benefits, but the effectiveness of wet ponds has not been validated. Because of the first-flush effect, it is generally believed that water quality would be improved if a detention basin were to capture the first flush and retain it for some period of time. The amount captured is generally expressed as an area-equivalent depth, D_e inches (e.g., 0.25 in. or 0.5 in.). The extended detention time, T_e hours, is often set at 12 or 24 hours.

The volume of first-flush storage V_e (ac-ft) is computed using the capture depth D_e and the drainage area:

$$V_e = D_e A / 12 \qquad\qquad (8.64)$$

The first-flush volume must be added to any dead storage volume, if present, to compute the extended detention storage volume. The sum of the dead storage (V_d) and the extended detention volume (V_e) is used with the stage-storage relationship to determine the extended detention flood elevation, E_o, which is also the invert elevation for the low-stage orifice when the low-stage flood volume is greater than the extended detention flood volume.

The above procedure indicates that the capture criterion is D_e area-inches, e.g., one-half inch for the entire watershed. An alternative philosophy suggest that during the early part of a storm only impervious areas will contribute direct runoff, and therefore, the design criterion should be an equivalent depth for just the impervious surfaces. For example, a 10-acre watershed that is 60% impervious would set the design criterion to capture one-half inches as 3 acre-inches, rather than 5 acre-inches needed for the criterion based on the entire drainage area. Where such a criterion is applied, Eq. (8.64) is rewritten as:

$$V_e = f D_e A / 12 \qquad\qquad (8.65)$$

where f is the fraction of imperviousness.

Arguments can be given for both criteria. If the storm intensity is high at the beginning of a storm, significant runoff can be produced from saturated pervious areas. Also, runoff from unconnected impervious areas, while counted when using the impervious-area-only criterion, may not appear as part of the first flush inflow. The better of the two criterion is not known, so it becomes more of a policy decision. For microwatersheds, the impervious-area-only criterion may be preferred, while for larger watersheds the total-area criterion may be more realistic.

In one sense, the design for extended detention becomes a three-stage riser system. However, the volume used for water quality control is actively part of the flood storage volume. The riser may include outlets for the extended detention, a low-stage orifice for control of high frequency events, and a high-stage weir or orifice for control of large volume storms. Under certain circumstances, the volume needed for the first-flush control may be greater than that for flood control, in which case the extended detention outlet also serves as the flood control outlet. For example, if the low-stage event has a small rainfall depth, the volume of flood storage may be small, such that the greater length of time for first-flush control storage may cause a need for a larger volume of storage.

The first-flush elevation, E_o, is established from the stage-storage relationship at the site and the volume, V_{de}, which is the sum of the first-flush volume and any dead storage.

The first-flush volume is released at a low discharge rate over a duration T_e (hours). The design discharge rate, q_e (ft^3/s), for the first-flush storage is computed:

$$q_e = AD_e / T_e \qquad (8.66a)$$

or the following if the volume depends on the imperviousness:

$$q_e = fAD_e / T_e \qquad (8.66b)$$

in which A is the contributing area (ac). For the units given, the conversion factors are algebraically about 1.0. For a 60-ac contributing area, the design discharge rate would be 2.5 ft^3/sec to detain 0.5 inch for 12 hours.

Generally, small orifices in the riser will serve as the outlets for the extended detention runoff. Thus, the required number of outlets (n) depends on the discharge through one hole, q_o, and the design discharge rate, q_e:

$$n = q_e / q_o \qquad (8.67)$$

where the discharge rate q_o through one orifice is:

$$q_o = C_o A_o [2g(E_o - E_d)]^{0.5} \qquad (8.68)$$

in which C_o is the orifice coefficient, usually 0.6; A_o is the area of a single hole of diameter D_o; E_o is the first-flush storage elevation; and E_d is the dead-storage elevation, which is also the elevation of the first-flush orifices.

For an extended-detention discharge computed with Eq. (8.64), the total orifice area A_o can be computed with the orifice equation:

$$A_e = \frac{q_e}{C_o [2g(E_o - E_d)]^{0.5}} \qquad (8.69)$$

in which E_o and E_d are the maximum and invert stage elevations for extended detention storage, respectively. The number of orifices needed in the riser is easily computed as the ratio of A_o to the area or one hole.

Once the extended detention storage is computed, the calculations for flood control storage design can be followed. The extended detention storage is part of the flood storage, but the flood control outlet may need to be adjusted for the presence of the extended detention orifices.

If a detention basin includes first-flush storage, then flow through the first-flush orifices must be accounted for when designed flood storage outlets. For a single-stage riser, the flow through the first-flush orifices would be:

$$q_{e1} = 4.82 A_e (E_1 - E_d)^{0.5} \qquad (8.70)$$

This would be used to adjust the design discharge for either a low-stage orifice of area A_o:

$$A_o = 2.2283 (q_{pb1} - q_{e1})/(E_1 - E_0)^{0.5} \qquad (8.71)$$

or a low-stage weir:

$$L_{w1} = (q_{pp1} - q_{e1})/\left[C_w (E_1 - E_o)^{1.5}\right] \qquad (8.72)$$

For a riser designed for high and low flood stages, the weir length for the high-stage event is computed by:

$$L_{w2} = (q_{pb2} - q_{o2} - q_{e2})/\left[C_w (E_2 - E_1)^{1.5}\right] \qquad (8.73)$$

The procedure is summarized in Worksheet 8.5.

8.5 WATER QUALITY IMPROVEMENT

Removal of pollutants can occur during the time that the water flows through the SWM basin. This improvement can have an immediate impact on downstream water quality. However, any water that is stored in the pond and gradually released, or is forced out during the next event, will most likely have different characteristics from the water that initially entered the pond. Time is an important parameter in allowing the progress of various processes and reactions. Longer times have the potential for better quality improvement.

8.5.1 Conventional Sedimentation Theory

Conventional sedimentation theory links particle sedimentation characteristics to the size and flow characteristics of the water passing through the basin. The theory, however, requires a number of simplifying assumptions that are more fitting for use in a water or wastewater treatment plant, where flows tend to be steady and controlled. We will develop the theory as a starting point and evaluate how it applies to SWM basins.

The analysis uses an ideal rectangular basin, as shown in Figure 8.9. Of interest is a sediment particle that enters at the top of the basin. The water flow carries the particle horizontally while it falls vertically, as described by Stokes Law. The particle reaches the bottom of the basin just at the exit. The time for this travel is designated as $\bar{t}$. During $\bar{t}$, the vertical distance traveled is H/V$_s$, and the horizontal distance is L/v$_f$, where V$_s$ and V$_f$ are, respectively, the Stokes settling velocity of the particle and the horizontal velocity of the flow moving the particle. Since the travel time is identical:

$$\bar{t} = \frac{H}{V_s} = \frac{L}{V_f} \qquad (8.74)$$

This equation is solved for the settling velocity and multiplied by the basin width, W. The product V$_f$HW is equal to the flow rate through the basin, Q; also, LW is equal to the bottom area of the basin, A. Thus:

$$V_s = \frac{V_f H}{L} = \frac{V_f HW}{LW} = \frac{Q}{A} \qquad (8.75)$$

The parameter Q/A is known as the loading. It has units of volume per time per area.

Any particle that has a settling velocity greater than or equal to Q/A will settle to the bottom of the basin before it reaches the end of the basin, and thus is considered as removed. Particles that have velocities (designated as V$_i$) less than Q/A are partially removed; those that enter near the bottom will settle, while those near the top will not and will be carried off with the effluent. The fraction of smaller particles (with velocity, V$_i$) removed, R$_i$, is equal to:

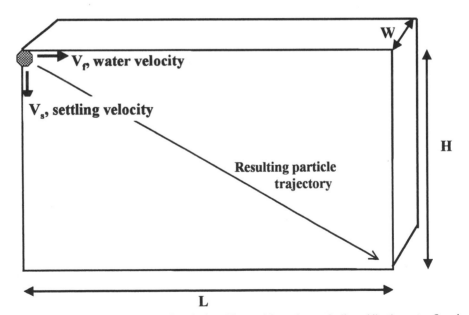

Figure 8.9. Diagram of ideal settling basin. The particle settles vertically, while the water flow is horizontal

Worksheet 8.5. Extended detention riser computations

1	Imperviousness, I	
2	First-flush depth, D_e (in.)	
3	First-flush duration, T_e (hr)	
4	First-flush volume, V_e (ac-ft)	
5	Dead-storage elevation, E_d (ft)	
6	Dead-storage volume, V_d (ac-ft)	
7	Volume, $V_{de} = V_d + V_e$	
8	First-flush elevation, E_o	
9	First-flush discharge rate, q_e (cfs)	
10	Orifice diameter, D_o (in.)	
11	Orifice area, $A_o = \pi D_o^2/4$	
12	Total orifice area, A_e	
13	Number of orifices, $n = A_e/A_o$	
14	Elevation, conduit centerline, E_c (ft)	
15	Tailwater elevation, E_t (ft)	
16	Weir coefficient	

Variable	Pre-development	Post-development
Drainage area, A (ac)		
Drainage area, A_m (mi^2)		
Time of Conc., t_c (hr)		
Curve number, CN		

OUTLET CHARACTERISTICS			
Var	Units	Value	Comment
n	—		
L	ft		
D	ft		Initial estimate
K_p	—		Table 8.2
K_p*L	ft		
$C_{\bullet}$			Eq. (8.31)
E_0	ft		Wet Pond
E_c	ft		Centerline
E_t	ft		Tailwater elev.
C_w	—		Weir coeff.

Step	Variable	Low stage	High stage
17	Rainfall, P (in.)		
18	I_a/P (pre-development) I_a/P (post-development)		
19	Runoff depth, Q_b (pre-dev.) (in.) Runoff depth, Q_a (post-dev.) (in.)		
20	Unit peak, q_{ub} (pre-dev.) (ft^3/s/mi^2/in.) q_{ua} (post-dev.) (ft^3/s/mi^2/in.)		
21	Peak discharge, q_{pb} (pre-dev.) (ft^3/s) q_{pa} (post-dev.) (ft^3/s)		
22	q_p ratio = $q_{pb}/q_{pa} = R_q$		
23	Storage ratio, R_v		
24	Depth of storage, d_s (in.)		
25	Volume of storage, V_s (ac-ft)		
26	Dead storage, V_d (ac-ft)		
27	Total storage, Vt (ac-ft)		
28	Elevation, E_j (ft)		
29	Diameter, D (ft)		
30	Extended detention discharge, q_e (cfs)		
31	Orifice width, W_o Orifice area, A_o (ft^2) Orifice height, H_o (ft) Orifice discharge, q_{o2} (ft^3/s)		
32	Low-stage weir length, L_{w1} (ft) Weir discharge, q_{o2} (ft^3/s)		
33	High-stage weir length, L_{w2} (ft)		
34	Invert elevation, E_i (ft)		

$$R_i = \frac{V_i}{Q/_A} \qquad (8.76)$$

Obviously, the smallest particles will be the least effectively removed.

Example 8.8   Water passes through a SWM basin at a flow rate of 14 cfs. The basin has a bottom area of 6000 ft². Using Stokes Law from Chapter 6, the settling velocities of 0.1 and 0.01 mm sediment particles are calculated.   A sediment specific gravity of 2.65 is used along with a temperature of 50°F.   For the 0.1-mm particle:

$$V_{0.1} = \frac{\left(\gamma_p - \gamma_w\right)D^2}{18\mu} = \frac{\left(2.65(62.4) - (62.4lb/ft^3)(3.28x10^{-4} ft)^2\right)}{18(2.74x10^{-5} lb - s/ft^2)} \qquad (8.77)$$

$$= 0.0225 ft/s$$

Similarly, for the 0.01-mm particle, $v_{0.01} = 2.25x10^{-4}$ ft/s.
    The loading is:

$$\frac{Q}{A} = \frac{14 ft^3/s}{6000 ft^2} = 2.33x10^{-3} ft/s \qquad (8.78)$$

The removals are then calculated. For the 0.1-mm particle:

$$R_{0.1} = \frac{V_{0.1}}{Q/_A} = \frac{0.0225 ft/s}{2.33x10^{-3} ft/s} = 9.66 \qquad (8.79)$$

Since the fraction is greater than 1, all of these particles should be removed.
    For the 0.01 mm sediment:

$$R_{0.01} = \frac{V_{0.01}}{Q/_A} = \frac{2.25x10^{-4} ft/s}{2.33x10^{-3} ft/s} = 0.0966 \qquad (8.80)$$

Only 9.7% of these smaller particles will be removed.

Many processes inherent to SWM basins will not allow the use of this simple analysis. Most important, the above analysis assumes steady-state conditions–that is, that all parameters are constant with respect to time. In practice, however, most all of the important parameters discussed above are variable in a SWM pond. Rainfall events produce runoff flows that vary from their beginning to their termination. Since they are designed for storage, the flow rate in is not equal to that out, and as a result, the basin volume is changing. Also, the sediment concentration and particle sizes in the runoff will not be constant and will vary depending on a variety of landscape and hydrologic conditions.

Nonetheless, the important point that results from the simple analysis is also somewhat intuitive. The efficiency of a SWM basin in sediment removal will depend on the area of the basin. The larger the basin, the longer that water will reside in the basin. This will provide more time for the sediment to settle and be removed.

8.5.2 Other Water Quality Issues in SWM Basins

Continued capture of sediment by a well-designed basin creates long-term issues that must be addressed. First, as sediment accumulates, the volume available for flow management will be reduced. As a result, the basin may become undersized and will not provide the desired benefits. Also, as this accumulation happens, velocities in the basin will likely increase; these increases can scour sediment from the bottom of the facility and carry it through the outfall, negating previous capture efficiencies.

Regular maintenance programs to remove trapped sediment should be established for SWM basins that accumulate significant sediment. Such maintenance will keep the basin operating at its design conditions. When disposing of captured sediment, it should be realized that several types of pollutants may be affiliated with the sediments in runoff and sediment quality should be checked. Disposal options will depend on sediment quality.

SWM ponds do not have provisions to capture dissolved pollutants, only those that are affiliated with the suspended solids. As a result, most dissolved pollutants will pass through these facilities and not receive treatment. When a permanent pool is present, however, slow chemical and biological processes may occur that can modify water quality. The quality can be improved, such as through biological degradation of organic carbon, or worsened, such as by the release of adsorbed compounds from settled sediments.

8.5.3 Case Study Information

Stormwater quality improvements in SWM basins have been documented in a few cases. Several ponds have been studied in North Carolina (Wu *et al.*1996). Averaging removals over several events, good to excellent removals were found for several water quality parameters. However, in evaluating individual events, a wide range in water quality improvement is noted. In some cases, complete retention occurred for a small event, which accordingly provides 100% pollutant removal. Also, several instances of "negative removals," where output loadings were larger than input, were noted for large events. TSS removals of 41-93% were found. Other removals noted include 22-87% Fe, 22-80% Zn, 29-53% TP, and 21-37% TKN (Table 8.3).

The ponds evaluated in Wu *et al.* had runoff surface-to-pond area ratios of 0.6 to 7.5. For TSS, Fe, and Zn, increased removals were found at higher ratios. A similar trend was apparent, but was less clear for TP and TKN. The presence of waterfowl in the ponds may have contributed to the discrepancies. Biological transformations could have also played a role.

Table 8.3. Percent annual removals of pollutants from detention ponds as reported by Borden *et al.* (1998) for basins A and B and *Wu et al.* (1996) for basins C through F.

Pollutant	A	B	C	D	E	F
TSS	56	20	41	62	79	93
VSS	32	30				
TOC	15	27				
TP	41	40	29	36	53	45
DP	54	15				
$NO_2 + NO_3$	16	66				
T-NH_3-N	2	(-64)				
TN	11	36				
TKN			22	21	37	32
Zn			22	32	51	80
Fe			22	52	66	87

A study of the removal of several pollutants from two wet detention ponds designed at 0.8 to 1% pond-to-runoff area demonstrated good removal for several pollutants (Table 8.3). With these two ponds, biological reactions were found to be very important in controlling the effluent pollutant levels. Again, on an event basis, in the situation of small storms, detention ponds can provide 100% pollutant removal by storing all of the runoff from a particular event.

In a third study, the pollutant removal efficiencies of two neighboring detention ponds in New Jersey were investigated (Bartone and Uchrin 1999). The bottom of one of the basins was concrete and stone; the other was vegetated with various wetland plants. Sampling was completed over four storm events. Pollutants examined included TSS, various nitrogen and phosphorus species, petroleum hydrocarbons, and fecal coliform and fecal streptococcus. Because the land use in the watersheds was high density residential, the input pollutant levels were low and in several cases, below detection limits. In the concrete basin, some reduction in concentrations of the nitrogen and phosphorus species was noted. Slight removals of fecal coliforms were also found (fecal streptococcus were below detection limits). However, in three of the four cases, the output TSS exceeded the input mass.

The surprise was in the results from the basin with the vegetated bottom. For nearly every water quality measurement made, the effluent was worse than the influent. In some cases, the quality degradation was marked, as in the case of fecal coliforms, where levels in the effluent were ten to 100 times higher. Apparently, the quality of the water held in this basin between storm events became very poor. During the event, this poor quality water is washed from the basin.

Overall, these studies suggest complexity in the use of SWM basins for water quality improvement. In general, some removal of pollutants is expected, especially for small events when the majority of the flow is retained. Also, the data of Wu et *al.* (1996) suggest that larger basins provide better quality improvement (which agrees with conventional sedimentation theory). Chemical and physical processes that occur within the ponds, especially between events, are poorly understood and can have a major impact on water quality leaving these facilities.

8.6 PROBLEMS

8.1 Develop in detail Eq. (8.4) from Eq. (8.1).

8.2 The riser on a retention pond will be designed to limit the discharge rate is 3.25 ft^3/s when the water surface elevation is 3ft above the center of the orifices in the riser. If each orifice has a diameter of 3 inches, how many orifices are required?

8.3 A detention-basin riser has eight 1-in. diameter orifices at the base of the riser. Compute the total discharge through the orifices for depths of 0.5, 1, 2, 3, and 4 ft. Plot the depth-discharge relationship. Assume an orifice coefficient of 0.6.

8.4 A detention-basin riser has a rectangular orifice 5 in. wide and 6 in. high. When the water level is 1.5 ft above the orifice, a discharge rate of 1.17 ft^3/sec is measured. Estimate the discharge coefficient.

8.5 A detention-basin riser will be designed to pass discharge of 20.4 ft^3/sec. Assume a discharge coefficient of 0.6 for the orifice, compute the area of the orifice if the design depth is 4 ft above the orifice.

8.6 With a detailed explanation for each step, derive the weir equation of Eq. (8.17).

8.7 Weirs and orifices are used to control rates of outflow from storage facilities. Which is the distinguishing difference between the two flow control measures?

8.8 A detention-basin riser uses a 2.75 ft weir ($C_w = 3.15$). Compute and plot the stage-discharge relationship for depths up to 5 ft.

8.9 Water can discharge from a controlled-outflow lake through two separate weirs; one with a length of 3 ft has an elevation of 127 ft and one with a length of 7 ft has an elevation of 129.5 ft. Both have a discharge coefficient of 3.0. Compute and plot the stage-discharge relationship for stages up to 6 ft above the low-elevation weir.

8.10 The design discharge for a detention-basin outlet is 139 ft^3/sec. Assuming a weir coefficient of 3, estimate the required weir length at a design depth of 4.5 ft.

8.11 Discharge measurements through a 5-ft weir were collected during four storm events, with the following values:

Storm	1	2	3	4
h (ft)	2.4	3.7	4.1	4.2
q (ft^3/sec)	55	102	126	127

Estimate the weir coefficient.

8.12 The design discharge for a detention basin outlet is 240 ft^3/sec. Assuming a weir coefficient of 3.1, estimate the required weir length at a design depth of 5.2 ft.

8.13 The pre- and post-development peak discharges are 33 and 51 ft^3/s, respectively. The post-development runoff depth from the 22-acre watershed is 1.7 inches. What is the required depth (in.) and storage (acre-ft) of a detention storage structure to reduce the post-development discharge into the basin to the pre-development discharge to flow out of the detention basin?

8.14 The NRCS curve numbers for pre- and post-development conditions of a 32-acre watershed are 66 and 78, respectively. The pre- and post-development times of concentration are 30 and 20 min, respectively. Assuming a 25-yr return period with a rainfall depth of 5.6 inches, estimate the depth (in.) and volume (acre-ft) of detention storage using the NRCS method. Use the NRCS rainfall-runoff equation to estimate depths of runoff.

8.15 The NRCS curve numbers for pre- and post-development conditions of a 54-acre watershed are 72 and 82, respectively. The pre- and post-development times of concentration are 50 and 35 min, respectively. Assuming a 50-yr return period with a rainfall depth of 6.4 inches, estimate the depth (in.) and volume (acre-ft) of detention storage using NRCS method. Use the NRCS rainfall-runoff equation to estimate depths of runoff.

For Problems 8.16 and 8.17, evaluate the characteristics of a single-stage riser with a weir for the data, characteristics, and planning method specified by the problem. Assume that $n = 0.013$, $L = 150$ ft.

8.16 Use the method and data of Problem 8.14. Derive the stage-storage relationship by assuming a detention basin with a rectangular bottom with an area of 0.9 acre and vertical sides. Use $E_o = 3$ ft; $E_c = 1.5$; $C_w = 3$.

8.17 Use the method and data of Problem 8.15. Derive the stage-storage relationship by assuming a detention basin with a rectangular bottom with a length of 300 ft, a width of 200 ft, and side slopes of 3:1 (h:v). Use $E_o = 2.5$ ft; $E_c = 1.5$ ft; $C_w = 3.1$.

8.18 The pre- and post-development discharges for the lower stage are 100 and 170 ft^3/s, respectively. For the upper stage, the values are 250 and 400 ft^3/s, respectively. The low-stage and high-stage post-development runoff depths are 1.5 and 2.8 inches, respectively. The stage-storage relationship for the site of the detention basin is $V_s = 1.65\, h^{1.2}$ where V_s [=] ac-ft and h [=] ft. The contributing drainage area is 140 ac. Assume the pond does not include dead storage. The C_* is 0.5, $E_c = -2.0$ ft, and $E_o = 0$ ft. Design a two-stage riser with a low-stage orifice with a width of 9.5 ft and a high-stage riser. Use a weir coefficient of 3.

8.19 Design a two-stage weir-weir riser for a 27-ac watershed for which the low-stage pre- and post-development peak discharges are 22 and 36 ft^3/s, respectively, and the high-stage values are 44 and 62 ft^3/s, respectively. The post-development runoff depths are 1.4 in. and 2.25 in., respectively. The dead storage is 0.5 ac-ft and $E_o = 1.6$ ft. The stage-storage relationship at the site is $V_s = 0.256\, h^{1.4}$ where V_s [=] ac-ft and h [=] ft. The weir coefficient is 3.1.

8.20 The peak flow to a stormwater detention pond is 120 ft^3/s. At this flow, the pond averages 4.5 ft deep, covering an area of 0.7 acres. What percentage of sediment particles 0.1 mm in size would settle in this basin under ideal sedimentation conditions. Assume a sediment particle specific gravity of 2.65.

8.21 A SWM basin is being designed for water quality improvement in addition to flow control. It is desired to have 90% of 0.15 mm sediment particles (S = 2.65) removed at a steady state flow of 85 ft^3/s. What should be the bottom area of the basin if ideal sedimentation occurs?

8.22 The suspended solids in runoff into a detention pond are dominated by three primary particles, with characteristics given below. Determine the fraction of each particle removed and the fraction of total suspended solids removal through the pond assuming ideal sedimentation. The average pond area is 1.2 acres with a steady state flow rate of 120 ft^3/s.

Sediment	Size (mm)	Specific Gravity	Concentration (mg/L)
Sand	0.09	2.65	100
Silt	0.01	2.8	40
Brake Dust	0.06	5.4	20

8-23 The mean TSS concentration entering a SWM basin is 120 mg/L and that leaving is 35 mg/L. The basin drainage area is 86 acres and receives 38 inches of rainfall per year, 75% of which becomes runoff and flows through the basin.
a) Determine the mass of sediment collected in the pond each year.

b) Assuming a 1 acre pond area and a bulk density of 1000 kg/m^3 for the settled sediment at the bottom of the basin, what should be the accumulated sediment depth in the pond after 20 years?

8.7 REFERENCES

Bartone, D.M. and Uchrin, C.G. (1999) "Comparison of Pollutant Removal Efficiency for Two Residential Storm Water Basins," *J. Environ. Engg., ASCE,* **125**, 674-677.

Borden, R.C., Dorn, J.L., Stillman, J.B., and Liehr, S.K. (1998) "Effect of In-Lake Water Quality on Pollutant Removal in Two Ponds," *J. Environ. Engg., ASCE,* **124**, 737-743.

Wu, J.S., Holman, R.E., and Dorney, J.R. (1996) "Systematic Evaluation of Pollutant Removal by Urban Wet Detention Ponds," *J. Environ. Engg., ASCE,* **122**, 983-988.

9

VEGETATIVE CONTROL METHODS

NOTATION

A   = drainage area
A   = cross-sectional area of flow
A_c = cross-sectional area
A_c = contributing surface area
A_h = total area of holes
A_s = surface area
A_l = cross-sectional area of a stem
C   = runoff coefficient
C   = particle concentration
C_d = discharge coefficient
C_w = weir coefficient
D   = storm duration
D   = height of check dam
D   = culvert diameter
D_e = flow depth
D_f = first-flush depth
D_h = diameter of outlet holes
D_p = pipe diameter
d   = flow depth
d_b = depth of flow on spreader
d_s = ponding depth
E_t = trap efficiency
f_c = ultimate infiltration rate
f_v = intensity of vegetation growth
H   = depth of bioretention pond
h_v = height of vegetation
i   = rainfall intensity
L   = length of buffer strip
L   = length of check dam pool

L = length of bioretention pond
L_w = weir length
m = mass of media volume for water
m_{si} = mass of suspended matter in inflow
m_{so} = mass of suspended matter in outflow
m_{sr} = mass of suspended matter retained
n = Manning's roughness
n = media porosity
n = number of stems
n_h = number of holes per foot
Q_c = runoff depth from contributing area
Q_o = outflow discharge
q = discharge rate
q_b = allowable discharge on level spreader
q_i = inflow discharge rate
q_o = outflow discharge rate
q_p = peak discharge
q_w = discharge over weir
R_h = hydraulic radius
S = longitudinal slope of check dam
S = slope of buffer
T = spread of flow
t_c = time of concentration
V = flow velocity
V_b = velocity on level spreader
V_c = volume of check dam
V_h = velocity of flow through holes
V_L = volume of infiltrated water per unit width
V_v = volume of vegetation
V_w = volume available for water storage
W = width of bioretention pond
W_b = width of level spreader
z = cross slope ratio
z = transport direction
ΔS = storage above the weir
Δt = time increment
α = collision efficiency factor
η = single collector collision efficiency

9.1 INTRODUCTION

Where volumes of runoff are large, the storage basin is often the necessary method of stormwater control. However, both off-stream and in-stream storage facilities have numerous disadvantages, which have led to the need for vegetative runoff control methods. Even when vegetated control methods can provide a partial solution, their use is still preferred because of their many advantages. The inclusion of vegetative control methods into drainage plans can reduce the amount of concentrated storage required, thereby lessening the disadvantages of storage facilities.

Cost is a major advantage of vegetative control. Both the construction and maintenance costs associated with vegetation control methods are lower than the costs associated with storage facilities. The maintenance of control methods based on vegetation are often minimal and even just part of the normal landscape maintenance responsibilities of the owner. For example, a grassed swale would require periodic mowing that would be part of the routine maintenance of a lot. Also, the vegetation or mulch in a bioretention facility may need to be replaced periodically. Such maintenance costs would be substantially less than those associated with the maintenance of a detention facility, especially the periodic dredging that is necessary to maintain the active storage volume.

Hydraulic modification is a second important benefit of vegetative control methods. Vegetation will modify the flow characteristics of the runoff, most noticeably the reduction of flow velocities and increasing both the natural storage and infiltration. Soils can be modified to increase infiltration rates and promote ground-water recharge that reduces the hydraulic loading discharged from the drainage area. This, in turn, reduces the inflow to headwater streams and increases dry-weather flow rates because of the release of ground water.

Vegetative control methods offer numerous aesthetic benefits. Grassed-lined swales and buffer strips are considered more beneficial aesthetically than paved drainageways. Their natural environment appearance is preferred to the man-made appearance of paved conveyance systems. They will generally be more acceptable to wildlife because they reflect the environment from which they evolved. Generally, vegetated drainageways will improve property values when compared with sites that included paved drainageways.

Vegetative controls also offer several water quality benefits. Overland flow on vegetated areas captures a greater portion of runoff and can provide for successful removal of pollutants and reduction of pollutant loads. The vegetation provides several benefits, including providing a rough surface to slow down flowing water. Flow through dense vegetation can provide filtering of suspended solids, as well as pollutants attached to the suspended matter. The height and density of the vegetation are primary determinants of the trap efficiency of a vegetated surface. The vegetation can also provide a mechanism for the uptake of pollutants, especially nutrients, but also other compounds. *Phytoremediation*, the use of vegetation to take up and degrade pollutants, has been studied and exploited for hazardous waste sites and highly polluted groundwaters. These same processes, with appropriate modifications, can be exploited to assist in the removal of captured pollutants in stormwater BMPs.

The action of the plants can provide other benefits. This can include the creation of a highly biologically active area near the root zone of a thriving plant population. These microorganisms can assist in pollutant breakdown. The vegetation can also provide physical benefits to the area by increasing soil media permeability, both directly and indirectly by encouraging a healthy ecosystem in the rhizosphere.

From a water quality perspective, when the runoff is infiltrated or evaporated, the pollutants that were in the water remain in the vegetated area and the soil media.

The pollutants, therefore, are removed from the runoff pathway and prevented from entering the streams and waterways.

The vegetation should be carefully managed in these treatment processes. Nutrients, carbon, and various pollutants may be accumulated in the biomass of grasses and other plants. However, to remove these pollutants from the stormwater management practice, the vegetation must be cut and removed. When grasses are cut and clippings are left in a filter strip or swale the clippings will decay, releasing previously-captured nitrogen, phosphorus, and other compounds back onto the soil surface. These pollutants can be mobilized by a rain event, causing a spike of these species in the effluent from the treatment area.

In this chapter, several vegetative control methods are introduced, including buffer strips, vegetative swales, and bioretention facilities. Sheet flow is the dominant hydrologic process on buffer strips. The flow on swales is concentrated, as opposed to sheet flow. Neither buffer strips nor swales involve permanent storage. Bioretention facilities provide for some storage, primarily initial abstraction storage, but they also share many of the advantages of the other vegetative controls methods.

Design procedures associated with vegetative control measures are provided. However, alternative design procedures are also available.

9.2   VEGETATED BUFFER STRIPS

The vegetated buffer strip, or filter strip, is a multipurpose alternative for both runoff volume and sediment control. Dense turf or vegetation judiciously placed across the path of surface runoff in a way that promotes sheet flow can reduce the velocity of flow, increase the likelihood of infiltration, and promote the trapping and settling of suspended matter. Such strips must be constructed in a way that encourages sheet flow and reduces the likelihood that concentrated flow will develop. Thus, buffer strips are most likely to be effective before the headwaters of first-order streams. A typical use would be along the perimeter of a parking lot. When properly sited and maintained, buffer strips serve as a protective cover and reduce the potential of erosion from exposed soils while promoting conservation practices. Buffer strip vegetation can range from well-maintained turf to wild grasses, shrubs, and trees, to natural forest. Each provides hydrologic and water quality benefits to a different degree. When properly sited, areas devoted to buffer zones will have positive aesthetic benefits and can be used for recreation.

Buffer strips are unlikely to be effective when not used in conjunction with other best management practices. Because of the necessity to maintain sheet flow conditions, buffer areas will be limited in size. Thus, the opportunity for infiltration and sedimentation is limited by the surface area of the buffer.

The design of a buffer strip will need to be coordinated with the runoff and erosion characteristics of the areas that drain onto the buffer strip. For example, the soil gradation characteristics of the inflowing sediment will influence the type of vegetation that should be used to achieve the design trap efficiency. The velocity of the inflow to the buffer will also influence the effectiveness of the buffer. The higher the inflow velocity, the longer the required buffer length. Accordingly, buffer strip slopes are limited to a maximum of about 5 to 8%.

The characteristics of the buffer site are also important considerations in design. The slope and roughness of the surface are primary determinants of the trap efficiency. The soil characteristics of the underlying soil will influence the amount of water that can potentially infiltrate. The nature of the cover material will also be an important factor in both the control of discharge rates from the outlet of the buffer area and the efficiency of sediment trapping..

Several mechanisms are operational within buffer strips to increase water quality. These include decreased flow rates and greater detention induced by the vegetation (larger value of Manning's n), both of which allow for greater sedimentation of particulates. Filtration of particulates by grass blades during the flow also reduces the suspended solids load. This process also reduces a number of toxics in the flow due to their affiliation with TSS. Finally, infiltration of a portion of the runoff can occur during buffer strip flow, decreasing the volume of the runoff discharged.

9.2.1   Performance Characteristics

The trap efficiency of suspended solids is the standard measure of the effectiveness of a buffer strip. The trap efficiency (E_t) is defined as the ratio of the mass of suspended material retained on the strip m_{sr} to the volume of material in the inflowing runoff m_{si}:

$$E_t = m_{sr} / m_{si} = \left(m_{si} - m_{so}\right)/ m_{si} = 1 - \left(m_{so} / m_{si}\right) \qquad (9.1)$$

in which m_{so} is the mass of suspended material in the outflow. Trap efficiency is a dimensionless value. A value of 1 indicates that all of the material in the inflow was trapped, while a value of 0 indicates that the outflow rate of material was equal to the rate of the material inflow. The inflows and outflows of Eq. (9.1) can be expressed either as masses or as rates of masses. With actual storms, the trap efficiency can vary greatly, and may include negative values, which result when material trapped during earlier storm events is washed out during a subsequent storm.

Ideally, the trap efficiency selected in the design of a buffer strip would be the long-term average of the trap efficiencies experienced during actual storms. Design methods are usually developed to achieve this, but the actual trap efficiency will depend on the type of storms experienced and the extent to which the buffer strip is maintained to the design conditions. The selection of the design trap efficiency depends on a number of factors and involves trade-offs in design objectives. Certainly cost is a factor, including the cost or value of the land that must be devoted to the buffer. The importance of preventing soil from entering the receiving stream should also be a consideration. For example, a relatively wide buffer strip would be desirable if the receiving stream was a prime trout habitat.

While vegetative buffer strips are designed primarily to control sediment and attached pollutants, they also reduce the volume of surface runoff that results from excess rainfall. A reduction in the surface runoff volume occurs as the vegetation impedes and retards the flow of water, allowing a portion of it to infiltrate into the soil. The rate of infiltration is a function of: (1) the condition of the vegetative cover including the height and density of the vegetation, (2) the infiltration properties of the underlying soil, (3) the rainfall intensity and duration, and (4) the antecedent soil moisture conditions. These factors act in an interrelated manner to influence the amount of water that infiltrates into a buffer strip.

The infiltration of surface water is a complex process that is time variant. During a storm event, the rate at which a soil may absorb water will decrease in an exponential manner until an ultimate infiltration rate is achieved. Most methods of estimating infiltration are based on empirical formulas that represent the results of field observations. Other methods are based on theoretical solutions of equations for porous media flow; however, many approximations have been developed. If high infiltration rates are desired, underlying soils may be modified by the addition of sand or organic material during buffer strip construction.

In addition to soil characteristics, the volume of water that infiltrates is also a function of the physical characteristics of the buffer strip. The ability of the vegetation to retard flow has the effect of decreasing the velocity of the runoff, which increases the detention time of the overland flow. The increased detention time increases the opportunity for infiltration. Differences in flow velocities can be significant for different types of vegetative cover.

Recognizing that infiltration losses may be small, the estimated reduction of the surface runoff from a vegetative buffer strip may be computed using the simplified model:

$$V_L = f_c DL / 12 \qquad\qquad (9.2)$$

in which V_L is the volume of infiltrated water per foot width of buffer strip in cubic feet per foot, f_c is the equilibrium or ultimate infiltration rate in inches per hour, D is the duration (hours) of the storm for which the rainfall excess rate exceeds f_c, and L is the length of buffer strip in feet. Values of the ultimate infiltration rate are given in Table 9.1 for various texture classes and soil types, where the soil types are classified using the NRCS hydrologic soil groups A, B, C, and D; the part of Table 9.1 used to estimate f_c will depend on the type of soil data available.

9.2.2 Buffer Strip Design

A vegetated buffer strip can be designed by specifying the type and characteristics of the vegetation, the length of the strip in the direction of the runoff flow, and the sediment trap efficiency, which is often set by policy. A design is generally based on the return period of the runoff, which is a primary determinant of the depth of flow. The vegetation should be maintained at a height that is at least as high as the design depth of runoff. The vegetation traps the suspended matter, so if the depth of runoff is greater than the depth of the vegetation, then the buffer strip will not achieve the desired trap efficiency.

The drainage area that contributes runoff to a buffer strip will influence the depth of runoff that enters the strip. However, once the flow is established on the strip, the shape, roughness, and hydraulic characteristics of the strip will define the flow depth and velocity. Thus, it is reasonable to use Manning's equation to define the velocity of the flow. The type of vegetation will determine the roughness coefficient. The slope of the buffer strip area may differ from the slope of the contributing area. The hydraulic radius will equal the depth of flow on the buffer strip, which actually depends on the density of the vegetal cover and the amount of runoff. As the vegetation density of the land cover increases, the depth of flow will need to increase to maintain the flow rate. Therefore, the depth of the inflow seems to be a reasonable approximation to the hydraulic radius of the flow on the buffer strip. This depth will depend on the return period of the design storm. If a temporally distributed model is used, the depth of flow will vary over the duration of the storm.

The following model can be used to estimate the required length of a buffer strip (L, ft) for a given runoff velocity (V, fps) and trap efficiency (E_t, %):

$$L = 0.0037\left(e^{0.11 E_t} - 1\right)\left(e^{2V} - 1\right) \qquad (9.3)$$

The model was developed with velocities between 0.25 ft/sec and 1.1 ft/sec and trap efficiencies between 75% and 95%. It may be appropriate to use a minimum buffer length of 20 ft even for relatively low trap efficiencies and low velocities. Using a minimum length reflects the fact that runoff from contributing areas may actually be at a higher velocity, which will require some distance to reach the equilibrium velocity established by the hydraulic conditions on the buffer. Buffer strips will be more effective if runoff enters from a level spreader area.

Equation (9.3) assumes a sandy soil. For soils with smaller particle sizes, a longer length will be needed. Stokes Law suggests that smaller particle sizes require greater times to settle. Since particles of silt are smaller than sand particles, increasing the buffer length will allow greater time for settlement. For medium silt soils, a 50% increase in buffer length is appropriate. For fine silt soils, a 100% increase in length is appropriate.

Table 9.1. Ultimate infiltration rates (f_c) versus soil texture class and soil groups

f_c (in./hr)	Soil class	f_c (in./hr)	Soil Group
5.0	Sand	0.40	A
1.5	Loamy sand	0.25	B
0.8	Sandy loam	0.10	C
0.4	Sandy clay loam	< 0.05	D
0.4	Loam		
0.25	Clay loam		
0.20	Silty loam		
0.15	Sandy clay		
0.15	Silt		
0.08	Silty clay loam		
0.04	Silty clay		
0.02	clay		

McCuen and Spiess (1995) derived the following model for computing the depth of flow, D_e, (ft), for shallow flow from small areas:

$$D_e = 0.000818 Li^{0.6} (nL)^{0.6} S^{-0.3} \qquad (9.4)$$

in which i is the rainfall intensity (in./hr), L is the length (ft) of the flowpath from the upper end of the flowpath, n is Manning's coefficient, and S is the slope of the flowpath (ft/ft). This expression can be used to estimate the hydraulic radius of the flow on the buffer strip.

Example 9.1. Runoff from a steeply sloped highway right-of-way transports a medium silt soil directly to a small upland stream. To reduce the amount of soil that enters the stream, a buffer strip is proposed. If conditions would allow the buffer to be located at a slope of 1.3%, what buffer length is necessary? Assume policy requires a 10-yr return period and a 6-min storm duration.

A short intense rainfall for the area would have a 10-yr rainfall intensity of 8 in./hr. Assuming that accumulation of runoff from the contributing highway would yield a hydraulic radius of 0.25 ft after the water drains onto the medium density buffer strip ($n = 0.2$), Manning's equation yields the following velocity:

$$V = \frac{1.49}{0.2} (0.25)^{2/3} (0.013)^{0.5} = 0.337 \text{ ft/sec} \qquad (9.5)$$

If the drainage policy requires a trap efficiency of 80%, then Eq. (9.3) yields a buffer length of:

$$L = 0.0037 \left(e^{0.11(80)} - 1\right)\left(e^{2(0.337)} - 1\right) = 23.6 \text{ ft} \qquad (9.6)$$

For a medium silt soil, this length should be doubled. Therefore, a buffer strip of 50 ft would be appropriate.

Example 9.2. An existing agricultural field is separated from a stream by a 45 ft buffer strip. The site is in a coastal area with a sandy soil. What trap efficiency can the state expect the buffer strip to provide?

An analysis suggests that runoff would have a velocity of 0.4 ft/sec on the buffer. Rearranging Eq. (9.3) to solve for the trap efficiency yields:

$$E_t = \frac{1}{0.11} \ln_e \left[1 + \frac{L}{0.0037\left(e^{2V} - 1\right)}\right] = \frac{1}{0.11} \ln_e \left[1 + \frac{45}{0.0037\left(e^{2(0.4)} - 1\right)}\right] = 83.7\% \qquad (9.7)$$

This trap efficiency assumes a sandy soil. If it were a medium silt soil, then the length used with Eq. (9.6) would need to be reduced by 50%, which would yield a trap efficiency of:

$$E_t = \frac{1}{0.11} \ln_e \left[1 + \frac{22.5}{0.0037\left(e^{2(0.4)} - 1\right)}\right] = 77.4\% \qquad (9.8)$$

The trap efficiency is reduced because the smaller silt particles have less opportunity for settling.

9.2.3   Design Sensitivity

Equation (9.3) is a useful model for estimating the length of a buffer strip needed to achieve a desired trap efficiency. A sensitivity analysis of Eq. (9.3) is then useful for (1) assessing the effect of increases in the trap efficiency on the required length; (2) evaluating the need for controlling the inflow velocity; and (3) examining the value of maintaining buffer strip roughness. At a particular site, the buffer strip designer may have some flexibility in setting the length and would, therefore, be interested in the effect on the trap efficiency. Knowing the sensitivity of the length to the flow velocity can support the need for controlling the inflow velocity, such as with a level spreader, and the value of maintaining buffer roughness, which is central to maintaining low flow velocities.

Equation (9.3) can be differentiated with respect to the two design variables, E_t and V, to assess the changes in L that will result from changes in E_t and V. Then depending on the values of E_t and V, the effect of changes can be assessed. The derivatives, or sensitivity functions, are:

$$\frac{\partial L}{\partial E_t} = 0.000407 e^{0.11 E_t} \left(e^{2V} - 1 \right) \tag{9.9}$$

$$\frac{\partial L}{\partial V} = 0.0074 e^{2V} \left(e^{0.11 E_t} - 1 \right) \tag{9.10}$$

The sensitivity functions show that changes in the length are dependent on both design variables.

The sensitivity functions are evaluated for selected velocities and trap efficiencies, with the results give in Table 9.2. For low velocities and trap efficiencies, the increase in the required lengths as either the required trap efficiency or velocity are increased is small. For example, at a velocity of 0.2 ft/s and a design trap efficiency of 60%, an increase in length of just 0.1 ft would increase the trap efficiency by 1%.

Since trap efficiency is an important design consideration, Eq. (9.3) can be algebraically transformed to solve for the trap efficiency:

$$E_t = 9.091 \ln_e \left(1 + \frac{270.3L}{e^{2V} - 1} \right) \tag{9.11}$$

Equation 9.11 can be differentiated to compute the sensitivity functions:

$$\frac{\partial E_t}{\partial L} = \frac{2457}{e^{2V} - 1 + 270.3L} \tag{9.12}$$

Table 9.2. Relative sensitivity of buffer strip length to the velocity (R_V) and trap efficiency (R_{Et})

V (ft/sec)	E_t (%)	L (ft)	$\frac{\partial L}{\partial E_t}$	$\frac{\partial L}{\partial V}$	R_{Et}	R_V
0.2	60	1	0.1	8	6.6	1.2
	80	12	1.3	73	8.8	1.2
	95	63	6.9	381	10.5	1.2
0.5	60	5	0.5	15	6.6	1.6
	80	42	4.6	133	8.8	1.6
	95	220	24.2	699	10.5	1.6
1.0	60	17	1.9	40	6.6	2.3
	80	157	17.3	363	8.8	2.3
	95	817	89.8	1889	10.5	2.3

Table 9.3. Relative sensitivity of buffer strip trap efficiency to the length (R_L) and velocity (R_V)

V (ft/sec)	L (ft)	E_t (%)	$\frac{\partial E_t}{\partial L}$	$\frac{\partial E_t}{\partial V}$	R_L	R_V
0.2	20	84.6	0.454	-55.1	0.108	-0.130
	50	92.9	0.182	-55.1	0.098	-0.119
	100	99.2	0.091	-55.1	0.092	-0.111
0.5	20	73.2	0.454	-28.8	0.124	-0.196
	50	81.5	0.182	-28.8	0.112	-0.176
	100	87.8	0.091	-28.8	0.104	-0.164
1.0	20	61.3	0.454	-21.0	0.148	-0.343
	50	69.6	0.182	-21.0	0.131	-0.302
	100	75.9	0.091	-21.0	0.120	-0.277

$$\frac{\partial E_t}{\partial V} = \frac{-4909e^{2V}}{270(e^{2V}-1)+(1/L)} \sim \frac{-18.18e^{2V}}{e^{2V}-1} \tag{9.13}$$

Values of the sensitivity functions are given in Table 9.3 for selected lengths and velocities. The negative values for velocity indicate that the trap efficiency decreases as the velocity increases.

Equations (9.9) and (9.10) are dimensioned and, therefore, depend on the units of both V and E_t. It is easier to assess sensitivities when they are presented in relative sensitivity form. The relative sensitivities of length with respect to E_t and V are:

$$R_{Et} = \frac{\partial L/L}{\partial E_t/E_t} = \left(\frac{\partial L}{\partial E_t}\right)\left(\frac{E_t}{L}\right) = \frac{0.11E_t e^{0.11E_t}}{\left(e^{0.11E_t}-1\right)} \sim 0.11E_t \qquad (9.14)$$

$$R_V = \frac{\partial L/L}{\partial V/V} = \left(\frac{\partial L}{\partial V}\right)\left(\frac{V}{L}\right) = \frac{2Ve^{2V}}{e^{2V}-1} \qquad (9.15)$$

Values for the relative sensitivities are also given in Table 9.2 for selected values of E_t and V. The relative sensitivities of L to E_t are larger than those for the velocity, which implies that decisions about establishing required lengths are more sensitive to the desired trap efficiency than to limits on velocities.

The relative sensitivity functions for trap efficiency are also of interest:

$$\frac{\partial E_t/E_t}{\partial L/L} = \frac{2457L}{E_t\left(e^{2V}-1+270.3L\right)} \qquad (9.16)$$

$$\frac{\partial E_t/E_t}{\partial V/V} = \frac{-4909e^{2V}V}{E_t\left(270\left(e^{2V}-1\right)+\left(1/L\right)\right)} \qquad (9.17)$$

The relative sensitivities for selected lengths and velocities are given in Table 9.3. The values for length suggest that an increase in the length of 1% increases the trap efficiency by about 0.1%. The effect of velocity is slightly greater, at least percentagewise. A 1% increase in the velocity causes a decrease in trap efficiency from 0.1% to about 0.3%. At least from a relative standpoint, it is more important to take design actions to limit velocities than policy actions to increase lengths.

Example 9.3. Assume that in Example 9.1 the length at the site available for the buffer was slightly more than the required 50 ft. Would it be more beneficial to increase the length of the buffer by 10 ft or to include a level spreader area that would reduce the average velocity entering the buffer by 20%?

The relative sensitivity values of Table 9.3 suggest values for R_L of about 0.105 and for R_V of about -0.15. Since 10 ft represents 20% of the length, then a 20% increase would increase the trap efficiency by 20% x (0.105) or about 2%. The 20% reduction in velocity with R_V = -0.15 suggests an increase of -0.15 x (-20) or about a 3% increase in trap efficiency. These sensitivity-based estimates suggest that it is better to incorporate the level spreader.

The same conclusion can be obtained by making the full calculations. A 60-ft buffer with a velocity of 0.337 ft/sec yields a trap efficiency of:

$$E_t = 9.091 \ln_e\left(1+\frac{270.3(30)}{e^{2(0.337)}-1}\right) = 82.2\% \qquad (9.18)$$

This is a 2% increase in trap efficiency. Note that the calculation was made with a 30 ft length because of the soil type. Equation (9.11) can also be used to show the effect of reducing the velocity by 20%, or 0.337(1-0.2) = 0.270 ft/sec:

$$E_t = 9.091 \ \ln_e\left(1 + \frac{270.3(25)}{e^{2(0.27)} - 1}\right) = 83.2\% \tag{9.19}$$

The actual increases in trap efficiency were accurately approximated by the sensitivity estimates.

9.2.4 Water Quality Performance

A few studies have evaluated water quality improvement for flow through vegetative buffer strips. Highway runoff quality data gathered by Wu *et al.* (1998) indicate that for small rain events, 50-84% of the TSS is removed by the adjacent grassy filter strip. At higher rainfall events, the runoff flow became deeper and the removal decreased from 20 to 35%. Similar results were obtained in another highway runoff treatment study (Yonge 2000). In this case, total suspended solids removal ranged from about 20 to 80%, with an average removal of 72% (average reduction of suspended solids from 41 to 6.7 mg/L). Total petroleum hydrocarbon removal was excellent, with most treated water having less than 1 mg/L TPH. With proper soil mixtures, infiltration of runoff can be an important water and pollutant attenuation pathway in vegetated buffer strips, reducing the volume of overland flow.

9.3   VEGETATED SWALES

Swales are upland concentrated flow channels that contain water only during storm events. They are designed to be vegetated, so if they are not properly maintained, they may not function as designed. Swales are generally sited to convey surface runoff past land parcels where it is undesirable to have water accumulate. In residential areas, they are often located between houses. Swales are also commonly placed alongside roadways to collect the sheet flow and direct the runoff to a collection, treatment, or discharge point (Figure 9.1). They are often the first concentrated flow path in the upland portion of the watershed and, therefore, are designed to drain sheet flow runoff. By directing the sheet flow runoff into concentrated flowpaths, the velocity is increased, which allows the surface runoff to be drained at a faster rate.

The design of a vegetated swale requires a minimum of inputs. The primary design variable is the discharge that the swale must pass. This can be computed using a peak discharge model such as the Rational method or the NRCS Graphical method. The discharge will depend on the design return period, the area and land cover of the contributing area, and the appropriate rainfall characterization necessary for the peak discharge model. The characteristics of the site where the swale is located are also important. This includes the longitudinal slope (S, ft/ft) of the swale site, the extent to which the site can be graded to control runoff, the sideslope z (i.e., z horizontal per unit vertical), and the roughness (n) of the vegetation to be used to line the swale.

The mechanisms of flow and treatment are similar to those of filter strips for the longitudinal flow; however, the flow is more concentrated and should be deeper in the swale. The swale may receive flow only at the inlet and convey it to an outlet, or, more commonly, it will also receive flow from the sideslopes. This sideslope flow is usually sheet flow, and the runoff flow behavior and sediment capture mechanisms are essentially those of grass filter strips.

9.3.1 Design Constraints

Once a reasonable design has been computed, it will be necessary to ensure that certain constraints are not violated. First, the velocity in the swale must not exceed the critical shear velocity of the vegetation and soil. This is necessary to ensure that the swale area will not erode if the grass cover thins or is worn out completely. Critical shear velocities are given in Table 9.4 for selected soil types. At points where a computed velocity exceeds the critical shear velocity, it is necessary to line the swale with riprap to prevent erosion within the swale area.

Figure 9.1. Vegetated swale for conveyance of highway runoff. This swale has been constructed to allow for water quality monitoring.

Second, the spread (T, ft) of the swale must be within the tolerance established by the physical surroundings. If a swale is located between two houses, obviously the spread cannot exceed the distance between the houses, and most likely, the swale should be designed so the maximum spread is some fraction, say 50%, of the maximum possible spread. For a vee-shaped swale, the spread is:

$$T = 2zd \qquad\qquad (9.20)$$

where d is the maximum depth of flow in the swale and z is the horizontal distance per unit vertical distance of the sideslope. For example, if the sideslope is 8 ft horizontal for each vertical foot (i.e., 12.5%), than z is equal to 8.

Third, the velocity of the water discharging from the lower end of the swale must be low enough to not cause erosion of the stream, treatment practice, or land area that receives the water. Riprap protection is generally required where the area around the outlet of the swale has the potential to experience erosion.

Fourth, the steepness of the sideslope is generally constrained. For swales of riprap, a maximum slope of 2-to-1 is generally the limit, while for grass-lined swales, a 4-to-1 limit is commonly imposed. Slopes steeper than these may be subject to failure.

Table 9.4. Maximum permissible velocities (ft/sec) in clear water (V_1) or in water transporting colloidal silts (V_2)

Material	V_1	V_2
Fine sand	1.50	2.50
Sandy load, noncolloidal	1.75	2.50
Silt loam, noncolloidal	2.00	3.00
Ordinary firm loam	2.50	3.50
Stiff clay, very colloidal	3.75	5.00
Alluvial silts, colloidal	3.75	5.00
Fine gravel	2.50	5.00
Coarse gravel	4.00	6.00

9.3.2   Design Procedure

Manning's equation is used as the basis for the design of a swale. Given the design discharge (q) and longitudinal slope (S), Manning's equation can be solved for the cross-sectional characteristics, assuming that the roughness (n) will be set by the intended vegetation and use of the completed swale:

$$AR_h^{2/3} = \frac{nq}{1.49\, S^{0.5}} \qquad\qquad (9.21)$$

For a vee-shaped swale, the area A and hydraulic radius R_h are:

$$A = zd^2 \tag{9.22}$$

$$R_h = 0.5\,zd\left(1 + z^2\right)^{-0.5} \tag{9.23}$$

The sideslope z is controlled by the conditions at the site, with steepness constraints considered. The depth d would depend on the amount of water entering the swale and the effect of swale conditions on the resulting velocity. Substituting Eqs. (9.22) and (9.23) into Eq. (9.21), the resulting value of d can be computed. Manning's equation can then be used to compute the velocity, which can then be compared with the critical shear velocity (Table 9.4). Equation (9.20) can then be used to compute the spread, which can then be evaluated for reasonableness for the site.

Example 9.4. A swale is proposed as a drainageway around the end of a strip mall. The design discharge from the parking lot is 1.3 ft³/sec, and the longitudinal slope of the swale area is 1.8%. The swale will be lined with riprap of a size that has a limiting velocity of 4 ft/sec and a roughness coefficient of 0.04. The allowable spread is 15 ft. The design sideslope and depth of flow must be determined.

Using Eq. (9.21), the following relates the known inputs and the design variables:

$$\frac{nq}{1.49 S^{0.5}} = \frac{0.04(1.3)}{1.49(0.018)^{0.5}} = 0.26 = zd\left[0.5zd\left(1 + z^2\right)^{-0.5}\right]^{2/3} \tag{9.24}$$

Equation (9.24) is solved by selecting a sideslope and then solving for d. For a sideslope of 5-to-1, a flow depth of 1.484 ft would result. This would yield a spread of 14.8 ft, which is below the 15 ft constraint imposed by the conditions at the site. The flow velocity is:

$$V = \frac{1.49}{0.04}\left[\frac{0.5(5)(1.484)}{\left(1 + 5^2\right)^{0.5}}\right]^{2/3}(0.018)^{0.5} = 4.04\,ft/\sec \tag{9.25}$$

This is an acceptable velocity given the allowable velocity of 4 fps (Table 9.4).

Example 9.5. An existing vee-shaped, grassed-lined swale has a longitudinal slope of 0.8%, sideslopes of 6 to 1, and a maximum allowable spread of 24 feet. The swale is located on a sandy loam soil. Additional development is taking place above the swale headwaters. What discharge can the development produce without requiring re-design of the swale?

Short grass has a roughness of 0.15. For a spread of 24 ft at a sideslope of 6 to 1, Eq. (9.20) gives an allowable depth of:

$$d = T/2z = 24/[2(6)] = 2 \text{ ft} \tag{9.26}$$

Equations (9.22) and (9.23) yield an area and a hydraulic radius of:

$$A = zd^2 = 6(2)^2 = 24 \text{ ft}^2 \tag{9.27}$$

$$R_h = 0.5(6)(2)(1 + 6^2)^{-0.5} = 0.986 \text{ ft} \tag{9.28}$$

Thus, the allowable discharge is:

$$q = \frac{1.49}{0.15}(24)(0.986)^{2/3}(0.008)^{0.5} = 21.1 \text{ ft}^3 / \sec \tag{9.29}$$

From the continuity equation, the velocity is:

$$V = q/A = 21.1 \text{ ft}^3/\text{s}/24 \text{ ft}^2 = 0.88 \text{ fps} \tag{9.30}$$

For a sandy loam soil, the critical velocity is 1.75 fps, which means that the swale would be stable under the computed discharge conditions.

9.3.3 Simple Water Quality Models

Limited water quality data have been collected on vegetated swale performance and water quality models are not well developed. Models have been used to described pollutant attenuation in swales, where the treatment efficiency is a function of the swale length.

For a linear model, the pollutant concentration is linearly dependent on the swale length:

$$C = C_0 - k_L L \tag{9.31}$$

where C and C_0 are the outlet and inlet pollutant concentrations (mg/L), respectively and L is the length (ft); k_L is an empirical parameter that describes the treatment efficiency per unit length and would be a function of flow velocity and vegetation characteristics. For very long swale lengths, values of $C_0 < 0$ are not physically possible, so computed values less than 0 should be set to 0.

The exponential model describes an exponential decrease in pollutant concentration with swale length:

$$C = C_0 \exp(-k_e L) \tag{9.32}$$

where k_e is the removal coefficient for the exponential model.

The models of Eqs (9.31) and (9.32) mechanistically consider only isolated flow that enters at the beginning of the swale and exits at the end. Additional sideslope flow would require a more complex model.

Example 9.6. The grass swale of Example 9.5 has a longitudinal length of 360 ft. The suspended solids removal coefficient for an exponential water quality model is 0.005 ft^{-1}. For an input TSS concentration of 56 mg/L the concentration leaving the swale can be calculated using Eq. (9.32):

$$C = (56\,\text{mg/L})\exp\!\left[-\left(0.005\,\text{ft}^{-1}\right)\!\left(360\,\text{ft}\right)\right] = 9.3\,\text{mg/L}$$

9.3.4 Water Quality Performance

A few studies have evaluated pollutant removals from swales, mostly from those parallel to highways. Barrett *et al.* (1998) noted that flow from a moderate-use highway that passed through a grassy swale had better quality than runoff from two highways with curb and gutter systems. Schueler *et al.* (1992) has reported pollutant removals in grassed swales of up to 70% TSS, 30% of total phosphorus, 25% of total nitrogen, and 50-90% for several metals. Additionally, removals of 67-93% of oil and grease and 65% of TSS and VSS were reported by Little *et al.* (1992).

Yu *et al.* (2001) compiled results from grassed swales from their work and that of others. Length is the most important parameter in ensuring good swale performance and a swale length of over 100 meters has proven to be successful in providing very good removal of suspended solids and other pollutants. Milder slopes are more effective in TSS removal, although the difference tends to be minor for very long swales. The authors recommend using slopes less than 3%.

Phosphorus removal data were much more scattered than TSS, apparently because of the likelihood of phosphorus being contributed by the vegetation in the swale. Zinc removal also increased in swale length, with removals approaching 100% within relatively short swale lengths. This should be attributed to some affiliation of the zinc with suspended solids.

The use of check dams in vegetated swales was found to have a significant affect in improving treatment efficiency (Yu *et al.* 2001). The check dams provided areas for the runoff to pool in the swale, permitting greater sedimentation of suspended solids and runoff infiltration. For a few small events, the check dams allowed the swale to completely retain all of the runoff. As a result, the hydraulic and pollutant loadings leaving the swale from this event were both zero.

9.4 BIORETENTION

Bioretention is an innovative stormwater management practice with great potential in improving water quality and managing flows. Bioretention systems are soil- and plant-based facilities employed to filter and treat runoff from developed areas. They are also commonly known as *rain gardens*. Typically, bioretention practices are placed at several points throughout a developed area. Bioretention systems are designed for water infiltration and evapotranspiration, along with pollutant removal by soil filtering, sorption mechanisms, microbial transformations, and other processes. Bioretention facilities can be small to collect excess runoff from a single lot or larger to address flows from parking lots or commercial building rooftops.

A bioretention facility (Figures 9.2 to 9.4) is generally designed to a size of about 5% of the runoff drainage area. A porous soil media is placed in the facility and covered with a thin layer of standard hardwood mulch. Various grasses, shrubs, and small trees are planted to promote evapotranspiration, maintain soil porosity, encourage biological activity, and promote uptake of some pollutants. Runoff from an impervious area is directed into the bioretention facility through a curb cut, grated drains, or other means. The facility is often surrounded by a berm that promotes ponding and increases the retention time. The water is allowed to pool for several inches and infiltrates through the plant/mulch/soil environment, which provides the treatment. Infiltrated water is usually collected in a perforated pipe and fed to a conventional storm drain or discharged directly to a stream. If surrounding soils are sufficiently permeable, however, the bioretention flow may be allowed to continue infiltrating to recharge ground-water supplies. Nonetheless, these situations must be looked at carefully because low infiltration rates may cause these systems to fail and can result in flooding of yards, parking lots, and walkways.

Bioretention is focused on smaller drainage areas to ensure capture of first flush runoff. Any pooling in the facility above the design value of about 15 cm should be directed to bypass treatment and to flow directly into a storm drain so as to minimize flooding concerns.

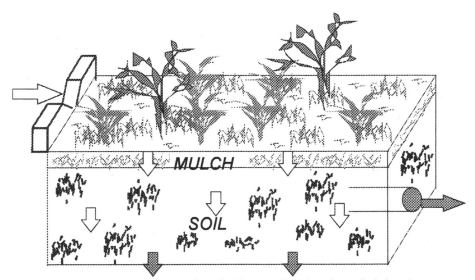

Figure 9.2. Diagram of bioretention cell with a controlled inlet and an underdrain outlet

9.4.1 Design Characteristics

Bioretention facilities are designed about 15-20 cm below grade to store and hold runoff that is conveyed to them (Figure 9.3 and 9.4). A swale overflow bypass or overflow drain is installed at this level. Any pooling above this level in the facility will flow into the overflow drain, which is connected with the underdrain to discharge into the storm drain or to a stream. Thus, very high flows will bypass the system and will receive minimal treatment. At the surface, a 3- to 6-cm layer of typical hardwood mulch is placed. Below the mulch will be the primary bioretention media. Older designs have used a porous soil, such as a sandy loam. Recent designs have used engineered mixtures of sand, topsoil, and leaf or similar compost as the primary medium. A common mix is 50% construction sand, 20-30% topsoil, and 20-30% leaf mulch (DER 2001). The clay content should be less than 10% to maintain high infiltration rates. Some recent data suggest that the sand plays a very important role in maintaining high permeability and should have d_{10} of at least 0.6 mm. The recommended media depth is 2.5 to 3 ft. A perforated PVC pipe, typically with a diameter of 4 inches, is placed along the length of the facility subbase at a slight slope to collect and drain the infiltrated runoff. The pipe is covered by a porous filter cloth to prevent clogging. This underdrain is connected to a typical storm drain at a manhole.

Various forms of vegetation should be established in a bioretention facility. The vegetation has several purposes, including promoting evapotranspiration, taking up nutrients, encouraging other biological activity that may cause degradation of captured pollutants, and providing a visually pleasant site. The plants that are selected for use in bioretention must be able to withstand periods of very wet and very dry conditions, be resistant to possible high concentrations of chemicals that may wash into the facilities, including oils and grease and salts from deicing, and withstand the temperature, wind, air pollution, and sunlight characteristic variations that may occur in an urban environment. A partial list of grasses, shrubs, and small trees that are recommended for bioretention is presented in Table 9.5. A more extensive list is given in DEP (2001).

9.4.2 Design Procedure

The design method presented here is based on the continuity of mass. The bioretention pond acts to intercept the first portion of the runoff, which would then represent an initial abstraction. The ponding area may need to be large enough to contain the entire volume of runoff from a small storm.

The design procedure assumes that a watershed area A_c (ft^2) drains to the bioretention facility, which has a surface area A_s (ft^2) and pond depth d_s (ft). The pond area is considered to be part of the contributing area. The runoff depth is assumed to be a design water quality volume that can be computed from a design first-flush depth, D_f, over the watershed area. Typically this value is ¼ to 1 inch.

9.4.2.1   Effect of Vegetation

Dense vegetation in a bioretention facility will decrease the storage volume available for water. If the pond is being created to store a certain portion of the runoff, then the design volume will need to account for the presence of the vegetation. The volume of vegetation will depend on the density of the plants, the volumetric characteristics of the submerged portion of the plants, and the height of the vegetation, h_v, or the ponding height, whichever is smaller. Only the height of the vegetation that is submerged should be included in the calculation of the volume of vegetation V_v. This would not be greater than the pond depth, d_s.

Figure 9.3. A bioretention facility that captures runoff from an office complex parking lot

Figure 9.4. Bioretention facility in a parking lot island strip

Table 9.5. Some selected vegetation recommended for use in bioretention (DER 2001).

Perennials for Saturated Soil	Drought-Tolerant Perennials	Groundcovers	Shrubs	Trees
Swamp Milkweed	Willowleaf Bluestar	Moss Sandwort	Red, Black, Purple Chokeberry	Red Maple
New England Aster	Big Bluestem	Wild Ginger	Inkberry	River Birch
Tickseed Sunflower	Columbine	Switch Grass	Creeping Juniiper	Bitternut Hickory
March Marigold	Butterflyweed	Leadwort	Groundsei Tree	Green Ash
Fringed Sedge	Wood Aster	River Oats	Beautyberry	Honeylocust
White, Rose Turtlehead	Blue Wood Aster	Green and Gold	Buttonbush	American Holly
Rose Mallow	Smooth or Blue Bird Aster	Hay-Scent Fern	Pepperbush	Black gum
Mist Flower	New England Aster	Bishop's Hat	Red Twig Dogwood	Swamp White Oak
Joe Pye Weed	Boltonia	Purple Lovegrass	Hearts-a-bustin'	Paw Paw
Boneset	River Oats	Red Fescue	Witch Hazel	Common Hackberry

The volume of vegetation is the product of the number of stems n, the mean cross-sectional area of a stem, A_1, and the submerged height of the vegetation, h_v:

$$V_v = nA_1 h_v \qquad (9.33)$$

in which h_v is the smaller of the actual height of the vegetation for submerged plants or the depth of the pond for vegetation that is not completely submerged. If the vegetation is uniformly distributed over the pond area, then the volume of vegetation could be approximated by:

$$V_v = f_v h_v A_s \qquad (9.34)$$

in which f_v is the fraction of the bioretention area taken up by vegetation, equivalent to nA_1/A_s. For a pond with a mature growth of vegetation, the density of growth f_v would vary with the type of vegetation. The volume of the basin that is available for the storage of water V_w is, thus, the difference between the physical volume of the pond $d_s A_s$ minus the volume of vegetation:

$$V_w = d_s A_s - V_v = d_s A_s - f_v h_v A_s = A_s \left(d_s - f_v h_v \right) \qquad (9.35)$$

If all of the vegetation is partially submerged, then $h_v = d_s$ and Eq. (9.35) reduces to

$$V_w = A_s d_s \left(1 - f_v \right) \qquad (9.36)$$

9.4.2.2 Storage Volume

The design of a bioretention pond volume will be based on the water quality volume concept. A runoff depth is considered to be representative of a first flush over the contributing drainage area, A_c. This depth, typically about ½ to 1 inch, is denoted as D_f. Thus, the water quality volume of inflow to the facility is $D_f A_c$. The total volume of storage in the facility V_t is equal to the depth d_s times the surface area A_s. The volume of storage equals the sum of the volume of water $D_f A_c$ and the volume of vegetation from Eq. (9.34). Therefore, the continuity of mass yields:

$$d_s A_s = D_f A_c + f_v h_v A_s \qquad (9.37)$$

If the depth of the pond is set by topography, vegetation constraints, lot layout practicality, or convention, then the surface area of the pond can be obtained by rearranging Eq. (9.37):

$$A_s = \frac{D_f A_c}{\left(d_s - f_v h_v \right)} \qquad (9.38)$$

If the height of the vegetation h_v exceeds the pond depth d_s, then Eq. (9.38) reduces to:

$$A_s = \frac{D_f A_c}{d_s \left(1 - f_v \right)} \qquad (9.39)$$

To use Eq. (9.38) or Eq (9.39) for designing the bioretention pond area, four inputs are required in addition to the design runoff depth. The runoff depth D_f will depend on rainfall characteristics and land use information, or possibly local regulations. The contributing area A_c must be delineated from a topographic map of the site. The height of the vegetation and the vegetation density f_v will depend on the type of vegetation that is most appropriate for the site.

Example 9.7. Consider the case of a bioretention facility with the pond designed to contain all of the runoff from a rooftop of a 28 ft by 52 ft house. The first ½ in. of runoff must be stored in the bioretention cell, which is equal to D_f. Assume that the type of vegetation to be used for the residential lot will have a height that exceeds the pond depth and a vegetation density of 20%. A design depth of 6 in. is assumed. Thus, Eq. (9.39) yields a surface area of:

$$A_s = \frac{0.5 \, in. \left[(28 \, ft)(52 \, ft)\right]}{(6 \, in.)(1-0.2)} = 152 \, ft^2 \qquad (9.40)$$

Therefore, the pond area would need to be about 8 ft by 19 ft. If a smaller first-flush depth were specified in the design policy, then a smaller pond could be used.

9.4.2.3 Infiltration

Rapid infiltration is critical to successful bioretention operation. Infiltrated water is exposed to the physical, chemical, and biological processes in the media that make bioretention an efficient water quality management device. Infiltration beyond the horizontal and bottom bioretention cell borders can allow recharge of groundwater, which can be very important in areas of high imperviousness. If the infiltration rate of the bioretention soil media is too low, water will quickly fill the storage pond volume and additional runoff will overflow into the bypass system, such that stormwater management benefits will not be realized.

· Bioretention soils should have infiltration rates from 1.5 to 4 inches per hour. Table 9.1 shows that this limits the media to a loamy sand or a sand. A mixture of sandy soil and sand, such as the 50/30/20 mix discussed earlier in Section 9.4.1, has been shown to perform well in balancing a high infiltration rate with good pollutant removal and should promote plant growth and survival (Hsieh and Davis 2005).

Example 9.8. Consider the case of a bioretention facility located in an island portion of a strip mall parking lot. Assume that the island is 24 ft by 10 ft and has a media depth of 2.5 ft. The contributing area is 0.25 acre, all impervious except for the island itself. The required water quality volume is calculated using a first flush depth D_f of ½ in. The corresponding water quality volume is:

$$V_c = (0.5 \, in./12 \, in./ \, ft)(0.25 ac)(43560 \, ft^2 / ac) = 454 \, ft^3 \qquad (9.41)$$

The vegetation density is 15%. Therefore, the storage volume is given by Eq. (9.36) for an 8 in. storage depth:

$$V_w = (10 \, ft)(24 \, ft)(8 \, in./12 \, in./ \, ft)(1-0.15) = 136 \, ft^3 \qquad (9.42)$$

Thus, 136/454, or 30% of the runoff volume can be stored. The remaining 318 ft³ corresponds to a depth in the bioretention cell of about:

$$D_b = \frac{318 \, ft^3}{(10 \, ft)(24 \, ft)(1-0.15)} = 1.56 \, ft = 18.7 \, in. \qquad (9.43)$$

An infiltration rate of 2 in./hr, will allow some of this runoff to be infiltrated, but the rest will bypass the bioretention treatment and discharge through the overflow drains. The actual amount will depend on the rate at which this ½-inch of runoff is produced.

9.4.2.4   Underdrain Design

The underdrain collects water that has infiltrated through the bioretention media, funneling the water to an existing stormdrain or other discharge point. Two parameters are critical to the design of an underdrain: the diameter of the pipe and the area of the drain holes per linear length. In addition, the slope of the underdrain and the location of the pipe must be set to ensure adequate drainage. The total area of the drain holes per length would depend on the supply rate of the water, which is a function of the infiltration rate of the soil used in the facility. Specifically, the ultimate infiltration capacity would serve as the design rate of inflow to the pipe.

For the design condition, the soil media area of the bioretention cell (see Fig. 9.5) will have length L, width W, and depth H. The pipe is laid at the bottom of the cell, generally along the center lengthwise. The pipe of diameter D_p is perforated with holes of diameter D_h and is laid at a slope S. The pipe has a roughness n; for plastic pipe, $n = 0.008$. The soil has an ultimate infiltration capacity f_c (see Table 9.1). Design conditions occur when surface ponding has just begun. Therefore, the soil is saturated, and infiltration is taking place at the ultimate capacity. Then the discharge to the pipe per unit length is:

$$q = \frac{WLf_c}{L} = Wf_c/12 \qquad (9.44)$$

The head on the water draining into the pipe holes is H. Thus, since the ground is assumed to be saturated, the velocity through the holes, V_h, is given by the orifice equation, with a discharge coefficient of 0.6. Thus, the area of the perforations (A_h, ft²) in the pipe per linear foot is obtained using Eq. (9.44) and the continuity equation:

$$A_h = q/C_d(2gH)^{0.5} = 0.2077qH^{-0.5} \qquad (9.45)$$

with q as ft³/s/ft and H in ft. The diameter of the holes would be set by the structural strength of the pipe. The holes should be uniformly spaced along the length. For a total hole area A_h, the diameter of each hole is:

$$D_h = (4A_h/\pi)^{0.5}/n_h \qquad (9.46)$$

in which n_h is the number of holes per linear foot.

The diameter of the pipe can be determined from Manning's equation. The unit discharge of Eq. (9.44) can be multiplied by the length L of the bioretention pond to obtain the design discharge Q. Thus, the required diameter is:

$$D_p = 1.333(Qn)^{0.375}S^{-0.1875} \qquad (9.47)$$

for Q as (ft³/s) and where S is the slope of the underdrain pipe (ft/ft).

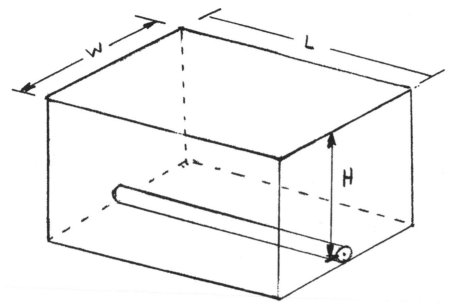

Figure 9.5. Schematic of a bioretention cell

Example 9.9. Assume that the bioretention cell of Example 9.8 is fitted with an underdrain and overlain with 3 feet of bioretention media with an ultimate infiltration capacity of 1.5 in./hr. The pipe, which is 10-feet long, would have a unit discharge from Eq (9.44) of:

$$q = (24\,ft)(1.5\,in./hr) = \frac{36\,ft\text{-}in.}{hr} = 0.000833\,ft^3/s/ft \qquad (9.48)$$

and using Eq (9.45) the area of the holes is:

$$A_h = 0.2077(0.000833)(3)^{-0.5} = 0.0001\,ft^2 \qquad (9.49)$$

If six holes are placed in the pipe per foot, then the minimum diameter of the holes, from Eq. (9.46) would be:

$$D_h = [4(0.0001)/\pi]^{0.5}/6 = 0.00188\,ft = 0.023\,in. \qquad (9.50)$$

Larger holes should be used, as this minimum diameter is not much larger than a medium coarse sand particle. The total discharge through the 10-ft long pipe is 0.00833 ft³/s. Equation (9.47) is used to compute the necessary diameter of the drain pipe when laid at a 1% slope:

$$D_p = 1.333[0.00833(0.008)]^{0.375}(0.01)^{-0.1875} = 0.098\,\text{ft} = 1.17\,\text{in.} \qquad (9.51)$$

Therefore, a diameter of 1.25 inches or more should be used.

9.4.3 Flow Routing

The primary pathway for the runoff is infiltration through the bioretention media. This infiltration and contact with the media removes pollutants through filtering and adsorption processes. Obviously, the greater that infiltration rate through the media, the more water will be treated by the media and less system bypass will occur. It is therefore important to maintain the infiltration rate as high as possible and practical. Some storage will occur in the ponding on the media surface, which will be adequate for very low magnitude events, but cannot manage volumes from large events, as seen in Example 9.8. Excess flow beyond that which can be infiltrated and stored will bypass the filtering media and discharge through the overflow outlet. Some settling of particulates may occur in bypassed flow, but generally the overflow will show little improvement of water quality.

A simple model of these flow conditions can be developed using a flow balance for the bioretention storage volume. Using Figure 9.5 as a guide, when the storage pond is not full, the flow balance in the pond volume can be described by:

$$\frac{dV}{dt} = Q_{in} - Q_{out} \qquad (9.52)$$

where V is the pond volume, Q_{in} is the input flow rate, and Q_{out} is the infiltration flow rate. For simple analysis, a constant infiltration rate, f_c, can be assumed:

$$Q_{out} = f_c A_s (1 - f_v) \qquad (9.53)$$

where f_v is the fraction of the area covered by vegetation, as discussed in Section 9.4.2.1. Eqs. (9.52) and (9.53) are used to increment the storage volume until it is full. When the pond is full and the input flow exceeds infiltration, the water balance is:

$$Q_{bypass} = Q_{in} - Q_{out} \qquad (9.54)$$

Any input flow greater than the infiltration rate will bypass the bioretention treatment.

The timing of the flow through the underdrain can be calculated by routing the flow through the media. The water approach velocity, V, is equal to the infiltration rate divided by the bioretention area. The travel velocity through the media pores, V_p, is faster than the approach velocity due to the presence of the soil media grains:

$$V_p = \frac{V}{n} \qquad (9.55)$$

where n is the media porosity.

Example 9.10. Column 2 of Table 9.6 shows the runoff hydrograph from a shopping center and parking lot during a small 3-hr storm. The flow goes into a bioretention cell, which has dimensions of 40 ft x 24 ft x 2.5 ft, with a vegetation density of 10% and a ponding depth of 6 in. The media infiltration rate is 2 in./hr with a porosity of 0.4. The available storage pond volume, Eq. (9.36), is:

$$V_w = (40\,ft)(24\,ft)(0.5\,ft)(1 - 0.1) = 432\,ft^3 \qquad (9.56)$$

The runoff inflow for each time increment is calculated as the mean flow during the increment. The infiltration rate cannot exceed the maximum value of 2 in./hr.

For each time increment, the integrated form of Eq (9.52) is used to determine the volume of stored water:

$$V_{t+1} = V_t + (Q_{in} - Q_{out})\Delta t \qquad (9.57)$$

When the stored water volume reaches the maximum permissible volume, additional storage cannot occur and Eq. (9.54) controls the flows. This occurs at a time of 2:21. The storage volume is full, and flow greater than the infiltration rate is bypassed.

The water velocity during infiltration through the bioretention media is (2 in./hr)/0.4 = 5 in./hr. For a 30-inch media depth, the travel time is (30 in.)/(5 in./hr)= 6 hours. The input and output hydrographs are shown in Figure 9.6. The delay created by the infiltration process is evident. The peak flow is reduced from 0.48 to 0.28 ft³/s (somewhat greater if we extrapolate the effluent to a peak). Input, treated, and bypass volumes can be calculated from the respective flows and time increments as in Chapter 7, using the trapezoidal rule (Columns 11-13). In all, 1872 ft³ of the total 2592 ft³ of runoff (72%) is infiltrated through the bioretention media; the remainder overflows.

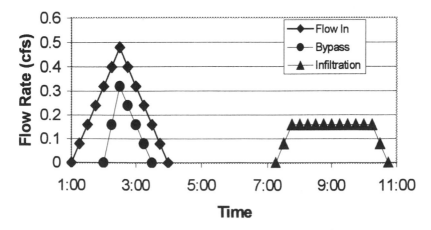

Figure 9.6. Inflow, bypass flow, infiltration outflow hydrographs for bioretention Example 9.9

Table 9.6. Bioretention flow information for Example 9.9

[1] Time	[2] Flow (ft³/s)	[3] Δt (s)	[4] Q_{in} (ft³/s)	[5] Q_{out} (ft³/s)	[6] $\Delta V = (Q_{in}-Q_{out})\Delta t$ (ft³)	[7] V_{pond} (ft³)	[8] D_{pond} (in.)	[9] Q_{bypass} (ft³/s)	[10] Time infil. water leaves	[11] V_{in} (ft³)	[12] V_{out} (ft³)	[13] V_{bypass} (ft³)
1:00	0					0	0					
1:15	0.08	900	0.04	0.04	0	0	0		7:08	36	36	0
1:30	0.16	900	0.12	0.12	0	0	0		7:23	108	108	0
1:45	0.24	900	0.20	0.16	36	36	0.5		7:38	180	144	0
2:00	0.32	900	0.28	0.16	108	144	2		7:53	252	144	0
2:15	0.40	900	0.36	0.16	180	324	4.5	0	8:08	324	144	0
2:21					108	432	6	0 then 0.28				
2:30	0.48	900	0.44	0.16		432	6	0.28	8:23			
2:45	0.40	900	0.44	0.16		432	6	0.28	8:38	396	144	144
3:00	0.32	900	0.44	0.16		432	6	0.20	8:53	396	144	252

Table 9.6. (cont.) Bioretention flow information for Example 9.9

[1] Time	[2] Flow (ft³/s)	[3] Δt (s)	[4] Q_{in} (ft³/s)	[5] Q_{out} (ft³/s)	[6] $\Delta V=(Q_{in}-Q_{out})\Delta t$ (ft³)	[7] V_{pond} (ft³)	[8] D_{pond} (in.)	[9] Q_{bypass} (ft³/s)	[10] Time infl. water leaves	[11] V_{in} (ft³)	[12] V_{out} (ft³)	[13] V_{bypass} (ft³)
3:00	0.32											
3:15	0.24	900	0.36	0.16		432	6	0.12	9:08	324	144	180
3:30	0.16	900	0.28	0.16		432	6	0.04	9:23	252	144	108
3:45	0.08	900	0.20	0.16	(−36)	396	5.5	0	9:38	180	144	36
4:00	0	900	0.12	0.16	(−108)	288	4		9:53	108	144	0
4:15				0.16	(−144)	144	2		10:08	36	144	
4:30				0.16	(−144)	0	0		10:23		144	0
Total										2592	1872	720
TSS mass, kg										8.81	0.32	2.08

The TSS concentration in the input flow is assumed constant (not a good assumption, but it will simplify the calculations) at 120 mg/L (3400 mg/ft^3). We can assume that TSS removal for infiltrated water is 95% and 15% for bypassed water. Since the TSS concentration is constant, the TSS mass is calculated as:

$$M = \int_0^T CQdt = C\int_0^T Qdt = CV \qquad (9.58)$$

The total influent volume is 2592 ft^3, of which 1872 ft^3 infiltrates through the media and 720 ft^3 is bypassed. At 120 mg/L (3400 mg/ft^3) the total TSS mass entering is 8.8 kg. With 5% TSS mass passing through the treatment (0.05)(3400 mg/L)(1872 ft^3) = 0.32 kg is carried with the treated water. Similarly, with 85% TSS mass in the bypass flow (0.85)(3400 mg/L)(720 ft^3) = 2.08 kg is released, for a total of 2.4 kg output. The overall mass removal is 73%.

It becomes obvious from this example that infiltration of the water through the media is extremely important for reducing peak flows, slowing the timing, and efficient pollutant removal. Maintaining high infiltration rates is critical for the successful operation of bioretention facilities.

9.4.4   Performance Characteristics

A number of bioretention facilities have been installed in the past few years. Research on bioretention performance with respect to water quality is beginning to provide a relatively clear understanding of the performance to be expected (Yu *et al.* 1999, Davis *et al.* 2001, 2003). Overall, the environmental benefits are many and significant.

9.4.4.1 Pollutant Removals

Heavy metals appear to be very efficiently removed through bioretention treatment. Figure 9.7 shows some data for removal of lead at different bioretention depths in pilot and field studies. Nearly all of the lead was accumulated at the surface of the bioretention media, likely in the organic mulch layer. Removals were greater than 95%, and effluent lead concentrations were approximately 2 μg/L. Very similar results have been noted for copper and zinc (Davis *et al.* 2003).

Figure 9.8 shows that removal of phosphorus through bioretention media. The removal is good, at about 70-85%. The trend for TKN removal is similar. Adsorption of these pollutants onto the bioretention media appears to be an important removal mechanism.

Little removal of nitrate has been found in bioretention systems. This is not surprising since nitrate is very mobile in soil media and minimal adsorption occurs. Removals of about 10% have been noted for some investigations, but in some studies, nitrate levels higher than input have been noted, most likely due to microbial conversion of previously captured organic nitrogen or ammonium to nitrate.

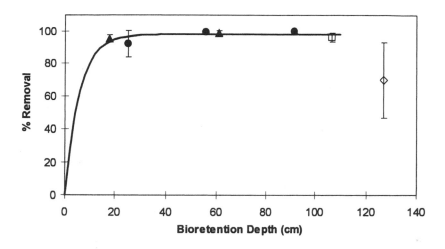

Figure 9. 7. Lead removal at different bioretention depths in pilot plant (closed symbols) and field experiments (open symbols). Error bars represent ± 1 standard deviation during experiment (adapted from Davis *et al.* 2003)

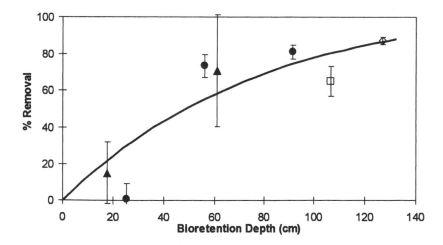

Figure 9. 8. Phosphorus removal at different bioretention depths removal at different bioretention depths in pilot plant (closed symbols) and field experiments (open symbols). Error bars represent ± 1 standard deviation during experiment (adapted from Davis *et al.* 2005)

9.4.4.2 Suspended Solids and Filtration Theory

Bioretention appears to be very effective in the removal of TSS and oils and grease. The media, especially a surface mulch layer, appear to be very effective in physically filtering TSS and sequestering oils from infiltrating runoff.

Filtration theory can be used for evaluating particulate, or colloid, capture during 1-dimensional transport through porous media. The theory is based on the equation:

$$\frac{dC}{dz} = \frac{-1.5(1-n)\alpha\eta}{d_c} \qquad (9.59)$$

where C is the particle concentration, z is the transport direction, n is the bed porosity, d_c is the diameter of a spherical collector (media particle), and η is the single collector collision efficiency. The collision efficiency factor, α, describes the "sticking" of the particles to the media and has a range from 0 to 1 (dimensionless). A value of 0 indicates that a particle will not stick when it strikes the surface of a media collector. Conversely, $\alpha = 1$ indicates that every collision results in particle attachment. The integrated form of Eq. (9.59) is:

$$\frac{C}{C_0} = \exp\left[\frac{-1.5(1-n)\alpha\eta L}{d_c}\right] \qquad (9.60)$$

where C_0 represents the input particle concentration and L is the filtration length (Logan *et al.* 1995). The collector efficiency η describes transport of the colloid from fluid streamlines to the media collector surface as the streamlines flow around the collector. The transport can occur via three mechanisms, diffusion through the flow streamlines to the collector surface, sedimentation from a streamline to the surface, and interception with a collector if a streamline is very close to the collector. The calculations of η for these three transport mechanisms are summarized in Table 9.6. Several dimensionless numbers are employed in these calculations. The value of η is also dimensionless.

Example 9.11. The average media size in a 2.5-ft deep bioretention facility is 0.2 mm at a porosity of 0.4. The input TSS level is 120 mg/L, dominated by 0.05 mm particles. With a sticking coefficient of 0.4 and a collector efficiency of 3×10^{-3}, Eq. (9.60) allows the calculation of the effluent TSS concentration as:

$$C = (120\,mg/L)\exp\left[\frac{-1.5(1-0.4)}{(0.2\,mm)}\left((0.4)(3\times10^{-3})(2.5\,ft)(305\,mm/ft)\right)\right] = 2\,mg/L \qquad (9.61)$$

Thus, 118 mg/L was trapped, which represents a trap efficiency of 98.3%.

Table 9.7. Parameters used in clean bed filtration model for calculation of collector efficiency, Eq. (9.59) (Logan *et al.* 1995).

$$\eta = 4A_s^{1/3}N_{Pe}^{-2/3} + A_s N_{Lo}^{-1/8}N_R^{15/8} + 0.00338A_s N_G^{-1.2}N_R^{-0.4}$$

| Diffusion | Interception | Sedimentation |

where: $\quad A_s = \dfrac{2(1-\gamma^5)}{2-3\gamma+3\gamma^5-2\gamma^6}$ and: $\quad \gamma = (1-n)^{1/3}$

Dimensionless numbers:

$$N_{Pe} = \dfrac{u^* d_c}{D_p} \qquad N_R = \dfrac{d_p}{d_c} \qquad N_G = \dfrac{U_p}{u^*} \qquad N_{Lo} = \dfrac{4H}{9\pi\mu d_p^2 U^*}$$

n = media bed porosity

U^* = pore velocity $= \dfrac{\text{approach velocity}}{n}$

H = Hamaker constant

d_c = collector (media) diameter

d_p = colloid (particle) diameter

D_p = colloid diffusion coef.

U_p = colloid settling velocity (found from Stokes Law, Chapter 6)

9.4.5 Maintenance and Sustainability

When a bioretention facility removes pollutants from runoff, the pollutants accumulate in the facility. The long-term fate of these pollutants must be considered. Continuously accumulating pollutants will cause the facility to fail, either by not removing additional pollutants (or exporting captured pollutants) or by clogging and not allowing water to infiltrate through the media. These factors must be considered in the design, operation, and maintenance of bioretention facilities. Novel advancements will exploit biological mechanisms to assist in pollutant degradation and removal from the treatment practice.

Phosphorus and heavy metals are expected to accumulate over time in the media and provisions must be made to address the long-term buildup of these materials. These pollutants may be accumulated in these facilities to a point where additional removal is not possible. Simple calculations have indicated that metals such as lead and zinc may accumulate to levels of concern in bioretention after 15 to 20 years of operation (Davis *et al.* 2003). Management decisions must be made as to the best ways to address these accumulations. Using mass balance considerations, accumulated phosphorus or metals may be removed from the bioretention media by scraping off the top mulch layer or a few cm of media every few years. The depth of material removed should be replaced by clean filter media. Most of the pollutants should be accumulated in this top layer. The layer materials will have to be disposed of in an appropriate manner. Another possibility is to use specialized vegetation, known as hyperaccumulators, which take up metals, phosphorus, or other pollutants into the biomass at high concentrations. The plants can be harvested yearly and properly disposed of, providing a pathway for captured-pollutant removal from the bioretention system. A third option is to plan on a major renovation of the facility after 15+ years. As more long-term research is completed on bioretention facilities, the proper solutions to these long-term issues will become clearer.

Example 9.12. Assume that the maximum value of accumulated lead permitted in a bioretention cell is 300 kg/ha. The cell has an area of 36 m^2 and a drainage area of 1000 m^2. The average yearly rainfall is 90 cm with 90% of rainfall becoming runoff. The EMC lead concentration is 75 µg/L. A mass balance analysis is employed to determine the lifetime of the cell based on allowable accumulated lead, assuming that all of the input lead is captured by the facility.

Lead input is found as:

$$\left(1000 \text{ m}^2\right)\left(\frac{90 \text{ cm}}{\text{year}}\right)(0.9)(75 \text{ µg} / \text{L})\left(\frac{1 \text{ m}}{100 \text{ cm}}\right)\left(\frac{1000 \text{ L}}{\text{m}^3}\right) = 6.08x10^7 \text{ µg Pb} / \text{yr} \qquad (9.62)$$

$$= 60.8 \text{ g Pb} / \text{yr}$$

The allowable lead is:

$$\left(\frac{300 \text{ kg}}{\text{ha}}\right)\left(\frac{1 \text{ ha}}{10^4 \text{ m}^2}\right)\left(36 \text{ m}^2\right) = 1.08 \text{ kg Pb} = 1080 \text{ g Pb} \qquad (9.63)$$

Therefore, the lifetime of the cell is:

$$\left(\frac{1080 \text{ g Pb}}{6.08 \text{ g Pb} / \text{yr}}\right) = 17.8 \text{ years} \qquad (9.64)$$

The allowable limit will be reached in just less than 18 years.

Since microbial transformations of nitrogen species are common, the fate of nitrogen compounds in a bioretention facility can be complex. Pathways may include ammonification, which is the conversion of organic nitrogen to ammonia. The nitrification of ammonia to nitrate is also likely under aerobic conditions. Therefore, while the capture of both organic nitrogen and ammonia is expected by bioretention media, the transformation of these species to other nitrogen forms can occur during the time between rainfall events. The result is leached nitrogen, in the form of nitrate, from the traditional bioretention system. Design modifications that create a carbon-rich anaerobic bioretention layer have been suggested to promote denitrification, which microbially converts nitrate to nitrogen gas, allowing it to be released to the atmosphere (Kim *et al.* 2003). This layer has a depth of 1 ft and media that consists of a mixture of sand and shredded newspaper.

Oil and grease pollutants can be biodegraded under appropriate environmental conditions. Some proof-of-concept studies on the capture of O/G by mulch layers, followed by the subsequent biodegradation of the captured O/G, have shown promising results (Hong *et al.*, 2005).

As a filtration practice, bioretention is very effective in the removal of suspended solids once the system matures. Bioretention cells will export TSS from the media for a few months after their installation. The long-term fate of captured particulate solids is unclear. Other infiltration facilities, such as sand filters, can become rapidly clogged due to TSS capture (see Section 10.6). Bioretention is expected to be more amenable to the integration of solids into the soil/mulch/plant matrix due to the additional chemical and biological processes expected to be present in these facilities.

The use of plants in bioretention can provide a number of benefits for addressing nutrient removal. During growth plants can take up captured nitrogen and phosphorus from the soil, which may prolong the useful life of the facilities. It must be remembered, however, that unless these plants are harvested, or they are somehow cut and removed, the nutrients will not be totally removed from the facility. If plants are allowed to die off or drop leaves over the fall and winter, these nutrients will be recycled back to the treatment practice, producing an added nutrient burden to the practice as the biomass decays.

A certain amount of periodic maintenance is necessary to maintain the aesthetics of bioretention facilities. Cutting of grasses and trimming plants will improve the look of the facility, but they can also have a major impact on the pollutant accumulation and removal. Layers of mulch may need periodic replacement as older mulch is degraded. The role of plants and mulch degradation in the accumulation and cycling of heavy metals and other toxics is not entirely clear at this point. Judicious use of removing plant and mulch organic matter from bioretention may provide a pathway for toxics removal from these facilities.

Example 9.13.   A bioretention facility receives 33 inches of runoff that contains 0.6 mg/L Zn from a 0.6-acre parking per year. Assume that a special breed of hyperaccumulator plant, *Thlaspi*, is grown in the facility to take up high levels of zinc into the plant biomass. The bioretention cell is 95% efficient in capturing Zn. To find the number of *Thlaspi* plants needed, the yearly accumulated Zn mass is calculated as:

$$M_{Zn} = \left(33/_{12}\,ft\right)\left(0.6\,ac\right)\left(43560\,ft^2/ac\right)\left(0.6\,mg/L\right)\left(28.3\,L/ft^3\right) = 1.22x10^6\,mg\,Zn \quad (9.65)$$

Thlaspi plants should be harvested each year; the plants average 200 grams of dry biomass with a Zn concentration of 2.0% of the dry weight.   Each plant will accumulate:

$$P_{Zn} = \left(200,000\,mg\right)\left(0.02\right) = 4000\,mg\,Zn\,per\,yr \quad (9.66)$$

Therefore, the required number of mature plants is:

$$n = \frac{M_{Zn}}{P_{Zn}} = \frac{1.22x10^6\,mg\,Zn}{4000\,mg\,Zn/plant} = 305\,plants \quad (9.67)$$

Approximately 305 *Thlaspi* plants must be planted and harvested each year so that Zn will not accumulate in the bioretention cell.

9.5   LEVEL SPREADERS

The effectiveness of some best management practices depends on the nature of the inflow to the facility. For example, a wide buffer strip will not trap much of the TSS entrained in the inflow if the inflow enters as concentrated flow rather than sheet flow. Similarly, infiltration trenches and sand traps require shallow depth sheet flow to be effective. Thus, the effectiveness of a design may require consideration of the topographic characteristics upgradient of the BMP.

A level spreader is a small area located immediately upgradient of a BMP facility that requires shallow-depth sheet flow as the inflow. The level spreader area must be graded so that it is relatively flat, which allows the inflowing water to spread out to a relatively uniform depth. An engineered weir or an earthen berm can be placed below the spreader area and upgradient of the buffer strip or infiltration trench. The weir is intended to provide greater control of the flow, both the discharge rate and the uniformity of the flow onto the BMP.

The goal of the design method is to determine the spreader area necessary to limit the flow velocity onto and the depth on the downgradient BMP. The length of the weir is assumed to equal the width of the BMP (W_b), as the objective is to spread the water uniformly over the downgradient BMP. To limit the discharge to the allowable rate, the depth (d_b) of flow and the velocity (V_b) on the buffer strip must be set. The depth of flow will need to be smaller than the height of the vegetation on the buffer and the velocity will need to be low enough that the objectives of the BMP are met. The velocity onto the BMP must be low enough to prevent erosion of the surface. Based on the continuity equation, the allowable flow rate (q_b) on the BMP can be approximated as:

$$q_b = V_b d_b W_b \tag{9.68}$$

The discharge over the weir q_w should not exceed the allowable discharge on the BMP (q_b). The area of the spreader surface (A_s) must be large enough to keep the depth of flow above the weir (h) low enough so that q_w does not exceed q_b. Equating the weir equation and the allowable flow rate of Eq. (9.68) yields the allowable depth above the weir on the level spreader area:

$$h = \left(V_b d_b / C_w\right)^{2/3} \tag{9.69}$$

in which C_w is the weir coefficient, with a value of 3 commonly used with the English system of units.

The continuity of mass equation can be used to relate the inflow rate q_i (ft^3/s) to the level spreader area, the discharge over the weir q_w, the storage above the weir ΔS, and the duration Δt over which this volume accumulates:

$$q_i - q_w = \Delta S / \Delta t = q_i - C_w L_w h^{3/2} \tag{9.70}$$

in which L_w is the weir length, which equals the width of the downgradient BMP. The storage ΔS of Eq. (9.70) is equal to the product of the surface area A_s and the depth of water above the weir, h. Solving for the surface area, A_s, yields:

$$A_s = \Delta t[q_i - V_b d_b W_b]/[V_b d_b / C_w]^{2/3} \qquad (9.71)$$

The time duration Δt can be approximated as the time at which the inflow exceeds the outflow discharge rate. Using the Rational method hydrograph with a time base equal to $2t_c$ and the time to peak of t_c, the duration Δt is given by:

$$\Delta t = 2t_c (1 - q_o / q_i) \qquad (9.72)$$

in which t_c is the time of concentration of the inflow hydrograph, q_o is the discharge over the weir, and q_i is the inflow discharge rate. Equation (9.68) can be used to estimate the outflow q_o from the level spreader area as q_b.

Example 9.14. A level spreader weir is needed to distribute sediment-laden flow across a 60-ft wide buffer strip. To achieve the desired trap efficiency, the flow velocity must not exceed 0.6 ft/sec at a depth of 0.2 ft. Flow into the spreader area is 8 cfs, and the inflow is based on a time of concentration of 20 minutes.

The maximum allowable flow over the weir is computed with Eq. (9.68):

$$q_b = q_o = (0.6\,\text{ft/s})(0.2\,\text{ft})(60\,\text{ft}) = 7.2\,\text{ft}^3/\text{s} \qquad (9.73)$$

At this discharge, the depth of flow above the weir is computed with Eq. (9.69):

$$h = ((0.6\,\text{ft/s})(0.2\,\text{ft})/3)^{2/3} = 0.117\,\text{ft} \qquad (9.74)$$

The time duration Δt is computed with Eq. (9.72):

$$\Delta t = 2(20\,\text{min})(60\,\text{s/min})(1 - 7.2/8.0) = 240\,\text{sec} \qquad (9.75)$$

Thus, the required area can be computed from Eq. (9.71):

$$A_s = 240(8 - 7.2)/(0.117)^{2/3} = 803\,\text{ft}^2 \qquad (9.76)$$

Since the weir is 60 ft long, then the spreader should be approximately 13.4 ft by 60 ft. This area should be graded such that it has minimal slope.

9.6   CHECK DAMS

Steeply sloped swales or narrow swales may provide little retardation of either flow rates or pollutant removal. Check dams are useful control structures that require little maintenance but have the potential to reduce flow rates in swales and increase trap efficiencies. A check dam is nothing more than a low-head weir, either machined or a compacted soil berm. Check dams are placed transversely across the swale, with the depth, often measured in inches rather than feet, largely influenced by topography.

The inclusion of a check dam into the flowpath of a swale serves to remove an initial abstraction from the runoff in terms of volume reduction. The volume behind a check dam is easily computed from geometric characteristics of the site:

$$V_c = zS^2L^3/3 = \frac{zD^3}{3S} \tag{9.77}$$

in which V_c is the volume of the check dam pool (ft^3), z is the cross-slope of the "vee" shaped swale, S is the longitudinal slope (ft/ft) of the swale, L is the length of the pool behind the dam (ft), and D is the height of the dam (ft) at the center of the "vee".

A full check dam provides little resistance to flow and thus has little effect on the travel time of flow through the check dam. However, some storage exists even after the volume below the top of the check dam has been filled. Therefore, the steady-state storage-routing time can be used as an approximation:

$$T_t = \text{volume/discharge rate} = V/q \tag{9.78}$$

Since the water spreads out over the surface of the pool behind the check dam and losses are associated with flow over the dam, the time computed by Eq. (9.78) should be greater than the uncontrolled travel time. Using the weir equation and the geometry of the pool behind the check dam, the following expression based on Eq. (9.75) can be used to estimate the steady-state time, T_t, in the pool:

$$T_t = \frac{z^{1.5}D^5}{19720C_wS^{2.5}q_p^{1.5}} \tag{9.79}$$

in which C_w is the weir coefficient ($C_w = 3$) and q_p is the peak discharge (cfs) into the pool.

Example 9.15. A check dam, 1-ft high at the centerline of a swale that parallels a highway has cross slopes of 8:1 (h:v) and a longitudinal slope of 1%. The discharge rate for which the check dam is designed is 2 ft^3/s. The volume in the pool behind the check dam is computed with Eq. (9.77):

$$V_c = \frac{zD^3}{3S} = \frac{8(1)^3}{3(0.01)} = 266.7 \text{ ft}^3 \tag{9.80}$$

The travel time through the pool and over the check dam is computed with Eq. (9.79):

$$T_t = \frac{8^{1.5}(1)^5}{19720(3)(0.01)^{2.5}(2)^{1.5}} = 13.52 \text{ min} \tag{9.81}$$

The travel time without the dam could be computed using Manning's equation and the velocity equation:

$$T_t = \frac{L}{V} = \frac{n\,D/S}{1.49\,R_h^{2/3}S^{1/2}} = \frac{0.15(1)/0.01}{1.49(0.4)^{2/3}(0.01)^{1/2}} = 185\,\text{sec} = 3.1\,\text{min} \qquad (9.82)$$

Because of the length and width of the pool, the travel time is extended by 10 minutes.

9.7 GREEN ROOFS

Green roofs are well established in Europe, but are gradually becoming more common in the U.S. A green roof consists primarily of a layer of growing mixture and vegetation placed on a building that has a flat or near-flat roof. Stormwater benefits are obtained through using a green roof, but additional benefits are found as well.

Green roofs consist of five layers that are placed over a traditional roof structure. The bottom layer is a watertight membrane that is sealed at all points on the roof. A thin plastic lattice or perforated system for drainage is placed above the membrane. A geotextile or filter cloth is placed over the drainage layer. A layer of growing media is placed above the cloth. This media supports the growth of the plants. The media layer can consist of a mixture of soil, sand, various forms of organic matter, such as peat and compost, and other materials. Finally, a layer of vegetation is established in the media.

A green roof can be classified as either intensive or extensive, the difference primarily being the thickness of the soil/planting mixture, and accordingly the type of vegetation that can be supported by the media layer. An intensive green roof has a media layer of 8 to 24 inches. Because of this thickness, the benefits of an intensive green roof are greater than that of an extensive roof. A greater diversity of vegetation is possible. Costs, which are related to the extra roof materials and the additional structural modifications due to the increased roof load, are the primary discouragement with extensive roofs. Since most green roofs being constructed in the U.S. are intensive, we will focus on these roofs.

An intensive green roof will have from 2 to 6 inches of lightweight growing media, consisting of soil and synthetic growing material. When wet, this media layer will add from 16 to 35 lb/ft^2 to the load carried by the roof, possibly necessitating some, but not major, structural modifications. With such a small layer of growing media, the vegetation is limited to small grasses and wildflowers. In the U.S., most of the green roofs are planted with sedums, a shallow root evergreen that has a high water uptake capacity. With proper plant selection, maintenance requirements for watering and fertilization should be low.

Stormwater benefits are derived from the vegetation and media layers. Rooftops are traditionally 100% impervious area with drains that direct the runoff beyond the structure. The media layer can hold up to a few inches of rainfall, depending on the media thickness and characteristics, removing this volume from the runoff pathway. The depth of flow held by the roof, S, is calculated as:

$$S = mD \tag{9.83}$$

where D is the media depth and m is the fraction of media volume that can be used for water holding capacity. This water holding capacity will also slow the flow from the roof. Water that is held by the media is transpired with the assistance of the plants. Water quality benefits may also be obtained, but information is sparse. Some removal of pollutants, such as nitrogen compounds, from the rainfall may occur. More important may be that the rainwater should not pick up any contamination from the roof material, as would be expected from a traditional roof. The green roof also should also reflect more sunlight than a traditional black asphalt roof, which can become very hot during bright sunny days. Therefore, the green roof should reduce the thermal load of the rooftop runoff and overall runoff temperature.

Example 9.16. During a school refurbishing project, a traditional roof has been replaced by an intensive green roof with 5 in. media thickness. Assuming that the media will hold 26% of its volume in water, the rainfall depth that can be held by the green roof is given by Eq. (9.83):

$$S = mD = (0.26)(5 \text{ in.}) = 1.3 \text{ in.} \tag{9.84}$$

If the total green roof area on the school is 300 ft by 100 ft, the rainfall volume on the roof is:

$$(300 \text{ ft})(100 \text{ ft})(1.3/12 \text{ ft}) = 3250 \text{ ft}^3 = 24{,}310 \text{ gal.} \tag{9.85}$$

In addition to stormwater benefits, the added layers provide insulation to lower heating and cooling costs for the building. The vegetation on the rooftop does not absorb heat to the same degree as traditional roofs, and they will help to minimize urban heat island effects. A vegetated roof can also provide aesthetic benefits.

9.8 PROBLEMS

9.1 Discuss the hydrologic processes important to runoff from vegetative control methods.

9.2 Discuss the physical and biological processes important to the functioning of vegetative control methods.

9.3 Discuss the concept of a buffer strip "storing" surface runoff when the site does not include a berm.

9.4 Discuss the situation where the sediment trap efficiency of a buffer strip for a storm event can be negative.

9.5 Would you expect the sediment trap efficiency of a buffer strip to change with the magnitude of the storm event? Discuss.

9.6 Sediment loads in runoff into a buffer zone q_{si} are obtained from sample measurements. Sediment loads in the outflow (q_{so}) are also measured for seven storm events. Compute the trap efficiency of the buffer strip for each storm and the average for all storms.

q_{si}	46	75	29	51	66	12	37
q_{so}	31	62	36	40	49	2	31

9.7 Six ft³/ft of runoff enters a buffer strip from a parking lot over a period of ½ hour. The buffer strip can infiltrate at a rate of 0.8 in./hr. If the buffer has length of 36 ft and a width of 400 ft, what volume of water is infiltrated? What percent of the inflow is retained in the buffer zone.

9.8 Design a buffer strip the is 60% effective in trapping fine silt soils. The vegetation has a roughness of 0.3. The site has a slope of 0.8% and the runoff has a design depth of 2 inches. The buffer strip is 550 ft wide.

9.9 What is the allowable velocity on a 30-ft-long buffer strip that needs to trap 85% of a fine sand.

9.10 Flow on a 20-ft-long buffer that is 400 ft wide is positioned to trap a medium sandy soil. The vegetation has a roughness of 0.25 and a design depth of 3 inches. The buffer has a slope of 1.3%. Determine the trap efficiency of the buffer.

9.11 A 42 ft-long area is available between a highway and a stream to use as a buffer strip. The runoff velocity through the buffer is expected to be 0.25 ft/s. Estimate the removal of zinc from the runoff flow, assuming that 50% of the Zn is attached to the soil (medium sand) and the other 50%, the dissolved portion, is not removed by the buffer strip.

9.12 If the swale of Example 9.5 could be lengthened by designing it as a meandering swale, such that the slope is 1.1% and the length remains the same, what is the improvement in trap efficiency?

9.13 Design a thick grass swale with 5-to-1 side slopes and a longitudinal slope of 1.7%. Use a vee-shaped cross section and assume that a flow of 2.1 ft³/s enters the upper end of the 75-ft long swale. Determine the flow depth and spread.

9.14 A Bermuda grass covered, vee-shaped swale at a 2.5% longitudinal slope and 7-to-1 side slopes drains runoff from a section of a parking lot. Determine the allowable discharge if the depth should not exceed 8 inches.

9.15 An existing swale has side slopes of 6 to 1 and 4 to 1 on the two sides. The swale is covered with a short grass and has a slope of 2%. If the spread is limited to 14 ft, what is the allowable depth and discharge?

9.16 If the peak discharge from the parking lot of Example 9.4 is reduced to 25 ft^3/s, what effect will this have on the design characteristics?

9.17 A paved right-triangular gutter ($s = 1.2\%$, $n = 0.02$) with a flow depth of 4 inches at the curb and a 12:1 cross slope drains to a grass-lined swale ($n = 0.09$) with a longitudinal slope of 1.1%. What is the depth, velocity, and spread of the flow in the swale?

9.18 Determine the sensitivity functions of the flow velocity in a longitudinal slope, the roughness, depth of flow, and cross-slope (z). Evaluate the sensitivity for the conditions: $s = 2.2\%$, $n = 0.04$, $z = 10$, and a depth of 5 inches.

9.19 A 1.6 ac portion of the shopping mall has a runoff coefficient C of 0.65, a design rainfall intensity of 4.5 in./hr, and a time of concentration of 12 minutes. What is the depth (in.) of direct runoff from the subwatershed?

9.20 The TSS concentration entering a 500 ft grassed swale is 245 mg/L. The concentration leaving the swale is 78 mg/L. Estimate the swale TSS removal coefficient using a linear water quality model.

9.21 Using the TSS data from Problem 9.20, determine the swale TSS removal coefficient using an exponential water quality model.

9.22 If the grass in a swale is replaced with a variety that grew denser and was allowed to grow taller, would you expect the exponential removal coefficient to increase or decrease? Explain.

9.23 An 80% removal of zinc is desired in a vegetated swale. If the exponential Zn removal coefficient is 4.5×10^{-3} ft^{-1}, what should be the length of the swale to meet the Zn water quality requirement?

9.24 The input zinc concentration to a swale is 650 µg/L. At the end of the 900-ft swale, the Zn concentration is reduced to 55 µg/L. How long should the swale be to produce a Zn concentration less than 20 µg/L?

9.25 A bioretention facility is being planned to treat runoff from 1-acre parking lot. Estimate the volume of media needed for the facility.

9.26 Vegetation in a bioretention facility has an average stem diameter of 0.25 in. per plant. The vegetation is planted at a density of 800 plants per square foot. If the vegetation is taller than the depth of water in the pond, what percentage of the volume is lost to vegetation?

9.27 Design a bioretention facility that has a vegetation density of 40% and will wholly contain one-half inch of rooftop runoff from a 22-ft-by-48-ft house. Assume a maximum depth of 8 inches.

9.28 Design a bioretention facility that will retain the first 0.25-inch from a 0.2-acre driveway. Vegetation has a density of 60%. The facility should have a maximum depth of 6 inches.

9.29 Ten samples of the number of stems in a 10 in.^{2} area has a mean of 60 and a standard deviation 12. The same samples have a mean stem diameter of 0.2 in. per plant with a standard deviation of 0.05 in. What is the distribution of plant volume if the plants have a height of 8 inches.

9.30 If the local policy for Example 9.7 had only required design for a ¼ inch depth, instead of ½ inch, how would this have changed the bioretention design area?

9.31 A plastic underdrain ($n = 0.012$) for a bioretention pond will be laid at a slope of 0.8%. If flow into the drain is expected to be 0.14 ft^3/s, what size drain (in.) is required?

9.32 An existing 1-in. plastic ($n = 0.015$) underdrain is laid at a slope of 1.8%. After an addition to the house from which rooftop runoff is directed to the bioretention pond, the discharge hydrograph through the bioretention is an isosceles triangle with a time base of 30 min and a peak discharge of 0.01 ft^3/s. What is the volume of the rock subgrade needed to store the water that will not drain during the peak area? Assume a 40% void ratio in the rock.

9.33 Use the runoff data from to route the event through a bioretention cell with 3 in/hr infiltration rate and porosity of 0.42. What fraction of the runoff volume infiltrates through the facility?

9.34 If the bioretention media of Example 9.11 are increased to an average size of 3 mm, estimate the output TSS concentration, assuming all other parameters are unchanged.

9.35 Bacteria, with an average size of 1 μm, enter a bioretention cell that has a media depth of 3 ft, average media size of 0.1 mm, and a porosity of 0.38. If the sticking coefficient is 0.6 and the collector efficiency is 10^{-4}, estimate the percent removal of the bacteria.

9.36 Show the development of Eq. (9.72).

9.37 A level spreader area will have an inflow rate of 4.6 ft³/s and will lie above a 65-ft wide infiltration pit that will be covered with grass (height = 2.5 in.). The allowable velocity of flow on the infiltration pit surface is 0.3 ft/s. The inflow is based on a time of concentration of 18 minutes. What design length is required?

9.38 Determine the allowable inflow rate to a 35-ft long, 80-ft wide level spreader if the buffer strip below the weir will have an allowable velocity of 0.25 fps at a depth of 0.2 ft. Assume the design storm used to compute the inflow hydrograph assumes a time of concentration of 26 minutes.

9.39 Derive an expression for the rate of change of the level spreader area to a unit change in the allowable velocity on the buffer strip.

9.40 If land development above the level spreader of Example 9.15 increased the discharge into the spreader to 8.9 cfs, what effect would this have on the flow velocity on the buffer strip?

9.41 Derive Eq. (9.77) for a check dam.

9.42 Derive an equation for the volume of a check dam for a semicircular swale.

9.43 A check dam is needed to control 80 ft³ of runoff in a swale laid at a slope of 0.8%. Design the dam for side slopes of 5 to 1.

9.44 An intensive green roof with a 6-inch media layer is constructed on a 50 ft x 100 ft commercial building. The layer will hold 20% of its volume in water. Determine the rainfall depth and water volume held by the layer.

9.45 In Problem 9-44, calculate the load on the roof (in lb/ft²) of the media, which has a bulk specific weight of 70 lb/ft³ and the held water.

9.9 REFERENCES

Barrett, M.E., Irish, Jr., L.B., Malinia, Jr., J.F., and Charbenuea, R.J. (1998) "Characterization of Highway Runoff in Austin, Texas, Area," *J. Environ. Engg., ASCE,* **124**, 131-137.

Davis, A.P., Shokouhian, M., Sharma, H. and Minami, C. (2001) "Laboratory Study of Biological Retention for Urban Storm Water Management," *Water Environ. Res.* **73**, 5-14.

Davis, A.P., Shokouhian, M., Sharma, H., Minami, C. and Winogradoff, D. (2003) "Water Quality Improvement through Bioretention: Lead, Copper, and Zinc Removal." *Water Environ. Res.* **75**, 73-82.

Davis, A.P., Shokouhian, M., Sharma, H., Minami, C. and Winogradoff, D. (2005) "Water Quality Improvement through Bioretention Media: Nitrogen and Phosphorus Removal." *Water Environ. Res.* **75**, under review.

Department of Environmental Protection Resources (DER) (2001) *Design Manual for Use of Bioretention in Stormwater Management*, Prince George's County (MD) Government, Department of Environmental Resources, Watershed Protection Branch.

Hsieh, C.-h and Davis, A.P. (2005) "Evaluation and Optimization of Bioretention Media for Treatment of Urban Storm Water Runoff *J. Environ. Eng., ASCE*, tentatively accepted for publication.

Hong, E., Seagran, E.A., and Davis, A.P. (2005) "Sustainable Oil and Grease Removal from Urban Stormwater Using Bioretention," *Water Environ. Res.* In press.

Kim, H., Seagren, E.A., and Davis, A.P. (2003) "Engineered Bioretention for Removal of Nitrate from Stormwater Runoff," *Water Environ. Res.*, **75**(4), 355-367.

Little, L.M., Horner, R.R., and Mar, B.W. (1992) "Assessment of Pollutant Loadings and Concentrations in Highway Stormwater Runoff," WA-RD-39.12.1, Washington State Dept. of Transp., Olympia, WA. Referenced in Barrett *et al.* (1998).

Logan, B.E., Jewett, D.G., Arnold, R.G., Bouwer, E.J., and O'Melia, C.R. (1995). "Clarification of Clean-Bed Filtration Models," *J. Environ. Engg., ASCE,* **121**, 869.873.

McCuen, R.H. and Spiess, J.M. (1995) "Assessment of Kinematic Water Time of Concentration," *J. Hydraulic Engineering, ASCE,* **121**(3): 256-266.

Schueler, T.R., Kumble, P.A., and Heraty, M.A. (1992) *A Current Assessment of Urban Best Management Practices: Techniques for Reducing Non-Point Source Pollution in the Coastal Zone.* Publication no. 92705, Metropolitan Washington Council of Governments, Washington, D.C.

Wu, J.S., Allan, C.J., Saunders, W.L., and Evett, J.B. (1998) "Characterization and Pollutant Loading Estimation for Highway Runoff," *J. Environ. Engg., ASCE,* **124**, 584-592.

Yonge, D. (2000) *Contaminant Detention in Highway Grass Filter Strips,* Washington State Transportation Commission Report WA-RD 474.1, Olympia, Washington.

Yu, S.L., Kuo, J.-T., Fassman, E.A., and Pan, H. (2001) "Field Test of Grassed-Swale Performance in Removing Runoff Pollution," *J. Wat. Res. Planning Mgmt., ASCE,* **127**(3), 168-171.

Yu, S.L., Zhang, X., Earles, A., and Sievers, M. (1999) "Field Testing of Ultra-urban BMP's," *Proceedings of the ASCE Environmental Engineering Conference.*

10

TRAPS, BASINS, AND FILTERS

NOTATION

A   = drainage area
A   = cross section area of sand filter
A_c = cross-sectional area
A_c = area contributing runoff to cistern
A_{cx} = cross-sectional area at distance x
A_j = projected area of jet
d   = depth of runoff into cistern
d   = depth of gravel fill
d   = required depth of trench or cistern
E   = USLE erosion rate
E_r = supply rate of soil
E_t = trap efficiency
E_{ti} = trap efficiency of particle size i
f_c = ultimate infiltration capacity
f_i = gradation fraction of particle size i
H   = height of sediment basin pool
Δh = vertical drop distance
k   = hydraulic conductivity
L   = horizontal length
L   = length of trench
L   = length of sand filter
L_e = total annual mass of TSS removed
L_m = annual solids loading
m   = maintenance interval
m_i = mass in inflow
m_o = mass in outflow
m_r = mass retained in basin
n   = porosity of gravel fill
n   = void ratio in cistern

Q   = flow through volume rate
Q   = average runoff depth
q   = sand filter flow rate
q_s = sediment load delivery rate
R_s = sediment deliver ratio
r   = half width of jet
S   = water volume
S   = sand filter storage
S_c = volume of cistern
S_s = volume of runoff into cistern or trench
T_d = time duration of direct runoff
t_D = duration of flow for sand filter
Δt = settling time
V   = horizontal velocity
V_s = settling velocity
V_t = volume of a sediment trap
V_x = horizontal velocity at distance x
v_c = gross volume of cistern
v_i = volume of soil in inflow
v_o = volume of soil in outflow
v_r = volume of soil retained
w   = width of trench
γ_s = specific weight of submerged soil
γ = specific weight of water
μ = viscosity of water

10.1 INTRODUCTION

Given the significance of suspended solids in water quality analysis and design, numerous types of stormwater management practices have been recommended for sediment control. Sediment control facilities are especially important near areas of exposed soils, such as construction sites. Where the duration of exposure is short, temporary management practices such as straw bales and silt fences are used. These practices are generally low in cost and easily installed. For larger projects, a small sediment trap may be necessary. Where sediment volumes are a continual problem, permanent facilities are necessary, but these need to be maintained in order to continue to be effective. A permanent sediment basin is often a reasonable control alternative. Wetlands use a forebay, a version of a sediment basin, to limit the amount of sediment that enters the vegetated portion of the wetland.

Sediment control methods take many forms, and universal agreement on definitions for each has not been reached. Sediment control fences and barriers made of straw bales are obviously temporary facilities. These barriers can be effective in sediment capture, but they must be properly installed and maintained. Classifying sediment traps and basins is more problematic. Some use the term *sediment trap* for facilities that control small drainage areas, say 2 to 10 acres, with the term *sediment basin* used for facilities that control larger areas. Others distinguish these practices based on the number of inflow points, with a facility where only one inflow point exists being designated a trap and a facility with more than one inflow point known as a sediment basin. Some distinguish between basins and traps on the basis of the permanency of the barrier, with the facility referred to as a basin only if it contains a permanent embankment. Herein, the distinction is made on the basis of the amount of conceptual development underlying the design method. Ideal settling theory is used to design a basin, with a simple continuity of mass employed for sediment trap design.

The design of permanent sediment basins and forebays are based on the principles that underlie setting basins used in drinking water and wastewater treatment facilities. Stokes Law, which was introduced in Section 6.9.3, is based on a force balance of particle weight, drag force, and buoyancy. It is the basis for determining the size characteristics of a sediment basin. Stokes Law assumes that each particle settles at a rate that is independent of other particles, so its accuracy may be less for cases of high sediment concentrations, where agglomeration of particles may occur.

10.2 SEDIMENT TRAPS

Sediment traps are intended for use in small drainage areas, such as construction sites. In situations where a trap is intended to be a permanent fixture, its use should be limited to contributing areas of 25 acres or less. Whether temporary or permanent, the removal and proper disposal of accumulated sediment is important. The maintenance of the basin is critical to its effectiveness as a trap. If accumulated sediment reduces the storage volume below the design volume, then the trap efficiency will be decreased and accumulated sediment may be washed out. The design of a sediment trap is based on a simple mass balance of expected sediment loadings. The design should also specify the frequency of maintenance that is reasonable to ensure that the available storage is adequate.

10.2.1 Trap Efficiency

Sediment traps collect the runoff flow, slowing the velocity. With a lower horizontal velocity, the time for the particles to settle is greater. Under these more quiescent conditions, the suspended sediment will settle to the bottom of the water flow. The sediment accumulates in the trap, leaving water with less sediment to flow out.

The trap efficiency (E_t) of a sediment trap is equal to the ratio of the mass of sediment in the inflow that is retained within the trap (m_r) to the total volume of soil in the inflow (m_i):

$$E_t = \frac{m_r}{m_i} = \frac{m_i - m_o}{m_i} = 1 - \frac{m_o}{m_i} \quad (10.1)$$

in which m_o is the volume of soil in the outflow. The last expression of Eq. (10.1) is typically used because trap efficiencies are usually estimated from grab samples taken from the outflow and inflow. The difference is assumed to be retained in the trap. The trap efficiency can vary considerably from storm to storm. Larger sediment fractions are expected to be retained during the smaller storms. During large storms, the mass of soil in the outflow may actually exceed the mass in the inflow due to washout, which according to Eq. (10.1) would indicate a negative trap efficiency. The average trap efficiency is the expected value over a large number of storms.

10.2.2   Design Procedure

The volume of a sediment trap is the primary design variable. The volume is generally expressed as a volume per unit time, such as acre-feet per year. If the trap is a temporary facility intended to be in place for only a portion of a year, then the volume can be reduced proportionally. If the distribution of rainfall is not uniform throughout the year, then this should be considered in designing the trap.

The mass balance used to estimate the required volume of a sediment trap (V_t, ac-ft/yr) is:

$$V_t = A\, E_r\, \left(1/\gamma_s\right) E_t \left(\frac{2000}{43560}\right) \tag{10.2}$$

in which A is the area (acres) that contributes to the trap, E_r is the supply rate of soil (tons/ac/yr), γ_s is the bulk specific weight of the captured soil (lb/ft^3), E_t is the trap efficiency fraction, and 2000/43560 are unit conversion factors.

The value of E_r in Eq. (10.2) is the amount of soil that is detached and transported to the site of the trap. It may be significantly less than the amount detached, as some detached material settles out as it is transported. The trap volume computed with Eq. (10.2) can be reduced by a fraction if it is believed that the inflow to the sediment trap is not well represented by E_r, the amount of soil eroded throughout the basin. For example, if the Universal Soil Loss Equation is used to estimate E_r, then the value reflects the amount of soil eroded rather than the amount that actually reaches the trap. The concept of a sediment delivery ratio could be used to define the fraction applied to Eq. (10.2). The following equations provide sediment delivery ratios (R_s) that can be multiplied by Eq. (10.2) when the estimated value of E_r is the amount of soil eroded rather than the sediment in the inflow to the trap:

$$R_s = e^{-0.04A} \text{ for sandy soils} \tag{10.3a}$$

$$R_s = e^{-0.01A} \text{ for silty soils} \tag{10.3b}$$

$$R_s = e^{-0.006A} \text{ for silty-clay soils} \tag{10.3c}$$

$$R_s = e^{-0.002A} \text{ for clayey soils} \tag{10.3d}$$

is which A is the area (acres) contributing flow to the trap.

The supply rate of soil can be estimated with an empirical model such as the Universal Soil Loss Equation, as discussed in Section 6.10, or empirically derived constants, such as those of Table 10.1. The adjustment factor of Eqs. (10.3) can be applied with the USLE but not to the values of Table 10.1.

The specific weight of the soil γ_s in Eq. (10.2) depends on the soil type. Table 10.2 gives the values of specific weights to use with Eq. (10.2).

When the value of E_r is given as an annual rate and the sediment trap is a temporary facility for a period much less than a year, the volume computed with Eq. (10.2) can be reduced to reflect the smaller expected volume of soil that is needed to trap. This can be approximated using the ratio of the average depths of rainfall expected during the time period that the trap is intended to function to the average annual rainfall depth.

Example 10.1. A 20-acre site with a silty soil is to be developed. The area will be exposed for about two months during which the rainfall is expected to be about 20% of the annual average rainfall. Local regulations require a trap efficiency of 95% for sediment in runoff. The storage volume in the sediment trap is calculated using Eq. (10.2).

For a construction site, Table 10.1 indicates a rate of 75 tons/ac/yr received at the site of the trap. Since the trap will function for less than a year, the rate is reduced to 20% of this value based on expected rainfall. For the silty soil, from Table 10.2, a specific weight of 75 lb/ft^3 is used. Equation (10.2) gives the following estimate of the required storage:

Table 10.1. Average annual erosion rates (E_r)

E_r (tons/ac/yr)	Land Use
0.25	Woods (good condition)
1.0	Suburban areas (residential with good grass cover)
2.0	Commercial/Industrial
5.0	Abandoned areas with developed cover
10.0	Cropland
75.0	Construction sites

Table 10.2. Specific weight γ_s of submerged soils

Soil	γ_s (lbs/ft^3)
Sand	100
Silt	75
Silty-Clay	65
Clay	50

$$V_t = AE_rE_t(2000/43560)/\gamma_s = 20[(0.2)(75)](0.95)(2000/43560)/75$$

$$= 0.174 \text{ ac-ft} = 7600 \text{ ft}^3$$

(10.4)

At an average 2-ft depth, this sediment storage would require an area of 3800 ft^2, which is about 0.44% of the total area of the construction site.

Example 10.2. A sediment trap is to be designed to prevent soil eroded from a 15-acre agricultural area that drains to a small wetland. Because the receiving area is a wetland, a 95% trap efficiency is required. The site, located in the middle of Ohio, has a soil with 30% sand, 50% silt, and 4% organic matter. The soil has a coarse granular structure and a moderate permeability. The area is at a 2% slope with a slope length of 900 ft. Corn is grown on the agricultural field. The volume of the trap is needed.

The USLE can be used to estimate the rate of erosion on the 15-acre watershed. The erosivity index is 150 (Figure 6.5), with a soil erodibility factor of 0.225 (Figure 6.6). The topographic factor is 0.39 (Table 6.12), with management and support-practice factors of 0.62 and 1.0, respectively (Tables 6.13 and 6.14). Thus, using Eq. (6.27), the USLE rate of erosion is:

$$E = 150(0.225)(0.39)(0.62)(1.0) = 8.2 \text{ tons/ac/yr.}$$

(10.5)

This erosion value is adjusted to a sediment trap input using Eq. (10.3b) for a silty soil:

$$R_s = e^{-0.01(15)} = 0.86$$

(10.6)

The specific weight, from Table 10.2, is 75 lb/ft^3. While the soil has 30% sand, which would suggest a lower sediment delivery ratio, the value of Eq. (10.6) will be applied with Eq. (10.2):

$$V_t = 15[(0.86)(8.2)](0.95)(2000/43560)/75$$

$$= 0.062 \text{ ac-ft} = 2680 \text{ ft}^3$$

(10.7)

At an average depth of 1.5 ft, a trap surface area of 1787 ft^2 (42 ft by 42 ft) would be needed.

10.2.3   Design Considerations

The effectiveness of a sediment trap can be increased by its proper siting. The following are some general considerations in locating a trap:

1. The trap should be located as close as possible to the source of the eroded material. This reduces the chance for deposition of soil in areas down gradient of the site from which the material was scoured.
2. If the sources of the soil are widely dispersed in the drainage area, locate individual traps near each source. In some cases, two or more traps can be more effective than one trap.

3. Site a trap so that it intercepts only soil-laden runoff. Flows from areas that do not generate much eroded material should be diverted around the trap. Clean water is more likely to pickup settled soil and carry it out of the trap. Larger volumes of water will also increase the flow velocity which reduces the effectiveness of the trap because the retention time is shorter.

4. If the trap is to be used on a construction site, it should be located so that it will not interfere with construction activity, yet be as close as possible to the exposed land surfaces.

The effectiveness of a trap will also depend on its proper construction. The following are some general considerations related to the construction of traps:

1. The natural terrain should be used to minimize construction disturbance and costs. Natural ridges can be used as a barrier or as a baffle. The use of natural depressions will minimize the need for excavation to create a trap. The soil disturbance associated with excavation can contribute to increased erosion.

2. The design should include an emergency spillway that will permit excess water to pass around the trap during large rainfall events.

3. If excavation is necessary, the side slopes should be no steeper than 3:1. Steep slopes are more likely to slough, which then contributes to the amount of loose soil.

4. All inflows should be situated so that they will freely drain even after sediment has accumulated to the design volume. The inlets should not be submerged or cause the backup of water such that areas are inundated.

5. All outlets should be protected to ensure that energy dissipation does not cause scouring. Riprap is a common means of protecting outlet areas. Gabions might be employed in critical areas.

Proper maintenance is also an important determinant of the effectiveness of a sediment trap. The following points should be considered:

1. The trap should be located so maintenance equipment will have adequate access for clean out.

2. The design plan should include provisions for the proper disposal of the accumulated soil.

The geometric characteristics of a sediment trap also influence its effectiveness. The following factors related to basin geometry should be considered in design:

1. Outlets should be located as far as possible from the inflow points. This increases the likelihood the material will settle and not reach the outlet. Where site conditions prevent having a reasonable distance between inflow and outflow points, baffles should be used to lengthen the travel time of the water in the trap. Small islands in the middle of the trap are used as barriers in larger traps.

2. Experience shows that the angle of dispersion of the inflow is approximately $30°$. Therefore, traps with a triangular shape with the inflow entering where the sides spread out at a $30°$ angle will minimize dead zones that are ineffective for settlement.

10.3   SEDIMENT BASINS

Sediment basins are generally permanent structures used to prevent scoured soil from entering streams, lakes, or other water bodies. In some cases, such as large, important projects, a sediment basin may be constructed with a short design life. The sediment basin may be converted to a detention basin (see Section 8.3) after a construction project has been completed and the drainage basin has been stabilized.

The design of a sediment basin uses the conventional sedimentation theory, as introduced in Section 8.5.1. The theory uses Stokes Law to describe particle settling velocity, which is linked to the horizontal particle velocity due to fluid movement. As given by Eq. (8.76), the trap efficiency, E_t, is calculated as:

$$E_t = \frac{V_s}{Q/A} = \frac{L}{H}\left(\frac{V_s}{V_x}\right) \tag{10.8}$$

where V_s is the particle settling velocity (ft/s, calculated by Stokes Law), Q (ft^3/s) is the basin input flow rate, and A (ft^2) is the bottom surface areas of the sediment basin. H (ft) is the average depth, L is the horizontal length of the basin, and V_x is the horizontal flow velocity (ft/s).

Generally, in a natural sediment basin if the inflow enters from a concentrated drainageway, such as a pipe or a small stream or gully, the flow will spread out both horizontally and vertically and steady, even flow does not result. Thus, the horizontal velocity at a distance x from the point of inflow, denoted as V_x, is given as a function of the varying cross sectional area A_{cx}:

$$V_x = Q/A_{cx} \tag{10.9}$$

The theory of jets indicates that a point discharge such as a pipe or small stream channel expands at a half angle of about 15°. Thus, the half width (r) and depth (r) would be a function of the distance x from the inflow point:

$$r = x \tan 15° \tag{10.10}$$

and the area A_{cx} at a distance x would be:

$$A_{cx} = 0.5 \, \pi \, x^2 (\tan 15°)^2 \tag{10.11}$$

Eq. (10.11) can be applied to Eq. (10.9) to find the limiting horizontal velocity. The projected area of the jet A_j is given by:

$$A_j = 2(1/2\,xr) = x^2 \tan 15° \tag{10.12}$$

At some point, the half distance or depth will extend to a boundary (i.e., basin side). Beyond that point, the rest of the full bottom area can be considered for sedimentation.

10.3.1 Design Procedure

In any particular case, the average discharge rate Q and the type of soil or particulate matter to be settled would be known. The setting velocity V_s could then be estimated using Stokes Law. The topography of the site would be used to select a reasonable value of the mean basin depth H. Local policy, which may vary depending on the nature of the receiving water body, would specify the expected trap efficiency, E_t. For a rectangular cross section, this leaves the length L and width W as the design variables.

$$LW = \frac{E_t Q}{V_s} \qquad (10.13)$$

Thus, any reasonable values of L and W that satisfy Eq. (10.13) should enable the expected trap efficiency to be met, subject to the concepts of Section 10.2.3.

Example 10.3. An existing sediment basin has a length of 240 ft, a width of 60 ft, a mean depth of 2.5 ft, and passes a flood flow rate of 355 ft³/sec. If suspended material in the inflow has a settling velocity of 0.02 ft/sec, the expected trap efficiency can be estimated. Rearranging Eq. (10.13), the trap efficiency is computed:

$$E_t = \frac{V_s LW}{Q} = \frac{(0.02 \text{ ft/s})(240 \text{ ft})(60 \text{ ft})}{(355 \text{ ft}^3/\text{s})} = 0.811 \qquad (10.14)$$

Therefore, the basin traps 81.1% of the incoming sediment.

Equation (10.13) can be used to compute the length necessary to trap 95% of the particles:

$$L = \frac{E_t Q}{V_s W} = \frac{0.95(355 \text{ ft}^3/\text{s})}{(0.02 \text{ ft/s})(60 \text{ ft})} = 281 \text{ ft} \qquad (10.15)$$

Therefore, retrofitting of an additional 41 ft length will be necessary.

Example 10.4. A sediment basin is needed to retain a medium sand ($d = 0.0012$ ft). The settling velocity is, therefore, computed using Stokes Law:

$$V_s = \frac{D^2(\gamma_s - \gamma)}{18\mu} = \frac{(0.0012)^2(165 - 62.4)\text{ lb}/\text{ft}^3}{18(2.5 \times 10^{-5} \text{ lb - s/ft}^2)} = 0.328 \text{ ft/s} \qquad (10.16)$$

The site will be excavated to a difference in depth of 1 ft between the inflow and outflow elevations. The length needed to trap 80% of the solids can be calculated for an inflow of 81.2 ft³/s and a width of 12 ft using Eq. (10.13):

$$L = \frac{E_t Q}{V_s W} = \frac{0.80(81.2 \text{ ft}^3/\text{s})}{(0.328 \text{ ft/s})(12 \text{ ft})} = 16.5 \text{ ft} \qquad (10.17)$$

The total active storage volume is 198 ft^3.

Example 10.5. A 28-ft long, 8-ft wide, 1.5 ft deep sediment basin receives runoff from a construction site. During one storm with an average runoff rate of 90 ft^3/s, the estimated soil eroded from the site was 6 tons, with 30%, 50%, and 20% being fine, medium, and coarse sand, respectively. The mean particle diameters for the three sands are given in Table 10.3. The settling velocity for each sand gradation is computed with Stokes Law for D in ft:

$$V_s = \frac{D^2(\gamma_s - \gamma)}{18\mu} = \frac{(102.6 \, lb/ft^3)D^2}{18(2.5 \times 10^{-5} \, lb\text{-}s/ft^2)} = 2.28 \times 10^5 \, D^2 \qquad (10.18)$$

The three settling velocity values are given in Table 10.3. The trap efficiency for each gradation is computed using Eq. (10.13):

$$E_t = \frac{V_s LW}{Q} = \frac{(28\,ft)(8\,ft)V_s}{(90\,ft^3/s)} = 2.489V_s \qquad (10.19)$$

The corresponding trap efficiency values are given in Table 10.3. The overall trap efficiency is a weighted average value:

$$E_t = \sum_{i=1}^{3} E_{ti} f_i = 0.204(0.3) + 0.567(0.5) + 1(0.2) = 0.5447 \qquad (10.20)$$

in which E_{ti} is the trap efficiency for sand gradation i and f_i is the fraction of the total in gradation i. With an overall trap efficiency of 54.5%, a load of 3.3 tons (of the 6-ton input) will settle into the dead storage zone of the basin. At 100 lb/ft^3 (Table 10.2), these 3.3 tons (6600 lbs) represents a volume of 66 ft^3. If the dead storage zone has a bottom area of 28 ft-by-8 ft, then the average depth of the deposited sand is 0.295 ft, or 3.5 in.

Table 10.3. Estimation of trap efficiency: Example 10.5

Sand	Fraction	Diameter (ft)	Settling velocity (fps)	Fraction trapped	Weight trapped (tons)
Fine	0.3	0.0006	0.0821	0.204	0.552
Medium	0.5	0.001	0.228	0.567	2.554
Coarse	0.2	0.003	2.052	5.107 (use 1)	1.800
					4.906

Example 10.6. Surface runoff from a driveway (25 ft x 12 ft) drains to a 14-ft-long, 8-ft-wide vegetated pond. For a typical storm, the runoff depth from the driveway during the intense portion of the storm is approximately 0.3 inches. As

the water is collected in the pond, the depth increases because the area of the pond is smaller that of the driveway. The depth is $300(0.3)/112 = 0.067$ ft. Particulate matter washed from the driveway has a specific weight of 150 lb/ft³ and a mean diameter of 0.0012 ft. When the pond in filled, the vegetation causes a roughness of 0.25. The site of the pond has a slope of 0.006 ft/ft.

The hydraulic radius of the flow is approximately equal to the flow depth of 0.067 ft. Therefore, Manning's equation yields the following horizontal velocity:

$$V_x = \frac{1.49}{0.25}(0.067\,\text{ft})^{2/3}(0.006)^{1/2} = 0.0762\,\text{ft/s} \tag{10.21}$$

The settling velocity is computed with Stokes Law:

$$V_s = \frac{(0.0012\,\text{ft})^2\,(150 - 62.4\,lb/\mathit{ft}^3)}{18(2.5\times10^{-5}\,\text{lb-s/ft}^2)} = 0.00769\,\text{ft/s} \tag{10.22}$$

Equation (10.8) is used to compute the trap efficiency:

$$E_t = \frac{14\,\text{ft}\,(0.00769\,\text{ft/s})}{0.067\,\text{ft}\,(0.0762\,\text{ft/s})} = 21.1 \tag{10.23}$$

Since the computed value is greater than 1, all of the particulate matter settles out. Equation (10.11) can be used to compute the distance over which the particles will settle out of the runoff:

$$L = \frac{E_t H V_x}{V_s} = \frac{1.0(0.067\,\text{ft})(0.00769\,\text{ft/s})}{0.0762\,\text{ft/s}} = 0.0068\,\text{ft} \tag{10.24}$$

Thus, the material is heavy enough to settle almost immediately.

10.3.2 Integrated Trap Efficiency

The computed trap efficiencies of Eq. (10.1) and (10.8) appear to be constants. In reality, the trap efficiency varies with the magnitudes of the inflow and outflow. For very small storms, a sediment basin may actually store all of the inflow runoff, which will yield a trap efficiency of 100%. For large storms, the runoff flow rates are higher; therefore, the horizontal velocities in the sediment basins will be higher, and trap efficiencies will decrease, as suggested by Eq. (10.8). Thus, the idea of a constant trap efficiency may not be conceptually realistic under dynamic flow conditions.

In the design of sediment basins for important projects, the variation of sediment delivery and trapping across storm frequencies should be addressed. As the return period of the storm increases, the mass of eroded material increases and the sediment basin trap efficiency decreases. But the expectation of eroded material decreases because high return period storms occur infrequently. An expected erosion volume and trap efficiency can be obtained by integrating both with the storm frequency.

The expected sediment outflow, with or without a sediment basin, is obtained by multiplying the sediment frequency curve by the exceedance frequency for runoff and the trap efficiency of the basin. For the expected sediment outflow where a basin is not in place, the trap efficiency is assumed to be 0. The computations require mean sediment rates for different exceedance frequencies, so the frequencies can be separated into ranges. Greater accuracy is achieved by using smaller frequency intervals.

Example 10.7. Table 10.4 includes a sediment frequency curve (column 1 vs. column 3) that is separated into 13 intervals, which range from 1% to 20%. The mean sediment rate for each frequency interval is given in column 3. For storms with an exceedance frequency of 98% or more, sediment is not produced because of the low rainfall impact and low runoff velocities. Multiplying the interval width (column 2) as a fraction by the mean sediment rate for that interval (column 3) gives the expected sediment rate for that interval (column 4). The sum of the expected rates for each interval gives the long-term expected sediment rate. In this case it is 58.6 tons/acre/year. Note that, while the sediment rate is high for the large events, their overall contribution is not dominating because the large storms occur infrequently. A greater proportion is generated from mid-sized storms that occur more often.

Table 10.4. Integration of long-term expected sediment rate for Example 10.6.

Exceedence frequency range (%)	Interval Δ (%)	Sediment frequency curve q_s (t/a/y)	Expected sediment (t/a/y)	Trap efficiency E_t	Expected basin outflow (t/a/y)
>99	1	0	0	1.00	0
98 – 99	1	0	0	1.00	0
96 – 98	2	2	0.04	1.00	0
90 – 96	6	7	0.42	1.00	0
80 – 90	10	13	1.30	0.96	0.05
60 – 80	20	24	4.80	0.90	0.48
40 – 60	20	40	8.00	0.81	1.52
20 – 40	20	62	12.40	0.70	3.72
10 – 20	10	106	10.60	0.55	4.77
4 – 10	6	152	9.12	0.34	6.02
2 – 4	2	215	4.30	0.11	3.83
1 – 2	1	308	3.08	0.05	2.93
< 1	1	451	4.51	0.02	4.42
			58.57		27.74

If the sediment enters a sediment basin and some of it is trapped, then the expected rate will be less depending on the trap efficiency of the basin. The integration method of Table 10.4 enables accounting for the variation of trap efficiency with storm size. The expected sediment of column 4 is multiplied by either the trap efficiency E_t for the frequency range (column 5) to compute the expected sediment mass trapped, or by $1 - E_t$ to compute the expected sediment mass transported in the outflow. Column 6 shows the expected sediment rate in the basin outflow for each frequency range, with the sum giving the expected long-term rate.

For the situation shown in Table 10.4, the overall trap efficiency is $1 - (27.74 / 58.57) = 0.526$ or 52.6%. This differs from the trap efficiency of the 2-yr event (50% exceedence frequency), which would be 81%. Generally, the long-term average sediment rate will be less than the rate for a 2-yr event because of the increasingly greater sediment rate with increasing storm size.

10.4 INFILTRATION TRENCHES

An infiltration trench is a useful alternative for reducing the surface runoff volume. An infiltration trench is a subsurface void filled with crushed stone and wrapped in a filter cloth. Water infiltrates into the trench from above, with a sod-covered, porous soil overlaying the crushed rock (Figure 10.1). Generally, the surface area directly above the trench is surrounded by a low berm in order to increase the retention time of the water on the surface. This provides for some ponding. The depth of the berm will be dependent on the intended use of the surface area. An infiltration trench will remain effective as long as the porosity of the overlying soil remains high.

In addition to residential areas, infiltration trenches are also applicable in agricultural areas. Although infiltration rates of cultivated land are often thought to be high, a combination of various agricultural practices may result in large quantities of polluted runoff. For example, the seasonal transitions of the crop growth cycle effects the volume of flows, with larger volumes of runoff occurring during periods of fallow land-use. Poor conservation practices coupled with soil textures that have a high runoff potential frequently create runoff problems that require control measures. Control measures that reduce the volume of runoff from agricultural fields have an additional benefit of reducing pollutant loadings that often accompany flows to receiving streams.

10.4.1 Considerations in Design and Construction

An infiltration trench may be described as a structural device for inducing infiltration into the surface soils, thus replenishing groundwater supplies. However, the primary goal of such a stormwater management device is more often to aid in the control of surface flows, rather than to serve principally as an artificial recharge device. Trenches are excavated and filled with crushed stone, which stabilizes the sidewalls and provides a significant increase in the storage capacity of the subsurface volume; the trench can also serve to filter sediment from the runoff. The volume of void spaces tends to clog with time; hence, a filter cloth can be placed around the stone and beneath a topsoil backfill. The filter cloth should prevent clogging and, thus, increase the design life of the infiltration trench. The infiltration characteristics of either the filter cloth or the topsoil backfill can be the limiting design factor; thus, it is important to consider their long-term characteristics in design. The topsoil overburden will prevent damage to the filter cloth and will also serve to filter pollutants transported in the stormwater runoff. Suspended solids in the runoff could clog the void spaces in the bottom and sidewalls of the trench.

10.4.2 Sizing of Storage-Trench Dimensions

Runoff from both impervious surfaces and surfaces with low infiltration capacities may be intercepted by an infiltration trench, which is preferably located in a low sloped area. The temporary storage and eventual percolation of storm runoff into the soil is the primary purpose of an infiltration trench. The desired dimensions of an infiltration trench will depend upon the volume of direct runoff for which control is needed and the characteristics of the watershed and soils. In general, the design of a trench must consider the limiting effects of the storage volume and the infiltration characteristics of both the topsoil backfill and the soil surrounding the infiltration trench. Quite often, the design assumes that the infiltration capacity of the surrounding soil is very small; this is a reasonable assumption because it would not be practical to install infiltration trenches in areas that have naturally high permeability soils.

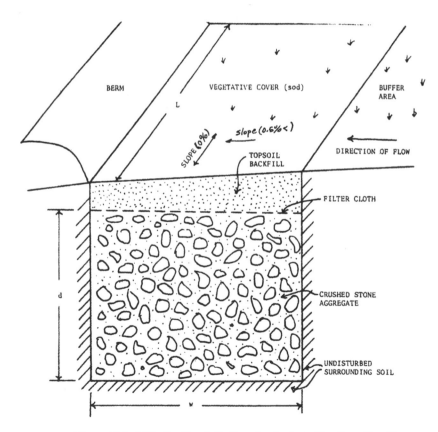

Figure 10.1. Infiltration Trench Configuration for Controlling Storm Runoff

The volume of available storage within an infiltration trench, which is the primary design variable, is a direct function of the porosity of the crushed stone or gravel fill material. The porosity, defined as the ratio of the volume of voids to the total volume of the infiltration trench, may be expressed as a percentage (or fraction) of the total volume; it is also called the percent voids. The void spaces are filled with water where available storage exists to detain surface flows. The maximum volume of storage occurs when the crushed stone is saturated, so that all available voids are occupied.

Various combinations of depths and widths of infiltration trenches may be used to achieve the necessary cross-sectional area of storage. The total volume of storage is computed by:

$$S_s = nHWL \tag{10.25}$$

in which S_s is the volume of available storage, H is the depth of gravel fill, W and L are the width and length of the trench, respectively, and n is the porosity of the gravel fill as a fraction.

The gravel backfill material is typically coarse aggregate from 3/8-inch to 1½-inch in size. The percent voids range in value from about 30 to 40 percent at dense compaction and loose compaction, respectively, for a gravel mixture. The void content of coarse aggregates may differ between gravels obtained from various regions throughout the country. Particular specifications of gravels are available upon request when purchasing fill material.

Storage of the direct surface runoff within the infiltration trench is accomplished by sizing the dimensions to accommodate the volume of water infiltrated from the surface flows. By simply equating the volume of storage with the volume of available surface runoff, the dimensions of the trench may be determined. The upland contributing area A_c will produce a runoff volume S_i, which equals the product of the average depth of runoff Q and A_c. The runoff depth is determined for a design rainfall and the land cover-soil complex for the contributing area. Q could be computed with either the direct runoff of either the Rational formula or the SCS rainfall-runoff equation (see Chapter 4). This choice depends on the policy statement for the locality. The contributing area is set by topography, although some land grading can be used to direct runoff to the location of the trench. The volume for the design storm represents the inflow, some of or all of which can be controlled.

The capacity of the overlying soil to infiltrate water through to the stone layer is generally not high enough to allow sufficient quantities of water to infiltrate without ponding. The velocity of runoff, even when the land surface above the trench is at a very low slope, would not allow much water to infiltrate without creating provisions for ponding. Thus, the depth of the berm that surrounds the surface area of the trench will be the primary factor in determining the volume of inflow that will be intercepted. A berm of depth d that surrounds the lower portion of the trench area will be able to contain a volume of dLW, where L and W are the length and width of the trench. If the objective is to control all of the inflow, the product of dLW would need to be equal to A_cQ. If A_cQ exceeds dLW, then the excess volume, $A_cQ - dLW$ will not be controlled. Neglecting the void space of the overburden soil, the void space within the rock must be sufficient to contain the water that infiltrates from the overlying surface area. Therefore, using Eq. (10.25):

$$S_s = nLWH = LWd \qquad\qquad (10.26)$$

Thus, the depth of the trench H must equal d/n.

This continuity of mass calculation assumes that the infiltration rate of the overburden soil will allow the water to enter the rock-filled storage area. If the water passes over the trench area too quickly to infiltrate, then a portion of the storage will not be used and the interception efficiency will be less than expected.

Example 10.8. Consider the case of a 0.5-acre residential lot that includes a large patio, which is made of an impervious material. Part of the back yard is to be devoted to an infiltration trench surrounded by a berm that is 10 ft wide. The trench area is landscaped such that the average person does not recognize the space as a BMP. Storm runoff from the patio will be directed towards a 20-ft by 10-ft infiltration trench area that will be surrounded by the berm. The trench is to be designed to contain the first 0.5-inch of runoff from the 1200 ft^2 contributing area. In this case, the design parameters are the depth of the berm and the depth of the trench, assuming a porosity of 0.35 for the crushed rock.

The contributing area will produce a runoff volume of 1200 ft^2 (0.5 in.)/(12 in./ft) = 50 ft^3 to contain the first 0.5-inch of runoff. Using Eq. (10.26), the height of the berm is calculated:

$$d = \frac{S_s}{LW} = \frac{50 \text{ ft}^3}{(20 \text{ ft})(10 \text{ ft})} = 0.25 \text{ ft} \qquad\qquad (10.27)$$

A 3-inch berm will be required. To store 3 inches of surface runoff for slow release to the ground water will require a trench depth of d/n = 0.25 ft/0.35 = 0.714 ft, or about 9 inches.

10.4.3 Sizing of Rate-Trench Dimensions

A trench surrounded by a berm is referred to as a *volume-trench* because the storage within the trench depends on the volume of water that can be stored behind the berm. For some sites, it may not be practical to include a berm in the stormwater management plan. The inclusion of a berm may be rejected for aesthetical reasons, such as if it were placed near the front of a residential lot. It may also not be practical on a field used for recreation. For volume-trench design, the overburden soil must have an infiltration capacity that would minimize the time of ponding. The property owner or user of the surface area may consider an excessive ponding time unacceptable.

Where berms are not practical, an alternative trench design can be used. When the volume of void space within the infiltration trench is a limiting factor, the infiltration rate through the topsoil overburden and filter cloth must also be considered in the sizing of the infiltration trench dimensions. This alternative is referred to as a *rate-trench* as the design characteristics are largely dependent on the ultimate infiltration capacity rate of the overburden soil. The stormwater must infiltrate through the topsoil overburden and into the trench prior to percolation into the surrounding soils. The total volume that enters an infiltration trench will depend on the ultimate infiltration capacity of the overlying topsoil backfill. The topsoil material should be selected such that the infiltration rate is high so storm runoff is more easily infiltrated into the trench. A sandy textured soil is recommended to fulfill the requirement of a high infiltration rate; a sandy soil is also capable of sustaining vegetative growth. A dense turf cover over the infiltration trench provides a leaf canopy that protects against the sealing of pores by sediments created upon rainfall impact. Also, vegetative root systems penetrate the soil and increase the infiltrating capacities by furnishing small channels between soil pores.

Without a berm, water will not pond on the surface above the trench. Thus, only the water that infiltrates during the duration of the direct runoff T_d (hr) will be managed by the trench. For an ultimate infiltration capacity f_c (in./hr) of the overburden soil, the volume of water for which trench storage is needed is:

$$S_s = LWf_cT_d \qquad (10.28)$$

in which S_s is the amount infiltrated and stored. Values of the ultimate infiltration capacity for various soil types are given in Table 9.1. As with a volume-trench, the depth of the trench depends on the porosity n of the rock fill. The required depth H (ft) is:

$$H = V_s /(nLW) = f_cT_d/n \qquad (10.29)$$

The filter cloth installed beneath the soil layer should have a water passage rate of at least f_c. To maximize the volume infiltrated, the surface area of the trench should be relatively flat, with a cross slope of 0% and a longitudinal slope much less than 0.5%.

Example 10.9. An infiltration trench is to be used on a residential lot to partially control runoff. The area available has a length of 10 ft and a width of 20 ft. The overburden soil will have an ultimate infiltration capacity of 1.3 in./hr, and the crushed rock in the trench will have a porosity of 0.35. The local drainage policy indicates use of a design storm of 0.8 in./hr intensity and a duration of direct runoff T_d of 2 hours. The land use of the contributing area A_c will produce an effective rainfall intensity of 0.3 in./hr. Therefore, the volume (ft³) of water produced during the design storm is:

$$S_i = i_eT_dA_c /12 = (0.3 \text{ in./hr})(2 \text{ hr})(A_c)/12 = 0.05A_c \qquad (10.30)$$

Equation (10.29) indicates a trench depth of:

$$H = f_cT_d/n = 1.3 \text{ in./hr}(2 \text{ hr})/0.35 = 7.43 \text{ in.} = 0.62 \text{ ft} \qquad (10.31)$$

and Eq. (10.28) yields a volume of storage for the water:

$$S_s = 10\,\text{ft}(20\,\text{ft})(1.3\,\text{in./hr})(2\,\text{hr})/12 = 43.33\,\text{ft}^3 \qquad (10.32)$$

Based on the volume of inflow computed using Eq. (10.30), this trench could contain the inflow from a contributing area of:

$$A_c = V_i / 0.05 = 43.33\ \text{ft}^3 / 0.05\,\text{ft} = 867\,\text{ft}^2 \qquad (10.33)$$

which represents an area of 30 ft by 29 ft. For a larger contributing area, runoff will exceed the infiltration trench capacity.

10.4.4   Siting of an Infiltration Trench

In addition to the factors previously identified, the effectiveness of an infiltration trench will depend on siting factors. The selection of a satisfactory location to implement an infiltration trench as a runoff control measure will necessitate a site investigation; information obtained pertaining to the inherent soil characteristics will reveal the suitability of the site environment to accommodate groundwater storage and recharge.   Hydrologic, topographic, and geologic investigations will provide the basis for a decision.

Much of the necessary information may be obtained from existing sources of data. County soil surveys that identify the soils and soil characteristics are available from the Natural Resources Conservation Service. The permeability is of particular interest because it reflects the rate at which the water moves through the soil. Additionally, the permeability may be measured on-site by various methods (e.g., percolation test and well pumping) and such site-specific data should be used whenever possible. For successful infiltration trench implementation, the individual soil horizons between the ground surface and the water table aquifer should not have any impermeable layers that restrict the downward movement of water.

The depth between the water level in the saturated aquifer and the soil surface should be determined to ensure that the trench will not rapidly become inundated due to a rise in the water table elevation during a large rain event. Areas with water table elevations that are seasonally high may require consideration of additional design requirements.   Coastal areas, for example, are frequently faced with groundwater tables that rise rapidly, which limit the ability of the ambient soil to adequately drain the stored stormwater runoff. Such circumstances may cause infiltration trenches to become ineffective in both the storage of stormwater and the recharge of ground water. Furthermore, the inundation of infiltration trenches by a rapidly rising water table may lead to the intrusion of soil particles into the trench from the surrounding soil, which could clog the available void space.   The placement of filter cloths along the bottom and sidewalls may be necessary to avert soil particle intrusion. However, high costs may prohibit such a requisite. Water table elevations should be several feet below the trench bottom to allow for adequate clearance as water levels rise. The actual distance will depend on soil characteristics.

10.4.5   Considerations in Selecting the Filter Cloth

The implementation of filter cloths as a separating layer between the topsoil backfill and the coarse aggregate fill should not be slighted. These fabrics are critical for the continued effective functioning of the infiltration trench. If the installation of a filter cloth is neglected, the topsoil overburden lacks the means of being held in place. Hence, the void space in the coarse aggregate fill will quickly clog, with the particles rendering the trench useless. Filter cloths are available with various specifications to address different geoenvironmental criteria. The selection of a filter cloth should be based upon: (1) permeability (i.e., size of pore openings), (2) strength of fabric, and (3) percent of open area. Filter cloth is being used more frequently to stabilize drain systems and care should be taken to ensure that it is not the limiting permeable material.

Recognizing that filter cloth restricts the movement of the topsoil material into the coarse aggregate fill, particles will tend to clog the pore openings within the filter cloth and reduce the permeability rate. This causes soil particles to form a mud-caked layer on top of the filter cloth, which requires periodic maintenance. Depending upon the depth of topsoil backfill, the accessibility of the filter cloth will vary. Maintenance schedules should be considered in infiltration trench design. Furthermore, the removal of the aggregate fill to be washed or replaced with new aggregate material may be required; however, proper maintenance of the filter cloth should make this an infrequent maintenance requirement.

10.4.6   Strategies for Increasing the Infiltration Potential

The total volume of surface runoff intercepted by an infiltration trench during a storm event is constrained by the volume of voids within the trench, the infiltration rate of the topsoil overburden, and the runoff collection surface area above the trench. Additionally, in the absence of a berm, the duration of the direct runoff, which from impervious surfaces is frequently short due to the low resistance associated with the relatively smooth surfaces found in developed areas. Because of all of these potential limiting factors, the fraction of the total surface runoff volume that is intercepted by a trench is often small. Therefore, a means of inducing greater infiltration would enhance the efficiency of the trench.

A number of different strategies may be used to effectively increase the volume of runoff entering the infiltration trench. To be effective, a strategy must overcome one of the constraints identified. For example, a vegetative buffer strip that is placed between the runoff generating surface and the site of the infiltration trench would reduce the flow velocity and thus increase the duration of flow across the surface area above the trench. Similarly, grading of the site so that the surface slope of the trench contact area is minimized should increase the contact time, and thus, increase the volume of water infiltrated. Also, a similar effect will be attained by increasing the surface roughness of the trench contact area.   An additional advantage of installing a vegetative buffer strip, increasing the surface roughness, or decreasing the slope is the filtration and removal of sediment particles from the surface flow as a pretreatment to the infiltration trench.

10.4.7 Water Quality Considerations

An infiltration trench has several components that contribute to water quality improvement. Foremost, infiltration through the overburden soil layer will filter suspended solids. Also, physicochemical interactions with the soil can remove other pollutants during infiltration. As the water infiltrates from the trench pore space through the native layers to the groundwater, more natural filtering and pollutant removal processes may be expected. As a stormwater management practice, any runoff and pollutants diverted to the infiltration trench will not be directly discharged to the local surface waters.

Of concern in infiltration trenches are very soluble pollutants that do not adsorb or otherwise interact with soils. These would include nitrate and chlorides from road salts. If the runoff contains high levels of these pollutants, they may be transported with the runoff to the groundwater.

10.4.8 Regulatory Considerations

Stormwater management policies and regulations play a critical role in the effectiveness of control methods, including infiltration trenches. Policies and regulations must provide control over the design, installation, and maintenance of the trenches. Failure at any of these aspects may lead to an ineffective infiltration trench. In order for a policy to be effective, it must provide for assurances that the design will properly coordinate the storage volume with both soil and site characteristics, as well as the characteristics of the filter cloth. Regulations must be sufficiently specific to manage site construction sequencing during the installation of the infiltration trench so that the soils in upland areas are not exposed; otherwise, excess eroded soil may rapidly clog the storage space within the infiltration trench. For maximum effectiveness, the infiltration trench should be installed after all of the land development has been completed and soils are stabilized. It should also be recognized that a poorly maintained infiltration trench will not function as the designer intended. Therefore, policies must provide control of maintenance, and regulations must provide for inspection and enforcement of the policies. Only when properly designed, installed, and maintained will an infiltration trench contribute to the control of stormwater runoff.

10.5 CISTERNS

A cistern is an aggregate-filled, subsurface storage facility that has a pipe inlet. Cisterns are sometimes referred to as dry wells. The pipe inlet usually drains runoff from rooftops, although they can also be employed as a control device for the first flush from a parking lot. They should only be used where the inflow does not contain considerable particulate matter that could clog the inflow pipe or the subsurface drainage areas. Where TSS levels are expected to be excessive, a surface filtering method should be included in the design plan to prevent cistern clogging. The filtering practice should be easily accessible for periodic maintenance.

Cisterns provide many environmental benefits. They provide for ground-water recharge, as their only outlet is by way of infiltration. Depending on the storage volume of the cistern relative to the volume of runoff from the design storm, the cistern will act minimally as an initial abstraction. If the relative volume of storage is large, then they may also reduce the peak discharge from the contributing drainage area. Water quality improvements are expected due to the action of infiltration through the soil to ground water.

Cisterns are most appropriate for small drainage areas. Of special concern when deciding to install a cistern is the underlying soil and the depth to the mean water table. The soil must have sufficient capacity to allow for infiltration between storms. If the water in the cistern cannot drain between storms, then all of the storage capacity will not be available for control of the next storm. Therefore, cisterns may not be appropriate in clayey soils. Sandy and loam soils are appropriate. Ultimate infiltration capacities of 0.25 in./hr or greater should underlie the cistern.

Areas subject to high water tables, even if seasonal, may not allow adequate water storage in the soil for a cistern to function as designed. The necessary distance between the bottom of the cistern and the mean water table elevation would depend on the storage capacity of the soil.

Cisterns are generally employed for the control of small volume rainfalls. Their capacity will be exceeded during the larger, less frequent storms, and therefore, they must be designed to include an outlet. The area around the inlet must be protected from scour, as the discharge velocities can be high. An overflow outlet should be directed away from the area of the cistern since overflow could reduce the effectiveness of the cistern itself.

The design of a cistern requires the specification of a design rainfall depth, P (in.), the drainage area to be controlled A_c (ft^2), and the void ratio (n), or porosity, of the crushed stone within the cistern. The rainfall depth would be for the duration and frequency specified in the local policy or drainage standards. Either the SCS method or the Rational formula could be used. Generally, cisterns are intended for the control of frequent storms, so the use of a 24-hr storm may not be appropriate. As an alternative to using a design storm, a depth of runoff could be specified for computing the required control volume. The product of the required depth (d) and the area (A_c) to be drained would set the volume of water to be stored in the cistern, i.e.,

$$S_s = dA_c \qquad\qquad (10.34)$$

The volume of the cistern (S_c) equals the volume of water divided by the void ratio (n):

$$S_c = dA_c / n \qquad\qquad (10.35)$$

Once the total volume is estimated, the length, width, and depth must be selected. The difficulty in setting these will depend on factors such as the depth to the mean annual water table, the location of lot lines, and the location of the overflow outlet. Buildings located near lot lines may require the cistern to be long and narrow to avoid extension beyond a property line.

Example 10.10. Assume that a cistern will be installed on a residential lot to contain the drainage from a 0.5-inch storm. The cistern must be designed to store the rooftop drainage from a 26 ft by 48 ft house. Separate inflow pipes are used to collect water from the front and back of the roof, with the inflow pipes directed to the same cistern. The total volume of water is:

$$V_s = 0.5 \text{ in.} (26)(48)/12 \text{ in./ft} = 52 \text{ ft}^3 \qquad (10.36)$$

Assuming a void ratio of 0.3, the required volume of the cistern is:

$$V_s = 52 ft^3 / 0.3 = 173 ft^3 \qquad (10.37)$$

Thus, a 4-ft deep, 7-ft long, and 6-ft wide cistern is installed.

10.6 SAND FILTERS

Sand filters are as constructed layers of sand through which runoff is directed. Sand filters are specifically targeted for the removal of suspended solids for water quality improvement, and their implementation is based on the mainstream success of rapid sand filters in the treatment of drinking water.

Stormwater sand filters are installed below the land surface and use a gravity head to drive the water through a layer of sand. The sand captures suspended solids by straining large materials at the sand layer surface and through attachment mechanisms throughout the filter depth. As the sand media captures suspended solids, the resistance to water flow increases. This decreases the flow through the filters and allows the buildup of a greater water head, diminishing treatment capacity. This decrease in flow must be considered in sand filter design.

10.6.1   Sand Filter Configurations

Sand filters are efficient for TSS removal, yet very susceptible to clogging. Additionally, the treatment of the water that flows through the filter media may be slow, necessitating the design of large filters. Because of these two concerns, sand filters are usually designed in tandem with a pretreatment pond or forebay storage volume. The forebay captures the water flow directly from the drainage area. This water detention allows some sedimentation of the larger suspended solids, thus introducing a pretreatment step for the removal of suspended solids to extend the design life of the sand filter. The storage of water also allows the flow into the filter to be moderated so that it is gradually fed to the filter for removal of finer particles.

Three different configurations of sand filters are shown in Figure 10.2. They differ in the type of storage in the pretreatment pond and if the filter basin follows or is within the pond. In Figure 10.2a, a pond or storage area is designed to hold the water quality volume to be treated. The water leaves the storage area through a control structure and is directed into the sand filter. The water head in the pretreatment storage area provides the head to the filter, as permanent storage is not built into the system. This configuration is known as the Austin sand filter, based on its extensive use in Austin, TX.

Figure 10.2b shows a similar filter design, with the primary difference being the presence of a permanent storage volume in the initial storage area. As runoff enters the storage, once the water level increases above the permanent pool, flow control devices allow water flow into the filter. Water head above the permanent pool provides the head for the filter.

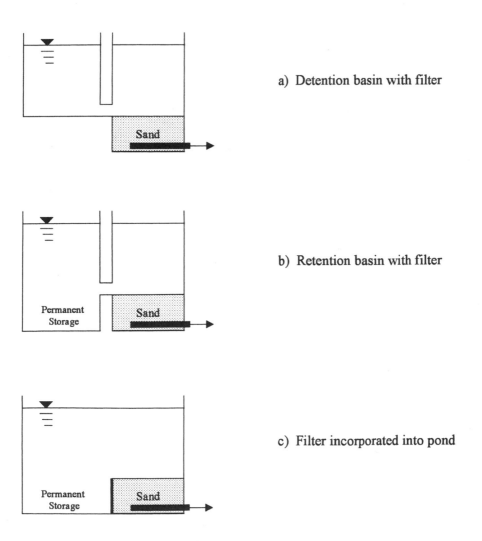

a) Detention basin with filter

b) Retention basin with filter

c) Filter incorporated into pond

Figure 10.2. Diagrams of stormwater sand filters. Adapted from Urbonas (1999)

The third arrangement is known as the Delaware filter (Figure 10.2c). In this case, the filter is integrated into the storage area itself. A permanent pond is created throughout the area where the filter is not present. Input flow accumulates over the permanent storage and the filter area, which provides the head for the filter. By integrating the filter into the storage area, some sedimentation will be taking place onto the surface of the sand filter, which can lead to premature clogging of the filter.

10.6.2 Design Procedure

The primary design parameter for the sand filter is the filter surface area. The greater the volume of water to be treated, the larger the filter area required. When designing a sand filter, both the hydraulic capacity and the suspended solids capture capacity must be evaluated.

The determination of filter area based on hydraulic capacity employs a modification of Darcy's Law:

$$q = kA\frac{dh}{dl} \tag{10.38}$$

where q is the flow rate, k is the hydraulic conductivity, A is the cross-sectional area, and dh/dl is the hydraulic gradient, for any consistent system of units. For sand filter operation, the volume of water (S) to be treated is infiltrated over a specific design time (t_D), producing the design flow rate:

$$q = \frac{S}{t_D} \tag{10.39}$$

The hydraulic gradient in the sand filter is calculated as the water head (h + L) divided by the depth of the sand media:

$$\frac{dh}{dl} = \frac{h+L}{L} = 1 + h/L \tag{10.40}$$

With these relationships, the design filter area is given by:

$$A = \frac{S}{kt_D(1 + h/L)} = \frac{S}{t_D i} \tag{10.41}$$

In the second part of Eq. (10.41), the water infiltration rate replaces the product of the hydraulic conductivity and hydraulic gradient.

Sand filters are generally designed to allow the captured water volume to be treated over a duration of about 0.5 to 1.5 days so as to provide reasonable storage, but to prevent creating opportunities for mosquito breeding. A value of 3.5 ft/day is a typical infiltration rate for clean sand, but 2.0 ft/day is more reasonable for a filter with some captured solids. Urbonas (1999) suggests an infiltration value of 2.5 in./hr for typical sand filters.

Urbonas (1999) also suggests that in sizing the filter area, an annual solids loading (e.g., lb/ft^2) should be considered. This leads to a second design equation:

$$A = \frac{L_a}{L_m m} \qquad (10.42)$$

where L_a is the total annual mass (weight) of TSS removed by the filter (lb), L_m is the design annual solids loading (lb/ft^2), and m is the maintenance interval (years) for the filter. Urbonas (1999) suggests a maximum of 0.36 lb/ft^2 for L_m. In cases where the calculated areas of Eqs. (10.41) and (10.42) differ, a compromise that considers that accuracy of the input parameters should be considered. After averaging the areas, the effect of averaging on the hydraulic and mass loading rates should be evaluated. Urbonas (1999) presents details on balancing the two loading criteria.

The value for L_a must be estimated considering the TSS production from the watershed and subtracting any TSS that is removed by the pretreatment sedimentation area and other practices. The capture mass by the filter alone is calculated, which can be found by assuming 95% TSS trap efficiency (Urbonas 1999) or by assuming that the TSS effluent from a sand filter is about 8 mg/L (Barrett 2003).

The sand layer should be several feet deep with a minimum depth of 1.5 feet. The elevation and available head can become an issue in design since a head of about 3 feet is maintained over the sand layer. In extreme cases, pumps may be necessary to move water from the filter underdrain up to the discharge point, thus increasing capital and maintenance costs. Clean sand sizes of 0.5 to 1.0 mm are recommended. Austin sand filters constructed in California had d_{50} (median media size) of 0.4 to 0.6 mm and uniformity coefficients of 1.6 to 2.0 (Barrett 2003).

10.6.3 Maintenance Considerations

Both the sand filter forebay and the filter itself will accumulate particulate matter during successful operation. Therefore, a maintenance program is critical to the continued performance of sand filters. The forebay must be cleaned at regular intervals to maintain the storage volume available for treatment and to prevent captured solids from being carried over to the filter during heavy rain events.

Unlike water treatment filters, sand filters that treat stormwater cannot be backwashed to removed captured solids. Over time, the filters clog because of the capture of solids. This decreases the flow through the filters, lengthens the time required for the treatment, and causes greater volumes to overflow or bypass the filter. Maintenance involves scraping or scoring the surface of the sand filter to break up any crusted layer formed from TSS capture. More complete maintenance may include scraping off a layer of crusted sand to expose the more permeable layers underneath. At some point in time, the entire sand media layer may need replacement. Special vacuum trucks can assist with these maintenance projects.

Example 10.11.   A sand filter is needed to treat runoff from a highly impervious commercial development that has a runoff coefficient of 0.9, a yearly rainfall of 38 in./yr, producing 32 inches of runoff per year, an EMC for TSS of 80 mg/L, and a drainage area of 2 acres. The system must treat the first inch of runoff.

Based on the criteria listed above, a typical sand filter layer depth of 2 ft is specified, which is greater than the minimum of 1.5 ft and will allow for some removal of top sand layers over time for maintenance. A 0.6-mm diameter sand will be used. The treated volume of runoff is calculated as:

$$V = (2 \text{ acres})(1 \text{ in. runoff})(0.9)(43560 \text{ ft}^2/\text{acre})/(12 \text{ in./ft}) = 6530 \text{ ft}^3 \qquad (10.43)$$

Therefore, to store and then treat the entire volume in 1 day with an average water head of 3 ft, using Eq. (10.41), the sand filter area is required is:

$$A = \frac{6530 \, ft^3}{(2 \, ft/\, day)(1 \, day)\dfrac{(3 \, ft + 2 \, ft)}{2 \, ft}} = 1310 \, ft^2 \qquad (10.44)$$

or with a 2.5 in./hr (5 ft/day) infiltration rate:

$$A = \frac{6530 \, ft^3}{(1 \, day)(5 \, ft/\, day)} = 1310 \, ft^2 \qquad (10.45)$$

Checking the solids loading criteria also, it is assumed that one-half of the runoff TSS concentration is removed by the pretreatment forebay, producing an EMC of 40 mg/L discharge onto the filter. With a TSS filter effluent of 8 mg/L, the yearly TSS accumulation in the filter is calculated using the simple method (see Section 6.4.1):

$$L = \frac{0.9(2 \, ac)(32 \, in. \; runoff)(40\text{-}8 \, mg/L \; TSS)(43560 \, ft^2/ac)(28.3 \, L/ft^3)}{(12 \, in./ft)(454,000 \, mg/lb)} \qquad (10.46)$$

$$= 469 \; lbs \; of \; TSS \; captured \; by \; the \; filter \; per \; year$$

Using the design filter loading of 0.36 lb/ft², along with a yearly scheduled maintenance, Eq. (10.45) gives:

$$A = \frac{469 \, lb/\, yr}{(0.36 lb/\, ft^2)(1/\, yr)} = 1300 \; ft^2 \qquad (10.47)$$

In this case, the areas calculated using both hydraulic and solids loading criteria are identical and can be used for design of the sand filter. A filter 32 ft by 42 ft would suffice.

10.6.4   Filter Performance

Data for sand filter water quality performance is summarized in Table 10.5. Sand filters are specifically designed for the removal of particulate matter and are very efficient in this regard. Also, since many pollutants are affiliated with TSS, their removal is accomplished simultaneously. This is seen in the removal of Total Pb and Total Zn, and to a limited extent Total Cu and Total P. The removal of dissolved metals is somewhat perplexing and suggests that chemical mechanisms may be playing a role in the pollutant removal. Nitrate production was found, indicating nitrification of captured TN.

Clark *et al.* (1998) studied different types of media and media mixtures to evaluate the performance for enhanced pollutant removal through a runoff filter. Sand was mixed with activated carbon, peat, compost, zeolite, cotton textile waste, and agrofiber to examine both infiltration and pollutant removal characteristics. The best infiltration characteristics were provided by the sand-activated carbon. The other mixtures became clogged with suspended solids more easily; the worst were the peat and compost mixtures. The organic media, however, provided the best removal of metals.

Clogging by suspended solids remains the most important parameter in considering filter design. Studies by Barrett (2003) indicated that cumulative suspended solids loadings between 5 and 7.5 kg/m^2 (1.0 to 1.5 lb/ft^2) over several months will result in filter clogging and infiltration rates that are unacceptable.

Table 10.5. Pollutant removal performance for sand filters

Pollutant	Percent removal Urbonas (1999) [a]	Barrett (2003) [b]
TSS	80-94	90
TP	50-75	39
Ortho-P	-	6
TN	30-50	22
Nitrate-N	-	(-74)
TKN	60-75	51
Total Cu	20-40	50
Dissolved Cu	-	6
Total Zn	80-90	80
Dissolved Zn	-	36
Total Pb	-	80
Dissolved Pb	-	39
TPH-oil	-	30
TPH-diesel	-	25
Fecal Coliform	-	65

[a] Most common range for reviewed data
[b] Reduction of EMC

10.7 PROBLEMS

10.1   Sampling of sediment inflow to and outflow from a sediment trap provides the following loads for 5 storm events. Compute the average trap efficiency for each storm event. Comment on the variation of trap efficiency with storm magnitude under the assumption that the inflow load increases with storm magnitude.

Storm	1	2	3	4	5
Inflow (ft^3)	27	16	32	8	11
Outflow (ft^3)	11	5	14	1	3

10.2   (a) The sediment inflow (I) and outflow (O) are related by the following model:

$$O = aI^b$$

where a and b are empirical coefficients. Derive an expression for the trap efficiency. (b) Use the following data to obtain least squares estimates of the coefficients a and b.

Storm	1	2	3	4
Inflow	50.2	126.0	39.8	79.4
Outflow	18.6	62.9	9.2	34.8

(c) For what inputs will the trap provide efficiencies of at least 75%?

10.3   A local policy requires a trap efficiency of 75%. The Universal Soil Loss Equation yields a sediment loss rate of 28 tons/ac/yr for a particular 5.6-ac watershed. What is the annual mass of soil transported out of a sediment trap at the outlet of the site?

10.4   Excess rainfall from a 28,000 ft^2 contributing area is directed to a berm-controlled area that overlies an infiltration trench. The overburden soil for the trench can infiltrate water at a rate of 0.3 in./hr. For a 6-hr storm that produces rainfall excess at a constant rate of 0.45 in./hr., what is the depth of water behind the berm at any time t during the 6-hr storm? The trench area is 1600 ft^2. What is the depth behind the berm at the end of the storm?

10.5   Estimate the volume of a sediment trap needed to trap 90% of eroded material from a 5.5-acre construction site where the dominant soil type is a silty-clay. What fraction of the construction site area must be devoted to the trap, if the trap has an average depth of 2 ft?

10.6   A sediment trap is designed without an emergency spillway, such that all runoff from the site passes through the trap. Stored water discharges from the trap along an 18-ft side. During one storm, the water depth reaches a stage 1.5 ft over the top of the discharge end of the trap. If the outfall side of the trap is not protected by riprap, is the integrity of the outfall in question?

10.7   (a) Based on Eq. (10.2), develop a design graph of the required volume of a sediment trap per acre, V_t/A, versus the soil supply rate (0 to 100 tons per acre per year) for a sandy soil. Show lines for trap efficiencies of 60, 75, and 90%.   (b) Use the graph to estimate the required volume of storage for an 8.5 acre watershed where the supply rate is 65 tons/ac/yr with a trap efficiency of 80%.

10.8   Estimates of soil detachment (E_d) are made throughout six watersheds over a six-month period. The sediment discharges (E_o) for the same period were also made at the outlets of the watersheds. Develop a relationship for the sediment delivery ratio R_s for the form of Eq. (10.3).

A(ac)	26	67	49	80	52	35
E_d	16	23	9.5	19	8.4	12
E_o	8.3	5.6	3.2	2.7	2.1	5.7

10.9   Sandy soil on abandoned land is detached at a rate of 4 tons/ac/yr. Determine the size of a sediment trap at the outlet of the 27-ac drainage area if it is necessary to trap two-thirds of the soil. The average depth of the site of the trap is 21 inches.

10.10   A 10.4-ac commercial site has a 375 ft^2 low spot with an average depth of 1.2 ft where sediment is trapped. The soil is predominantly clay. What trap efficiency is expected?

10.11   Using flow through concepts that underlie a settling tank, derive a general expression that shows the variation of the trap efficiency for spherical particles as a function of particle diameter.

10.12   Design a sediment basin to contain 75% of soil with a diameter 0.0008 ft and a specific weight of 125 lb/ft³. Use Stokes Law to compute the settling velocity. Topography limits the elevation drop between the inflow and outflow points to 2.5 ft. The average inflow rate is 11.4 ft³/s.

10.13   Water moves horizontally through a 32-ft-long sediment basin at a velocity of 0.25 ft/s. The inflow enters the basin at the top of the storage area. If the soil is mostly very fine sand, what is the average vertical distance a particle will drop as it passes through the basin?

10.14   Water is discharged at rate Q from a pipe into a sediment basin that expands with a half-angle of 12°. The basin has a constant depth D. Derive an expression for the horizontal velocity as a function of the distance from the pipe outfall, X.

10.15   The flow rate into a sediment basin is 29 ft³/s. The basin is 62-ft long, 18-ft wide, and 1.5 ft deep. If the sediment in the inflow has a settling velocity of 0.015 fps, what is the expected trap efficiency?

10.16 A sediment basin is required to retain 75% of the fine sand ($d = 0.0004$ ft) that enters with the flow ($Q = 43$ ft^3/s). An area that is 15 ft wide is available, which would yield an average depth of 16 inches. What length is required?

10.17 The average runoff from an agricultural field of 37 cfs includes 4 tons of sediment during a storm. The gradation curve indicates 15% very find sand, 45% find sand, 30% medium sand, and 10% coarse sand. If the sediment basin is 25 ft long, 7 ft wide, and 2.5 ft deep, what is the trap efficiency?

10.18 A bioretention cell is planned to receive runoff of 0.4 ft^3/s for a duration of 30 minutes. The contributing area is 600 ft^2. The runoff will include particulate matter ($\gamma = 150$ lb/ft^3) with a mean diameter of 0.015 inch. The vegetation has a roughness of 0.45, and the pond has a slope of 0.5%. What size retention basin is needed to trap 80% of the particulate matter?

10.19 What would the trap efficiency be in Example 10.7 if the sediment frequency curve (column 3 of Table 10.4) were: {0, 0, 2, 5, 10, 20, 33, 50, 87, 128, 175, 247, 328} tons/ac/yr.

10.20 If an infiltration trench was filled with spherical rock 1 inch in diameter and placed in layers with the rocks center above center, what is the porosity? State assumptions clearly.

10.21 An area of 32,000 ft^2 contributes direct runoff of 0.6 in. to the site of an infiltration trench. If a berm of 3 inches surrounds the trench area, what area must be allocated to the trench in order to trap the first 0.25-inch of the runoff? What volume (ft^3) of water will bypass the trench area?

10.22 A 6000 ft^2 area contributes 0.3 in. of runoff during a design storm. (a) What berm depth is needed to control all of the runoff if the surface area of the infiltration trench is 0.35 ft^2? (b) If a storm has an average depth of 0.55 inches, what volume (ft^3) of runoff will overflow the berm of part (a)?

10.23 An area of 1500 ft^2 contributes runoff to a 180 ft^2 area surrounded by a berm. The trench within the berm is intended to control the first 0.25-inch of runoff. What surface area over the infiltration trench is required?

10.24 A trench with a surface area of 160 ft^2 has overlying soil with an infiltration capacity of 0.8 in./hr. The crushed rock in the trench has a porosity of 0.33. The design storm has an intensity of 1 in./hr and a duration of 1 hour. What area size is required to control the runoff if the depth is 4 inches?

10.25 An infiltration trench must be designed to capture 0.1 in. of direct runoff from a 34,000 ft^2 area. The overburden soil has an infiltration capacity of 0.55 in./hr. The design storm has a rainfall excess of 0.6 in./hr for 3 hours. What surface area is needed for the trench?

10.26 Discuss cistern design from the standpoint of the continuity of mass.

10.27 What is the effect on the design of cisterns for locations where water tables are high?

10.28 A cistern is used to collect runoff from a 22 ft × 52 ft rooftop. The soil that underlies a cistern has an ultimate infiltration capacity of 0.5 in./hr. For a 6-hr storm with an intensity of 0.5 in./hr. If a cistern has a bottom area of 12 ft², what proportion of the constant-intensity storm will infiltrate during the storm? Assume vertical infiltration only.

10.29 A house with a roof area of 32 ft by 64 ft uses two cisterns to collect runoff. The design rainfall depth is 0.6 inch. Assuming a void ratio of 0.33, what volume cisterns are required?

10.30 A cistern with a total volume of 150 ft³ and a void ratio of 0.34 collects runoff from a 40 ft-by-20 ft impervious surface. Assuming all rainfall contributes to runoff, what is the intensity of a 30-minute storm that will cause the capacity of the cistern to be exceeded?

10.31 A sand filter is proposed for runoff treatment from a urbanized business area. The drainage area is 1.5 acres with a runoff coefficient of 0.8. The first ¾ in of runoff must be treated. Approximately 35 inches of runoff is generated per year with a TSS of 100 mg/L. What area of sand filter is required?

10.32 The runoff from the parking lot of a new basketball arena will be treated via a sand filter. The lot is 100% impervious and covers 1.75 acres. The first inch of runoff must be stored and treated. The runoff TSS is 75 mg/L and 40 inches of runoff is generated per year. Determine the sand filter area required.

10.33 Several sand filters are to be constructed around a large 10.acre paved parking lot. It is expected that each filter will have an area of 400 to 500 ft². An average of 32 inches of runoff is generated per year, containing 92 mg/L TSS. How many sand filters are required for the parking lot?

10.34 A sand filter has an area of 500 ft². It treats 36 inches per year of runoff over 0.5 acres of drainage area (c = 0.87). The average TSS reduction is from 68 to 8 mg/L. How long can the filter be expected to operate under these conditions before excessive clogging occurs?

10.35 A sand filter receives runoff from a 1-acre parking lot, reducing the TSS concentration from 89 to 12 mg/L. The average runoff is 29 inches per year. The filter area is 1000 ft². How long is the filter expected to operate before excessive clogging will occur?

10.36 A sand filter treats runoff from a commercial area, 1 acre, 82% impervious. An average of 42 in. of runoff is generated per year with the runoff quality given in the table below. Using removal data from Barrett in Table 10.5, determine the mass of each pollutant removed by the filter over one year.

Pollutant	Runoff Concentration
TSS (mg/L)	94
TP (mg/L)	0.85
TN (mg/L)	2.3
Total Cu (μg/L)	67
Total Zn (μg/L)	650
Total Pb (μg/L)	43
Fecal Coliform (MPN/100 mL)	9800

10.37 The runoff from the parking lot of a new basketball arena is treated via a 900 ft^2 sand filter. The lot is 100% impervious and covers 1.75 acres. Approximately 40 inches of runoff is generated per year, containing 90 μg/L Total Cu, 40 μg/L Total Pb, and 570 μg/L Total Zn. If the removal of these metals through the filter is 75%, determine the total yearly mass removal for the three metals.

10.38 Table 10.5 shows a "negative removal" for nitrate. Discuss what this means. What is happening to nitrogen compounds in the sand filter?

10.8 REFERENCES

Barrett, M.E. (2003) "Performance, Cost, and Maintenance Requirements of Austin Sand Filters," *J. WaterRes. Plan. Mngmt, ASCE*, **129**, 234-242.

Clark, S.E., Pitt, R., Brown, P., and Field, R. (1998) "Treatment by Filtration of Stormwater Runoff Prior to Groundwater Recharge," *WEFTEC 98*, Orlando FA.

Urbonas, B.R. (1999) "Design of a Sand Filter for Stormwater Quality Enhancement," *Water Environ. Res.*, **71**, 102-113.

11

WETLANDS

NOTATION

A_c = cross-sectional area of wetland
C_d = orifice coefficient
D = pipe diameter
D_d = depth of dead storage
D_i = diameter of inflow pipe
d = particle diameter
d_h = distance moved horizontally
d_v = vertical drop of sediment particle
E_d = dead storage water surface elevation
E_f = trap efficiency
ET = evapotranspiration
E_t = evapotranspiration rate
g = gravity
H = forebay height
h = representative stem length
h_i = height of invert
I = inflow
I_c = retardance index
i = infiltration rate
k = consumptive use coefficient
L = forebay length
L_d = length of dry period
L_d = length of weir (dry season)
L_w = length of weir (wet season)
M = average stem density
n = Manning's coefficient
O = outflow
P = rainfall depth
P = wetland perimeter

p = percentage of daily daylight hours
Q = inflow
q_i = discharge rate into wetland
R_h = hydraulic radius
R_{qd} = dry season peak discharge ratio
R_{qw} = wet season peak discharge ratio
S = specific gravity of water
S = storage
S = slope of flowpath
S_s = specific gravity of soil
ΔS = change in storage
T = mean daily temperature
T_d = detention time
t = time
t_r = retention time
Δt = change in time
V = velocity of flow
V_a = active storage volume
V_h = horizontal velocity
V_{max} = storage at maximum flood stage
V_o = volume of sediment in outflow
V_s = volume of storage
V_s = settling velocity
V_{sd} = dry season storage volume
V_{st} = volume of sediment stored
V_{sw} = wet season storage volume
v_i = volume of sediment in inflow
W = forebay width
υ = viscosity of water

11.1 INTRODUCTION

The word *wetland* means different things to different people. This is a result of the wide array of benefits that wetlands offer. A large wetland intended primarily for wildlife habitat enhancement is much different than a small wetland constructed for water quality improvement in a residential area. Of course, these diverse wetland types have similarities, but they have obvious differences. This chapter will deal with the similarities.

Storage is the primary hydrologic process on which wetland design depends. Two types of storage are important: active and dead. Active storage is that where an outlet controls the duration of time in which the water stays within the wetland boundaries. Dead storage represents the volume in which water can leave only by evaporanspiration or infiltration. Both types of storage play important roles in the functioning of a wetland to meet the design objectives. Active storage can be divided into two components: that provided in a forebay and storage in the main cell of the total wetland. Generally, the purpose of having a forebay is to provide a site where easily settleable solids can be deposited. The forebay can then be easily accessed for the removal of the solids. The removal of the solids in the forebay will extend the life of the main cell of the wetland, a site that is generally more difficult to maintain because of access problems for equipment.

11.2 WATER BUDGET

A wetland can be considered as a system. Typically, the continuity of mass is used to define a water budget of a system:

$$\sum I_i - \sum O_i = \sum \frac{dS_i}{dt} \tag{11.1}$$

in which the first term represents the sum of all inflows to the system, the second term represents the sum of all outflows from the system, and the term on the right side of the equal sign is the sum of the storages within the system per unit time. Inflows to a wetland would include rain that falls directly on the wetland, surface runoff that can be both sheet flow and flow concentrated in streams, tidal overflows along the perimeter of the wetland, and ground-water inflows, which includes both interflow during and immediately following storm events and seepage due to a rising water table. Outflows include evaporation from the water surface or soil surfaces within the perimeter of the wetland, transpiration from vegetation within the confines of the wetland, seepage from surface water into the ground, concentrated overflows into streams that drain the wetland, sheetflow that discharges water over a berm that may surround part of the wetland, and tidal outflows. The storage is generally considered to be the amount of surface water contained within the perimeter of the wetland at any time, and t is the time period for which measurements are made.

Generally, inflows and outflows are measured in units with dimensions of length cubed over time (L^3/T), such as ft^3/sec or m^3/s. The values of storage and time would then appropriately be recorded with dimensions of length cubed (L^3) and time (T), respectively, with units such as ft^3 or m^3 for S and seconds or hours for time. If the surface area of the wetland is large and remains relatively constant, the inflows and outflows may be recorded as depths, with units such as ft or m. These are assumed to apply uniformily over the entire area of the wetland. It is sometimes convenient to multiply Eq. 11.1 by the time period, which is expressed as finite time Δt and divided by the area A of the wetland, which yields:

$$\sum d_{Ii} \Delta t - \sum d_{Oi} \Delta t = \sum d_{si} \tag{11.2}$$

in which d_I, d_O, and d_S are depths of inflows, outflows, and storages, respectively, and the subscript i refers to the time increment.

The time period Δt used for accounting will depend on the time interval either at which measurements are made or at which decisions are needed. For example, if stream inflow is a major contributor of water to the wetland and the stream is gaged with readings every 15 minutes, then computations may be made at a 15-minute increment. For a natural wetland where only monthly values are needed, computations may be made on a monthly basis. Where water quality is the main interest and grab-sample measurements are made at irregular intervals, water balance computations are much more involved and largely depend on the availability of a model.

Example 11.1. Table 11.1 provides a monthly water balance for an 11-acre wetland that drains a 224-ac watershed. The average monthly rainfall onto the wetland was taken to be the long-term mean monthly rainfall from a Weather Bureau climatic atlas. The total inflow was estimated from a water yield model of both surface and subsurface inflows. Another model was used to estimate the total outflows. Evapotranspiration estimates for the wetland were obtained from local pan evaporation data, a regional pan coefficient, and other weather station data. The rain incident to the wetland (P), the streamflow (Q), the evapotranspiration (ET), the infiltration (I), and the outflow (O) were expressed in common units, i.e., ft³/day. For a larger wetland, units of acre-feet/day may be more appropriate. The water balance of Eq. (11.1) is specifically stated for this example as:

$$P + Q - ET + I - O = \Delta S \qquad (11.3)$$

in which ΔS is the change in mean monthly storage, which is also expressed in ft³/day. The calculations account for the differing number of days per month. For some months, the water level decreased by as much as 0.06 in./day or increased by as much as 0.04 in./day. The net change over the year was 0. However, in any one year, the net change in storage could be positive, i.e., a rise in the water surface elevation, or negative. In unusually wet years, the storage would increase, while in unusually dry years, the water level would fall below normal.

11.3 STORAGE ACCUMULATION METHOD

For wetlands that are insensitive to short duration storms, the storage accumulation method can be used to establish the necessary outlet facility characteristics, such as the weir length. The design method is based on the continuity of mass equation:

$$I - O = \frac{\Delta S}{\Delta t} \qquad (11.4)$$

in which I and O are the inflow and outflow discharges (ft³/s), and ΔS is the change in volume (ft³) that occurs during time period Δt (seconds). Equation (11.4) can be rearranged to compute the change in storage during Δt:

$$\Delta S = \Delta t (I - O) \qquad (11.5)$$

Since Eq. (11.5) is a linear function, the accumulated storage at time t is given by:

$$S_t = S_{t-1} + \Delta t (I - O) \qquad (11.6)$$

Table 11.1. Monthly water balance of a wetland

Month	Rainfall (in.)	Inflow (ft³/day)	Inflow (ft³/day)	ET (in./day)	ET (ft³/day)	Infiltration (ft³/day)	Outflow (ft³/day)	Outflow (in./day)	Storage (ft³/day)	Storage (in./day)
Jan	5.2	6700	38000	0.01	400	0	42800	1.07	1500	0.04
Feb	4.8	6800	35000	0.02	800	200	39600	0.99	1200	0.03
Mar	4.3	5500	28000	0.05	2000	600	29500	0.74	1400	0.04
Apr	3.9	5200	21000	0.09	3600	1100	20700	0.51	800	0.02
May	2.9	3700	15000	0.14	5600	1700	11200	0.28	200	0.01
Jun	2.1	2800	12000	0.21	8400	1900	5400	0.14	-900	-0.02
Jul	1.7	2200	10000	0.24	9600	2100	1900	0.05	-140	-0.04
Aug	1.4	1800	8000	0.20	8000	2000	2100	0.05	-2300	-0.06
Sep	1.9	2500	17000	0.17	6800	1700	12100	0.30	-1100	-0.03
Oct	2.4	3100	25000	0.12	4800	1500	22300	0.56	-500	-0.01
Nov	3.6	4800	30000	0.07	2800	900	31100	0.78	0	0
Dec	4.4	5700	34000	0.03	1200	200	37200	0.93	1100	0.03
	38.6									

If the accumulated storage exceeds the storage capacity of the wetland, then the excess is discharged through an emergency overflow spillway and the accumulated storage is set equal to the maximum. If the computed storage at anytime is less than 0, it should be set to zero.

The outflow rate O is controlled by the outflow structure, which can be a weir or an orifice, or both. The outflow may be a target discharge that is necessary for the quality of aquatic habitat in the stream below the wetland. In this case, it is a minimum target discharge, and greater discharge rates can be permitted. The target discharge may also vary seasonally throughout the year. This would require an outlet that could be manually controlled.

The time interval selected for computations can affect the accuracy of the design storage. If the interval is too large, then the peaks and troughs of the time series of storage will be smoothed too much, with the required storage being underestimated. A small interval increases the computational effort, but computers make this a minor problem.

Assuming that a gauged record of streamflow is not available at the site, then it will be necessary to use a nearly raingage record to generate runoff for the period. If daily rainfall depths are available, then a model such as the SCS rainfall-runoff equation could be used to compute the daily inflow rates. Equation (11.6) can then be used to compute the temporal sequence of accumulated storage.

Figure 11.1 shows a storage vs. time graph. For the case shown, the initial storage S_0 is not zero, as it would be when the berm is first formed. If the wetland is constructed during a wet period of the year, then storage would increase until the start of the dry season. At time t_1 in Figure 11.1, the storage capacity is reached, and water passes through both the principal and emergency spillways. At time t_2 the outflow rate exceeds the inflow rate and storage is depleted below the maximum allowable storage S_n. At time t_3, the storage is depleted and drainage ceases to discharge from the wetland. Between times t_3 and t_4, the inflow is zero or not great enough to offset infiltration and evapotranspiration losses. Thus, the outflow is zero. At time t_4, the inflow is sufficient to replace storage and provide some discharge into the downgradient stream.

In the design phase, the system is modeled with allowance for difference storage configurations and outlet facility characteristics. The effect of variation in these design parameters is assessed to select the combination that best meets the design objectives. For each combination, the number of emergency spillway overflows and the number of times that downstream flow requirements are not met would be considered, along with other objectives, in selecting the final design.

11.4 RESIDENCE TIME IN WETLANDS

Time is an important aspect of wetland design. A number of time parameters must be considered. Where the inflow contains significant amounts of soil and settleable solids, the settling time for the average particle is a standard time parameter. The duration of time over which a specified portion of the inflow occurs is often used as a standard of comparison, where the inflow is based on a design storm of some return period, or frequency. For example, the design policy may aim for storing 50% of the flow for a 24-hour, 2-yr storm. Note that the design involves a volume, a duration, and a frequency. Volume-duration-frequency information is a common hydrologic design requirement. Time is also a factor in the design of the dead storage. The design criteria may want to limit the duration over which the pond stays dry during periods of low inflow or drought.

Time is an important design parameter because it relates to other design parameters, such as the trap efficiency, the volume and peak reduction ratios, and the ease of maintenance. If a forebay is built into a wetland, the detention time of the forebay will be direct determinant of the its trap efficiency. The greater the detention time, the greater the proportion of solids that will settle in the forebay. This is also true of the vegetated portion of the wetland. If retention of water to reduce the peak discharge is a primary design criterion, then the relationship between retention time and peak reduction must be known.

Time is important in the attenuation of flood peak discharges. It is a primary factor in the trap efficiency of the wetland. Where maintenance is a primary consideration, the design will affect the time between scheduled maintenance. This is especially true in the design of a forebay intended to limit the introduction of settleable solids to the vegetated portion of the wetland. Where it is important to maintain wet-pond dead storage, the expected pond depletion time due to evapotranspiration and infiltration is a design factor.

The standard method of representing residence time T_d in a storage facility is the ratio of the total active storage volume V_s to the discharge rate:

$$T_d = V_s / Q \qquad (11.7)$$

where Q is the steady-state rate of flow through the wetland. This may be valid for very large reservoirs, but it is unlikely to provide accurate mean residence times for small retention facilities such as wetlands. Equation (11.7) can be derived from the continuity of mass when the outflow is insensitive to the inflow. The continuity of mass is:

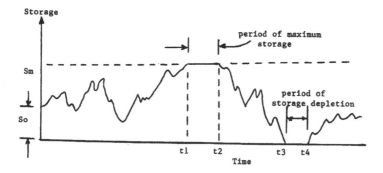

Figure 11.1. Storage versus time with initial storage, S_0, and maximum allowable storage, S_m.

$$I - O = \frac{dV_s}{dt} \tag{11.8}$$

If the inflow is ignored, then Eq. (11.8) can be solved for the time, which for the numerical approximation is:

$$\Delta t = \frac{\Delta V_s}{-Q} \tag{11.9}$$

In this case, the incremental volume ΔV_s is the active storage (ft^3) for the storm event and Q would be the average outflow (ft^3/s). The negative sign in Eq. (11.9) only indicates that the time increment Δt would decrease as the outflow rate increases, so the minus sign can be ignored.

A second measure of the average residence time in a wetland can be computed if inflow and outflow hydrographs are available. The time between the centers of mass of the two hydrographs would reflect the residence time in the wetland. An estimate could be made for an existing wetland if measured hydrographs are available. In the design of a wetland, the hydrographs would need to be computed with a model, which is generally a design storm hydrograph model.

Example 11.2. When using the method based on the centers of mass, it is necessary to have representative hydrographs. For wetland design, measured hydrographs would obviously not be available, so it would need to be based on model generated hydrographs. The inflow hydrograph reflects the runoff from the upstream contributing area while the outflow hydrograph is based on a hydrologic routing of the inflow hydrograph using the geometric and physical characteristics of the proposed wetland.

The inflow and outflow hydrographs for a small wetland are given in columns 2 and 6 of Table 11.2, respectively. A variable time increment is used for both hydrographs. The trapezoidal rule can be used to compute the volumes under the hydrographs, with a volume of 156,240 ft^3, or 3.587 ac-ft. Therefore, the center of mass of each hydrograph occurs when a volume of 78,120 ft^3 has passed either the inflow or outflow point. This volume is most easily identified by computing the cumulative hydrograph, which is given in columns 4 and 8 of Table 11.2. For the inflow hydrograph, the center of mass occurs at 12.86 hr. The center of mass of the outflow hydrograph occurs at 22.05 hr. Therefore, the retention time is 9.19 hours.

11.5 PERMANENT POND DEPTH ESTIMATION

The design depth of dead storage will be determined in part by water requirements for the vegetation and in part by the loss rates, which includes both evapotranspiration and infiltration losses. If the expected rates of infiltration and evapotranspiration are known for the design conditions, the time duration during which the dead storage would drain or evaporate can be computed. For a design depth of dead storage (D_d, ft), infiltration rate i (ft/day), and rate of evapotranspiration E_t (ft/day), the drain time Δt (days) during a dry period is:

Table 11.2 Estimation of wetland retention time for design-storm inflow hydrograph and routed outflow hydrograph.

(1) Storm time (hrs)	(2) Inflow rate (ft³/s)	(3) Inflow volume (ft³)	(4) Cumulative inflow (ft³)	(5) Storm time (hrs)	(6) Outflow rate (ft³/s)	(7) Outflow volume (ft³)	(8) Cumulative outflow (ft³)
12.0	0		0	16.0	0		0
		5400				900	
12.2	15		5400	16.5	1		900
		20160				2700	
12.4	41		25560	17.0	2		3600
		27720				5400	
12.6	36		53280	17.5	4		9000
		20880				8100	
12.8	22		74160	18.0	5		17100
		13320				9000	
13.0	15		87480	18.5	5		26100
		9360				9900	
13.2	11		96840	19.0	6		36000
		7200				9900	
13.4	9		104040	19.5	5		45900
		6120				9000	
13.6	8		110160	20.0	5		54900
		5400				8100	
13.8	7		115560	20.5	4		63000
		4680				7200	
14.0	6		120240	21.5	4		70200
		4320				7200	
14.2	6		124560	22.0	4		77400
		3960				7200	
14.4	5		128520	22.5	4		84600
		3600				6300	
14.6	5		132120	23.0	3		90900
		3240				5400	
14.8	4		135360	23.5	3		96300
		2880				5400	
15.0	4		138240	24.0	3		101700
		6300				5400	
15.5	3		144540	24.5	3		107100
		4500				5400	
16.0	2		149040	25.0	3		112500
		5400				10800	
17.0	1		154440	26.0	3		123300
		1800				9000	
18.0	0		156240	27.0	2		132300
						7200	
				28.0	2		139500
						5400	
				29.0	1		144900
						3600	
				30.0	1		148500
						3600	
				31.0	1		152100
						2700	
				32.0	<1		154800
						1440	
				33.0			156240

$$\Delta t = \frac{D_d}{E_t + i} \qquad (11.10)$$

If an estimate of the expected length of a dry period (L_d, days) can be made, then the period during which the wetland will not have a pond can be made as $L_d - \Delta t$. The effect of such a dry period on the health of the vegetation can be assessed.

The drain time of the wetland can be designed to minimize the possibility of a damaging dry period by controlling the infiltration rate. Including a soil with a limited infiltration rate as the subbase of the wetland can increase the design drain time of Eq. (11.10).

Example 11.3.   A 20-ac wetland is planned for an area subject to periods of near-zero rainfall during the summer months. The vegetation to be used in the wetland should have a minimum of 3 inches of water. The soil that exists at the site has an ultimate infiltration capacity of 0.015 in./hr or 10.8 in./month. Evapotranspiration for the vegetation can be estimated using the Blaney-Criddle method:

$$E_t = kpT / 788 \qquad (11.11)$$

in which E_t is the mean monthly evapotranspiration (inches/month) for a month with a mean temperature $T(°F)$ and a percentage of daily daylight hours p; and k is the consumptive use coefficient. For the vegetation planned for the site, the consumptive use factor is 0.9. What depth of dead storage is required to maintain the minimum pond level in the absence of rainfall during a two-month period?

Assume the Weather Service data indicates the following values for p and T during the design period:

Month	p	T	k	E_t
1	57	81	0.9	5.27
2	50	77	0.9	4.40

Therefore, the ET for the vegetated wetland during the design period is 9.67 in. Thus, the infiltration and ET losses during the 2-month period is 31.27 in., or 2.61 ft, with 69.1% due to the infiltration losses. Therefore, a wet-pond depth of 34.3 inches is required in order to maintain a 3-inch wet pond during the 2-month dry period. Such a depth may not be compatible with the vegetation on the topography. If the wetland was constructed with a clay liner below the wet pond area, then a wet pond depth of 13 inches would suffice.

11.6   ACTIVE STORAGE DESIGN

The active storage of a constructed wetland may be designed for multi-purpose use. Flood control is often an important objective. The volume of active storage depends on rainfall characteristics, watershed characteristics, the allowable outflow, and site parameters. The design volume would depend on the return period and duration of the design rainfall. High return period designs will require large volumes. Short duration design storms will require relatively low storage volumes. The design volume of storage will depend on factors that govern both the peak discharge and the volume of the runoff inflow. This includes the runoff curve number of the contributing area, the drainage area, and the time of concentration of the inflow. If the control of settleable solids is an objective, then the trap efficiency will be another design criterion.

The design of the active storage volume and layout varies with the configuration of the system and the location of the system. If the wetland includes a forebay, the storage volume within the forebay area is part of the total active storage volume. In-stream and off-stream wetlands require different design methods. The latter type of wetland must have a constructed outlet control facility to maintain the design volume. It is generally more accurate to size the volume and outlet control structure simultaneously.

Active storage is generally estimated with a model. Simplified methods only require peak discharge estimates for the inflow and design outflow. More complex methods use a design-storm approach that involves hydrograph generation with a unit hydrograph model and a routing component that hydrologically routes the inflow hydrograph through the wetland. The method presented in Section 8.3 is an example of a simplified method.

11.7 FOREBAY DESIGN

Excessive volumes of sediment, especially when pollutants are attached to the soil particles, can be detrimental to the health of a wetland. Sediment accumulation within a wetland reduces both the volume of storage available for water and the rate that water can infiltrate to the ground-water table. It can also reduce water clarity, thus reducing the aesthetic quality of the wetland, and possibly damage the vegetation.

To reduce the volume of soil material that flows into the primary portion of a wetland, forebays can be included in the design of a wetland. A forebay is a portion of the wetland located near the place where most of the surface inflow enters the wetland. It is intentionally located and designed to maximize the sediment trap efficiency. Considerable sedimentation can occur in a properly designed forebay. Forebays should be situated where equipment can have easy access for periodic dredging. It would be necessary to maintain a certain design volume that is free of accumulated sediment so that the volume will be sufficient to maintain the design trap efficiency.

Stokes Law can be used as the basis for forebay design. The Law is commonly used in the design of sedimentation tanks that are part of wastewater treatment systems. Stokes Law defines the settling velocity (V_s, ft/s) as a function of the mean particle diameter (d, ft), the specific gravities of the soil (S_s) and water (S), and the kinematic viscosity of the water (v, ft²/s):

$$V_s = gd^2(S_s - S)/(18v) \qquad (11.12)$$

Stokes Law is derived by balancing the drag force on a soil particle as it falls through a liquid medium, the weight of the particle, and the buoyant force on the particle. Since the specific gravities and the viscosity are generally considered to be constants, Stokes Law relates the settling velocity directly to the particle diameter.

The trap efficiency (E_f) of a forebay is defined as the proportion of sediment in the inflow that is stored within the forebay:

$$E_f = \frac{m_{st}}{m_i} = 1 - \frac{m_o}{m_i} \tag{11.13}$$

in which m_{st} is the mass of sediment retained in the forebay, and m_i and m_o are the sediment masses in the inflow and outflow, respectively. Grab samples of the inflow and outflow are often used to estimate m_o and m_i. The trap efficiency of a forebay is a function of the volume of storage, the gradation characteristics of the sediment, the geometric characteristics of the forebay, the inflow rate of the water, and the hydraulic characteristics of the forebay outlet.

Inputs for the design of a forebay include: (1) the inflow discharge rate Q; (2) the allowable depth H of the forebay, which is set by the topography of the site and with some allowance for excavation; (3) the settling velocity V_s of the sediment particles of diameter d, which is computed with Stokes Law; (4) the forebay width W, which is set either by the site topography or by the hydraulic theory that limits the rate at which the inflow spreads out; and (5) the trap efficiency E_f, which is set by either policy or the aquatic constraints placed on the acceptable loading rates of sediment in the wetland. These design variables are related by three constraints. First, the continuity equation relates the average horizontal velocity of the water V_h, as well as the suspended sediment, to the discharge and cross-sectional area:

$$Q = \text{Area} * \text{Velocity} = WHV_h \tag{11.14}$$

Second, during any time interval Δt, a particle of sediment dropping at velocity V_s will drop a distance $d_v = V_s\Delta t$ and move horizontally a distance $d_h = V_h\Delta t$. Therefore,

$$\Delta t = \frac{d_h}{V_h} = \frac{d_v}{V_s} \tag{11.15}$$

Third, if a particle moves horizontally at velocity V_h and vertically at velocity V_s, then the particle will drop distance D_v by the end of the forebay of length L. If d_h is the length of the forebay, $d_h = L$, and d_v is the greatest height from which a sediment particle will fall in time Δt, $d_v = D_v$, then $D_v = L(V_s/V_h)$. If the soil particles are uniformly distributed vertically within the inflow, then a proportion D_v/H will be trapped:

$$E_f = D_v / H \tag{11.16}$$

Solving Eq. (11.15) for V_h and substituting it into Eq. (11.14) yields:

$$Q = WHd_h V_s / d_v \qquad (11.17)$$

Solving Eq. (11.16) for D_v and assuming that d_v equals D_v when d_h equals the length of the forebay L, substitution into Eq. (11.17) yields:

$$L = QE_f / WV_s \qquad (11.18)$$

Equation (11.18) indicates that the length of the forebay is directly related to the design inflow Q and the necessary trap efficiency E_f and indirectly related to the forebay width W and the settling velocity V_s.

The efficiency of Eq. (11.16) assumes that the inflow to the wetland occurs uniformly over the entire depth of the wetland. It is quite likely that the inflow will enter at the top of the wetland through a pipe on a shallow surface stream. In such cases, Eq. (11.16) does not apply.

The inflow to a wetland forebay enters through a pipe of diameter D_i or stream of depth D_i. Assuming that the sediment is uniformly distributed across the diameter of the inflow pipe D_i and that mixing does not occur as the flow and sediment move through the forebay, then the proportion of particles that fall to the bottom prior to the end of the forebay can be represented by D_d/D_i. The trap efficiency (E_f) is given by:

$$E_f = \begin{cases} 1 & \text{if } L > H V_h / V_s \\ 0 & \text{if } L < (H\text{-}D_i)(V_h / V_s) \\ 1 - (H/D_i) + (L/D_i)(V_s/V_h) & \text{otherwise} \end{cases} \qquad (11.19)$$

Example 11.4. A 20-ac residential area that overlays a sandy soil produces a discharge of 40 ft³/sec. The runoff enters a 9-ft wide area where a forebay to a constructed wetland will be located. The forebay is needed to trap 90% of the medium sand that is suspended in the inflow. Stokes Law is used to compute the settling velocity of the sandy soil (mean diameter = 0.00115 ft):

$$V_s = \frac{32.2 \text{ ft/sec}^2 \ (0.00115 \text{ ft})^2 (2.65 - 1)}{18(1.2 \times 10^{-5} \text{ ft}^2/\text{sec})} = 0.325 \text{ ft/sec}$$

The forebay length is then computed using Eq. (11.18):

$$L = \frac{40 \text{ ft}^3/s \ (0.9)}{9 \text{ ft} \ (0.325 \text{ ft/sec})} = 12.3 \text{ ft} \qquad (11.20)$$

11.8 IN-STREAM WETLANDS

Some rivers contain natural lowland areas where relatively heavy vegetation has developed. Where topography permits, wetlands can be constructed in the river system. Such in-stream or riverine wetlands can provide numerous benefits, from flood control to habitat enhancement, from water quality control to aesthetical enhancement.   Modifications to natural wetland areas can increase their effectiveness.

The primary design variables for an in-stream wetland are the cross-sectional area characteristics and the velocity of flow. The flow rate into the wetland would depend on the hydrology of the upstream contributing area. The density and depth of the wetland vegetation will influence the effectiveness of the wetland to reduce flowrates and to remove suspended solids from the flow. The greater the retardance of the flow by the vegetation, the larger the storage volume needed to contain the inflow to some design criterion.

The design of an in-stream wetland area uses the continuity equation and Manning's equation. The flow rate q_i into the wetland can be determined for the contributing watershed using either a peak discharge method such as the SCS Graphical method or a unit hydrograph design-storm model such as HEC-1 or the SCS TR-20 computer program. Once the velocity (V, ft/s) through the wetland is determined with the method described below, the cross sectional area (A_c, ft^2) of the wetland is derived from the continuity equation:

$$A_c = q_i / V \qquad (11.21)$$

The velocity of flow in the wetland area is computed with Manning's equation:

$$V = \frac{1.49}{n} R_n^{2/3} S^{1/2} \qquad (11.22)$$

The hydraulic radius is the ratio of the cross-sectional area A_c to the wetland perimeter P. Since the velocity depends on the cross-sectional area by Eq. (11.21) and the cross-sectional area depends on the velocity by Eq. (11.22) the design process is iterative. Manning's equation depends on the hydraulic gradient (S, ft/ft), which is usually approximated by the slope of the channel bed.

Assuming that the wetland area is heavily vegetated, tables for Manning's roughness coefficient (n) found typically in text books are not applicable. The values of n in Table 11.3 can be used for flow depths greater than 1 ft. For flow depths less than 1 foot, the value of n will depend on the density and stem length of the vegetation, the velocity of flow, and the hydraulic radius (SCS, 1992). Since the velocity depends on n, the procedure is iterative.   SCS (1992) provides the following equation for predicting the retardance index, I_c:

$$I_c = 2.5h^{1/3} M^{1/6} \qquad (11.23)$$

Table 11.3. Manning's n values for in-stream wetlands for depths greater than 1 ft.

n	Stream Conditions
	Fairly regular section, full stage
0.030 – 0.035	No pools, no vegetation
0.035 – 0.040	Some grass and weeds, no brush
0.040 – 0.045	Some weeds and stones, light brush
0.050 – 0.070	Some weeks; heavy brush on banks
0.045 – 0.055	Some pools and shoals, clean
0.040 – 0.050	Gravel or cobble bottom, novegetation
0.060 – 0.085	Sluggish sections, weedy, deep pools
0.070 – 0.100	Dense weeds and reeds
	Irregular sections, slight channel meander increase above by 0.01 to 0.02
	Sections with lower stage and very flat slopes, increase above by 0.005 to 0.010

Table 11.4. Estimation of Manning's roughness coefficient (n) for in-stream wetlands for flow depths less than 1 ft as a function of the velocity V, the hydraulic radius R_h, and the retardance index I_c

$\hat{n} = c_1 e^{c_2 x} X^{-c_3}$ where $X = V * R_h$

Applicable range of I_c	c_1	c_2	c_3	[a] lower limit on X	[b] Upper limit on X
$8.5 \le I_c$	0.27	0.046	0.81	0.6	20
$6.5 \le I_c < 8.5$	0.15	0.042	0.70	0.4	20
$5.0 \le I_c < 6.5$	0.085	0.039	0.65	0.2	10
$3.5 \le I_c < 5.0$	0.053	0.074	0.55	0.1	10
$I_c < 3.5$	0.034	0.053	0.38	0.1	10

[a] If X is less than the lower limit, use the lower limit of X
[b] If X is greater than the upper limit, use the upper limit of X

in which h is a representative stem length (ft) and M is the average stem density (stems per ft^2). Manning's roughness coefficient can then be determined using I_c, V, and R_h as input to one of the equations of Table 11.4.

The procedure for designing the cross section of an in-stream wetland requires initial estimates of the following: values of h and M to compute I_c with Eq. 11.23; the channel slope in the wetland; the inflow discharge rate q_i; an approximate width of the wetland (W); and the velocity of flow in the wetland. Given these input values, the following procedure is used:

(1) Compute the cross-sectional area: $A_c = q_i / V$

(2) For the computed A_c and width, estimate the depth of flow and hydraulic radius.

(3) If the depth is greater than 1 ft, use Table 11.3 to obtain an estimate of the roughness coefficient; otherwise, use the product of the hydraulic radius and the velocity with Table 11.4 to estimate n

(4) Compute the velocity with Manning's equation (Eq. 11.22)

(5) If the velocity of step 4 is different than the assumed velocity of step (1), repeat steps (1) to (5).

The length of the wetland portion of the stream will need to be sufficiently long to allow the flow from the upstream drainage area to become distributed across the cross section. If longitudinal sections of the wetland area have different cross-sectional characteristics, then separate calculations will need to be made for each section.

Example 11.5. Assume that a 2000-ft section of a stream has a slope of 0.2% and a natural width of about 150 ft. The discharge rate from the upstream portion of the watershed is 50 cfs for the design flood. The section immediately upstream has a width of 20 ft and a depth of 1 ft, which yields a velocity of 2.5 ft/s. The vegetation planned for the wetland will have a density of 800 stems per square foot and stem lengths of 3 ft.

Equation (11.23) can be used to compute the retardance index:

$$I_c = 2.5(3\,ft)^{1/3}(800)^{1/6} = 11.0 \qquad (11.24)$$

For an assumed velocity of 0.4 fps within the wetland, the continuity equation yields a cross-sectional area of 125 ft². For a width of 150 ft, the depth is 0.833 ft. For a depth less than 1 ft, Table 11.4 is used to obtain a roughness coefficient. The product of the velocity and the hydraulic radius of 0.824 ft is 0.33. This yields a value of 0.10 for n. Manning's equation is used to compute the velocity:

$$V = \frac{1.49}{0.10}(0.824)^{2/3}(0.002)^{1/2} = 0.586\,ft/s \qquad (11.25)$$

As this differs from the assumed value of 0.4 fps, repeating the above steps yields a cross-sectional area of 85.4 ft², a depth of 0.57 ft, a hydraulic radius of 0.565 ft and a revised velocity of

$$V = \frac{1.49}{0.1}(0.565)^{2/3}(0.002)^{1/2} = 0.455\,ft/s \qquad (11.26)$$

The iteration should be continued. After several iterations, the velocity is 0.503 ft/sec. Thus, the velocity in the wetland portion of the river is 20% of that in the upstream section and the cross-sectional area of the wetland is 5 times as large as that of the upstream section. With the lower horizontal velocity and the buffering of the vegetal growth in the wetland, suspended material is more likely to be trapped in the wetland.

Because of the iterative solution procedure, it is quite possible that the solution will not converge if a poor initial estimate of the velocity is selected. For example, in this example, the solution would not converge if the initial estimate of the velocity was less than 0.33 ft/s or greater than 0.96 ft/s. If a solution is not converging after several iterations, then another initial estimate should be selected.

11.9 OFF-STREAM WETLANDS

Off-stream wetlands are often areas of natural depressions. Additionally, the topography around a natural depression can be changed to enhance the storage-discharge characteristics of the site. With off-stream wetlands, a considerable portion of the total available storage is often dead storage. Therefore, the infiltration characteristics of the underlying soil, the location of the water table, and the evapotranspiration potential of the site become primary determinants of the hydrologic effectiveness of the site as a wetland. If the average dead-storage depth of the site is small, then the wetland will function primarily as a first-flush storage area. Once the storage volume of the depressions is filled, the inflow will essentially pass directly out of the wetland area, with only a very slight attenuation of the discharge rates. Form the standpoint of the runoff hydrograph, the water stored in the dead-storage portion will act much like an initial abstraction. On the recession side of the surface runoff hydrograph, the water will discharge at a slow rate because of the shallow depths over a large surface area.

The first-flush volume, or dead-storage volume, can be computed from a detailed topographic map of the depressed area. The stage-storage relationship for storage above the invert of the outflow basin is also necessary to evaluate the storage-discharge characteristics of the site. It is generally assumed that for large depressional wetlands the outflow is insensitive to the inflow because of the relatively large surface area and shallow depths. The outflow characteristics of the wetland can be changed by modifying the invert elevation and length of the controlling berm. A two- or multi-stage berm can be designed to provide greater control of the outflow.

If the volume of storage is relatively large and the outflow is insensitive to the inflow, then the retention time (t_r, hours) can be approximated by:

$$t_r = V_a / q \qquad (11.27)$$

in which V_a is the active storage volume (ft^3) and q is the discharge rate (ft^3/hr). As water flows into and out of the wetland, the discharge rate and the active storage volume will both change. Therefore, the value given by Eq. (11.17) yields an approximate value. The difference in the centers of mass of inflow and outflow can also be used to estimate the average retention time.

Off-stream wetlands are generally considered as an option where the natural topography encourages their use. In low-sloped areas with natural depressions, the primary design variable would be the length of the outflow weir, which is often a berm that is vegetated. The length of the berm is set either to limit the outflow rate to control downstream inundation or to produce a retention time that will allow sufficient settling of solids suspended in the inflow.

To design the length of the berm, either a simple routing procedure based on the continuity of mass or the storage-indication method can be used (McCuen 2005). The inflow hydrograph would usually be developed by a design storm, unit hydrograph method. Alternatively, a triangular hydrograph that uses the Rational method to define the peak and the time of concentration as the time to peak could be used. The weir equation could be used to compute the outflow discharge and the stage-storage relationship could be used to compute the stage needed for the weir equation. The weir length could be modified until the design objective is met.

Example 11.6. A 40-ac watershed that is partially farmland, with the remainder being open space and forest cover, drains to a depressional area before discharging into a stream. The depression will store about 0.2 ac-ft of dead storage. Above the dead storage area, the storage-discharge relationship is:

$$V_s = 175,000 + 22,500h^2 \qquad (11.28)$$

in which V_s is the volume (ft^3) and h is the depth (ft). The outflow area could easily be modified to include a berm that would act as a broad-crested weir with a length from 10 to 30 feet. The vegetated berm would have a discharge coefficient of 2.6.

The site will be evaluated for the minimum and maximum weir lengths of 10 and 30 ft. The variables of interest for each design would be the maximum outflow rate, the maximum volume of active storage, the maximum depth of storage, and the retention time. The Rational formula yields a peak discharge of 30 cfs for the 40-ac drainage area, and the computed time of concentration is 0.5 hr. Therefore, the design inflow hydrograph is an isosceles triangle with a time base of 1 hr and a peak of 30 cfs. This yields an inflow volume of 54,000 ft, or 1.25 ac-ft. The dead storage will act as an initial abstraction and remove 0.2 ac-ft, or 16% of the inflow volume. Water will begin to accumulate as active storage after 17 minutes, which is the time required for the dead storage to fill up.

The following procedure was used to route the runoff through the active storage area of the depressional wetland. The volume of inflow during a time increment of 0.1 hour was added to the active storage volume, which was then used to compute the depth of active storage with the stage-storage relationship of Eq. (11.28). The depth of active storage was used with the weir equation to compute the outflow rate. This was used to compute the volume of outflow for the time period, which was then subtracted from the active storage volume.

The peak outflow rate would vary with the weir length. It is 2.88 cfs for the 10-ft weir and 7.19 cfs for the 30-ft weir. The maximum depth of storage was 0.23 ft for the 10-ft weir and 0.204 ft for the 30-ft weir. The maximum active storage volumes were 0.929 ac-ft and 0.782 ac-ft for the 10-ft and 30-ft weir lengths, respectively. The longer weir length allows the water to discharge more quickly, so the storage volume is less than the storage for the shorter weir length. Using Eq. (11.27), the retention times were computed to be 3.91 hr and 1.32 hr for the 10-ft and 30-ft weir lengths, respectively. For a volume of inflow of 54,000 ft^3, the time of the center of mass would occur where the volume was 27,000 ft^3. Using the difference in the times to the centers of mass the retention times were 4.35 hr and 1.45 hr for the 10-ft and 30-ft weir lengths. Using the centers of mass provides longer retention times because of the accounting for the dead storage abstraction.

The design length could then be selected based on the criterion that is most important to the use of the depressional area as a wetland. The length of the berm could then be used to compute the effects of the wetland on the inflow, both the discharge and the suspended solids.

11.10   OUTLET CONTROL STRUCTURES

Off-stream wetlands provide storage space for collecting surface runoff and interflow from areas that have high water tables. Numerous design methods for sizing outlet control structures have been proposed. Most are variations based on either the weir equation or the orifice equation. Some methods allow for the control of the water level in the wetland, while other outlet configurations are uncontrolled.

11.10.1   Uncontrolled Outlets

The outlet for an off-stream facility can be either a weir or an orifice. For example, a culvert through a berm can be used to limit the discharge. The two simplest cases will be illustrated here, the case for an unsubmerged inlet and outlet and the case where the inlet is submerged and the outlet is unsubmerged. Solutions for others cases are available in texts on hydraulic routing.

If both inlet and outlet are unsubmerged, then Manning's equation can be used to compute the discharge rate from the wetland. If a circular pipe is used, then Manning's equation can be transformed to solve for the diameter of the pipe:

$$D = 1.333(qn)^{3/8} S^{-3/16} \qquad (11.29)$$

in which q is the design discharge (ft^3/s), n is Manning's roughness coefficient, S is the slope of the pipe (ft/ft), and D is the pipe diameter (ft).

Once the inlet becomes submerged, the pipe acts as an orifice. Submergence of the pipe outlet actually requires a depth that is about 20% greater than the pipe diameter. The orifice equation assumes that the height h is measured from the center of the pipe. It is usually desirable to measure the height of the water surface above the invert h_i rather than the center of the culvert. The relationship between the height above the center (h), the pipe diameter (D), and the height above the invert is:

$$h + \frac{D}{2} = h_i \qquad (11.30)$$

Substituting Eq. (11.30) into the orifice equation yields:

$$q = C_d \left(\frac{\pi D^2}{4} \right) [2g(h_i - 0.5D)]^{0.5} \qquad (11.31)$$

Since the discharge is known, Eq. (11.31) can be solved iteratively for D. Equation (11.31) can also be rearranged to an equation in terms of D:

$$D^5 - 2h_i D^4 + \frac{16q^2}{C_d^2 \pi^2 g} = 0 \tag{11.32}$$

Equation (11.32) would need to be solved iteratively for the diameter D.

Example 11.7. A constructed wetland with a 15-inch dead storage depth needs an overflow outlet to pass storm runoff into the down-gradient receiving stream. The design discharge is 0.7 ft^3/s. The proposed design will pass the flow through a circular culvert without submergence at either the inlet or outlet. The corrugated metal pipe will be laid at a slope of 0.008 ft/ft and have a roughness coefficient of 0.024.

Using Eq. (11.29) the pipe diameter is:

$$D = 1.333[0.7(0.024)]^{3/8}(0.008)^{-3/16} = 0.712 \text{ ft} \tag{11.33}$$

which is an 8.5 inch pipe. If a 6-in commercially available pipe is used, then the design discharge will not pass but the maximum storage will be larger. The estimated diameter can be rounded up to a 9-in. pipe, which will pass a discharge slightly greater than the design discharge.

Example 11.8. A wetland that will have a design discharge rate of 18 ft^3/s is proposed. Due to its proximity to a roadway, the maximum depth is 4.5 ft. The pipe is sized assuming a discharge coefficient of 0.6.

Using Eq. (11.32) the following function is set by the design conditions:

$$f(D) = D^5 - 2(4.5)D^4 + \frac{16(18)^2}{(0.6)^2 \pi^2 (32.2)} = D^5 - 9D^4 + 45.31 = 0 \tag{11.34}$$

The polynomial can be solved iteratively, as follows:

D	3	2	1.5	1.57	1.571
f(D)	-441	-66	7.34	0.17	0.06

Therefore, either an 18-inch or 21-inch pipe could be selected. If the 18–inch pipe is used, the design flow will not be met and the water surface elevation will exceed the design depth of 4.5 ft. Stage-storage information is needed to determine the exact increase in height above the 4.5-ft design value. If the 21-inch pipe is selected, then the flow can exceed the design discharge of 18 cfs, which may cause problems down gradient from the wetland. A rating curve of the downgradient stream is necessary to determine the extent of this problem. In any case, the selected design would need additional information to determine the effect of the pipe size.

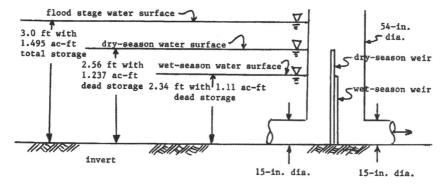

Figure 11.2. Schematic of a two-season outlet-control structure.

11.10.2   Controlled Outlets

Where it would be beneficial to be able to vary the water surface elevation of the wetland seasonally, an uncontrolled outlet could not be used. Figure 11.2 shows an outlet facility that enables the flood storage water surface elevation to be the same for all seasons, but simultaneously allow for variation in the seasonal dead storage water surface elevation. The dead storage water surface elevation can be controlled by inserting or removing boards in the vertical tube. Such controlled outlets would enable the dead storage to be increased at the start of the dry season and then reduce the dead storage at the start of a wet season. In terms of active storage, the controlled outlet could be used to minimize the land area devoted to the wetland, while meeting all of the objectives associated with the wetland. The controlled outlet allows the same design water surface elevation to be used for both the wet and dry seasons.

Example 11.9. A seasonal-dependent design storm model predicts discharges for a 24-acre developed watershed of 7.6 cfs for the dry season and 14.0 cfs for the wet season. The runoff depths from the contributing watershed are 0.4 in. and 0.7 in., respectively. The constructed wetland is intended to control downstream flow rates to 3.0 cfs and 7.0 cfs for the dry and wet seasons, respectively. A design objective is to have the active storage water surface elevation at 3 feet above the outlet pipe invert for both seasons. The stage-storage relationship is:

$$V_s = 0.4h^{1.2} \tag{11.35}$$

in which V_s is the storage (ac-ft) and h is the pool depth (ft) above the invert. What are: (1) the water surface elevations for the dead storage during both seasons; (2) the lengths of the outlet weirs for both seasons; (3) the riser pipe diameter; and (4) the outlet pipe diameter, D.

The TR-55 storage volume method can be used to estimate the active storage volumes. The ratios of the peak discharges for the dry and wet seasons, respectively, are:

$$R_{qd} = \frac{3.0\ \text{cfs}}{7.6\ \text{cfs}} = 0.395 \tag{11.36}$$

$$R_{qw} = \frac{7.0\ \text{cfs}}{14.0\ \text{cfs}} = 0.500 \tag{11.37}$$

Using Figure 8.4, the storage ratios for the dry and wet seasons are 0.323 and 0.275, respectively. Thus, the flood storage volumes are:

$$V_{sd} = 0.323(0.4\ \text{in.})(24\ \text{ac})/12\ \text{in./ft} = 0.2584\ \text{ac-ft} \tag{11.38}$$

$$V_{sw} = 0.275(0.7\ \text{in.})(24\ \text{ac})/12\ \text{in./ft} = 0.3850\ \text{ac-ft} \tag{11.39}$$

The total storage at the flood-stage water surface elevation of 3 ft is:

$$V_{max} = 0.4(3\ \text{ft})^{1.2} = 1.495\ \text{ac-ft} \tag{11.40}$$

Therefore, the dry- and wet-season dead storages are:

$$V_{dd} = 1.495 - 0.258 = 1.237\ \text{ac-ft} \tag{11.41}$$

$$V_{dw} = 1.495 - 0.385 = 1.110\ \text{ac-ft} \tag{11.42}$$

These correspond to dead-storage water surface elevations of:

$$E_d = (V_{dd}/0.4)^{5/6} = 2.562\ \text{ft} \tag{11.43}$$

$$E_w = (V_{dw}/0.4)^{5/6} = 2.341\ \text{ft} \tag{11.44}$$

Thus, the active-storage heads above the dry-season and wet-season weir inverts are:

$$h_{id} = 3.0 - 2.562 = 0.438\ \text{ft} \tag{11.45}$$

$$h_{iw} = 3.0 - 2.341 = 0.659\ \text{ft} \tag{11.46}$$

At a maximum flood stage of 3 ft, the weir lengths needed to pass the design downstream flow rates are computed using the weir equation:

$$L_d = \frac{q}{C_w h_{id}^{1.5}} = \frac{3.0}{3(0.438)^{1.5}} = 3.45\ \text{ft} = 41.4\ \text{in.} \tag{11.47}$$

$$L_w = \frac{q}{C_w h_{iw}^{1.5}} = \frac{7.0}{3(0.659)^{1.5}} = 4.36\,\text{ft} = 52.3\,\text{in}. \qquad (11.48)$$

Since the barrel of the riser will need to be slightly larger than the larger weir length, a 54-in. diameter riser barrel should be used. The diameter of the outflow pipe is computed using the larger discharge and Eq. (11.32):

$$D^5 - 6D^4 - \frac{16(7)^2}{(0.6)^2 \pi^2 g} = 0 = D^5 - 6D^4 + 6.853$$

Solving for D yields a diameter of 1.09 ft. Therefore, a 15-in. piper should be used.

Figure 11.3 shows the wetland and riser design parameters. During the dry season, the 3.45 ft weir controls the outflow and maintains a dead storage of 1.237 ac-ft, which can be decreased by evapotranspiration and infiltration. For the wet season, the dry-season weir is removed, and the 4.36-ft weir controls the outflow. Less dead storage is needed during the wet season. For both seasons, the flood stage is 3 ft above the invert of the outlet pipe.

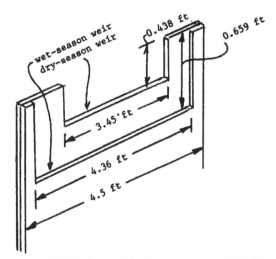

Figure 11.3. Weir characteristics for a two-season outlet facility.

11.11   WATER QUALITY IN WETLANDS

Fully functional wetlands have many provisions to improve the quality of incoming runoff. As discussed in previous sections, sedimentation due to storage can play an important role in the removal of suspended particulates and the pollutants affiliated with them. Residence time in the wetland can also encourage several chemical processes. Most important, though, are the myriad of biological reactions that can be promoted in an active wetland ecosystem.

A large body of work has investigated wetlands as treatment zones for municipal wastewater, industrial wastewater, and agricultural runoff (e.g., Kadlec and Knight 1996). Information on urban runoff is much more scarce, but the same concepts that are important for these other applications can be applied here.

The presence of flora and fauna create a complex, but effective environment for pollutant management. Wetlands can be considered as complex ecosystems with interdependent communities. Microorganisms such as bacteria, fungi, and algae will thrive under appropriate environmental conditions. These organisms can metabolize carbon and nitrogen pollutants. Similarly, the vegetation can play a direct role in pollutant management by taking up and mineralizing various compounds, but also indirectly, by supporting pollutant degrading microbial populations. Both invertebrate and vertebrate animals will exist in the wetland, but their impact on water quality is expected to be minimal.

Biological degradation processes are important for addressing organic and nitrogen pollutants. Bacteria and other microorganisms can degrade organic pollutants via the aerobic decomposition reaction:

$$\text{Organic matter} + O_2 \quad \leftrightarrow \quad CO_2 + H_2O + \text{cells} \tag{11.49}$$

This reaction was discussed in Chapter 2. The rate at which this reaction takes place depends on the type of organic matter, the concentration of oxygen available, the number of degrading organisms, and other environmental factors such as pH and temperature.

11.11.1 Modeling Water Quality Improvement

The simplest way to model the degradation process is through a zero-order or a first-order reaction. In a zero-order process, the rate of reaction is constant and independent of the concentration of reactant present.

$$\text{Rate} = \frac{dC}{dt} = \text{constant} = k_0 \tag{11.50}$$

The value of the rate will depend on a number of environmental conditions, as described above. The units of k_0 are mass per volume per time, i.e., mg/L-min. According to Eq. (11.50), the conversion will depend on the residence time in the wetland. Residence time calculations have been discussed in Section 11.4. Using Eq. (11.50) with a wetland residence time allows the calculation of pollutant concentration leaving a wetland:

$$C = C_0 - k_0 \, \Delta t \tag{11.51}$$

where C_0 is the concentration of the pollutant input to the wetland.

Under steady-state conditions, that is where the input and output flows are identical and constant, and where the input concentration is constant, the output concentration is constant. When one or all of these parameters change with time, the concentration leaving the wetland is not constant and the calculation of must be done incrementally.

Example 11.10. The removal of phosphorus through a wetland is described using a zero-order reaction, with a rate constant equal to 0.004 mg/L-min. The flow into and out of the wetland is 4000 L/min and the effluent phosphorus concentration is desired. The wetland has an average depth of 1 m and a surface area of 8000 m². The input phosphorus concentration is 10 mg/L. First, the wetland volume is found as the product of the area and depth, giving 8000 m³. The average time of the water in the wetland is the volume divided by the average flow rate:

$$\Delta t = \frac{V}{Q} = \frac{8000 \ m^3}{4m^3 / \min} = 2000 \ \min \qquad (11.52)$$

Using this answer with Eq. (11.51) provides the output phosphorus concentration:

$$C = C_0 - k_0 \Delta t = 10\frac{mg}{L} - \left(0.004 \frac{mg}{L - \min}\right)(2000 \ \min) = 2\frac{mg}{L} \qquad (11.53)$$

A first-order reaction is another common reaction order. In this case, the reaction rate is proportional to the concentration of reactant present:

$$\text{Rate} = \frac{dC}{dt} = k_1 C \qquad (11.54)$$

The rate constant, k_1, has units of time^{-1}. This equation can be integrated to:

$$C = C_0 \exp(-k_1 \Delta t) \qquad (11.55)$$

where, again, the Δt is the time increment for the pollutant in the wetland "reactor."

Example 11.11. The removal of nitrogen in a wetland is described by a first-order process with a rate constant, $k_1 = 0.001$ min^{-1}. The input total N concentration is 4.8 mg/L and the desired total N output concentration is 1.5 mg/L at an average flow rate of 600 L/min. The average wetland depth is 0.8 m. In order to determine the area of wetland required, Eq. (11.55) is rearranged and used to solve for Δt.

$$\Delta t = \frac{-\ln C/C_0}{k_1} = \frac{-\ln\left(\frac{1.5 \ mg/L}{4.8 \ mg/L}\right)}{0.001 \min^{-1}} = 1160 \ \min \qquad (11.56)$$

The volume is related to the flow rate and time by:

$$V = Q \Delta t = \left(0.6 \ m^3 / \min\right)(1160 \ \min) = 692 \ m^3 \qquad (11.57)$$

Therefore, the wetland area is given by:

$$A = \frac{V}{H} = \frac{692\,m^3}{0.8\,m} = 870\,m^2 \tag{11.58}$$

For more dynamic conditions, Eqs. (11.51) or (11.55) can be applied in incremental steps.

Example 11.12. As in Example 11.11, the removal of nitrogen through a wetland is described using a first-order reaction, with a rate constant equal to 0.001 min^{-1}. Again, the wetland has an average volume of 692 m^3. The wetland flow and input nitrogen concentration is given as follows.

Time increment	Input and output flow rate (L/min)	Input N conc. (mg/L)	Δt (min)	Output N conc. (mg/L)
1	800	4	865	1.68
2	1100	8	629	4.27
3	600	3	1153	0.95
4	200	1	3460	0.03

The average residence time of the water in the wetland for each increment is found as the volume divided by the average flow rate:

$$\Delta t_1 = \frac{V}{Q_1} = \frac{692\,m^3}{0.8\,m^3/min} = 865\,min \tag{11.59}$$

Using this answer with Eq. (11.55) provides the output phosphorus concentration for each increment. For example, for the first increment:

$$C = C_0 \exp(-k_0 \Delta t) = 4\tfrac{mg}{L}\exp\left(-0.001\,min^{-1} \bullet 865\,min\right) = 1.68\frac{mg}{L} \tag{11.60}$$

These calculations are repeated for the remaining time increments to provide the dynamic effluent nitrogen concentrations.

11.11.2 Wetland Design for Water Quality

Some recent work has presented design information for water quality improvement for constructed wetlands for treating urban and highway runoff (Shutes *et al.* 1999). Foremost, as with many natural treatment systems, pretreatment is very important. A well-maintained sediment trap preceding the wetland is critical for the removal of many input suspended solids so as to prevent the accumulation of solids in the wetland itself. Optimum performance has been noted with a wetland design area that is 2-3% of the drainage area. A length-to-width ratio of 1:1 to 1:2 has been suggested. The wetland surface should include 0.44 m of pea gravel under 0.15 m of soil, for a total media depth of 0.6 m. The soil supports the wetland vegetation and associated microbiological communities. The pea gravel allows for subsurface flow of water. This subsurface area is an active biological zone and provides many important processes for pollutant removal. An impermeable clay layer below the gravel may be necessary to keep water in the wetland for supporting the various ecological processes under long periods of dry weather. The wetland should have a slope of about 1% to promote the water movement.

Ideally, the wetland should be sized to hold a 10-year storm event, although land area may not allow this. The hydraulic retention time should be at least 30 minutes for the maximum expected flow. Hydraulic retention times of 3-5 hours should be expected for annual storms and 10-15 hour retention times will provide optimum wetland performance. The maximum time should be limited to 24 hours to minimize stagnation and other problems associated with standing water.

Hydraulic loadings less than 1 m^3/day/m^2 should be maintained, and flow velocities should be below 0.3 to 0.5 m/s. Velocities greater than 0.7 m/s can cause damage to wetland vegetation.

Suggested vegetation includes reedmace (*Typha latifolia*) and common reed (*Phragmites australis*) (Shutes *et al.* 1999). Regardless, it is important that the wetland vegetation blend in with the natural surroundings and diverse plant communities may provide more effective water treatment than wetlands containing only a few plant species (Karathanasis *et al.* 2003). The wetland planting designs should consider placements to minimize flow short-circuiting.

11.11.3 Water Quality Performance

Because of the complex ecological communities established in working wetlands, a number of pollutant removal mechanisms are operational in a wetland. Physical processes dominate for the removal of suspended solids. Chemical processes such as adsorption and precipitation are important for the removal of inorganic compounds such as heavy metals and phosphorus, as well as hydrophobic organics, such as oils. Importantly, biological reactions can reduce BOD, nitrogen, and phosphorus levels in runoff waters.

A compilation of data on stormwater wetlands has shown that pollutant removals can be satisfactorily described using a first-order plug flow model using an areal rate constant, k_a:

$$\frac{C}{C_0} = \exp\left(-k_a\middle/q\right) \tag{11.61}$$

Here, q is the hydraulic loading to the wetland, flow rate per unit area (Q/A). Pollutant removal rate constants for total phosphorus, ammonia, and nitrate in stormwater wetlands are similar to those found for wastewater treatment wetlands (Carleton *et al.* 2001). The average value of k_a for total phosphorus was 11.3±17.6 m/yr. Those for ammonia and nitrate were 5.0±5.3 and 9.9±16.4 m/yr, respectively. The variability of pollutant removal in wetlands is high and in some cases, negative removals (pollutant concentration increases) were found.

Percent pollutant removals in stormwater management wetlands were also regressed against the wetland:watershed area ratio (AR). Although much scatter in the data was noted, these regressions provide a preliminary estimate for wetland performance for design or analysis (Table 11.5). For example, for 31 measurements, TSS removal averaged 43%, but ranged from (−300) to 95%.

Example 11.13. Using Eq. (11.61), the removal of total phosphorus (TP) in a stormwater wetland can be estimated by employing the noted phosphorus rate constant of 11.3 m/yr. The input TP is 3.2 mg/L. The yearly rainfall is 120 cm with a runoff coefficient of 0.75 over the watershed area, which is 40,000 m^2. The wetland area is 590 m^2, producing a hydraulic loading to the wetland of:

$$q = \left(1.2\,\dfrac{m}{yr}\right)(0.75)\left(\dfrac{40,000\ m^2}{590\ m^2}\right) = 61.0\,\dfrac{m}{yr} \qquad (11.62)$$

Using a rearranged form of Eq. (11.61):

$$C = C_0 \exp\left(\dfrac{-k_a}{q}\right) = (3.2\,\text{mg/L})\ \exp\left(\dfrac{-11.3\,\text{m/yr}}{61.0\,\text{m/yr}}\right) = 2.7\,\text{mg/L} \qquad (11.63)$$

Example 11.14. Using the regression equations of Table 11.5, the TP and TSS removal can be estimated for the same wetland as examined in Example 11.13. The ratio of wetland area to watershed area is calculated as:

$$AR = \left(\dfrac{590\,m^2}{40,000\,m^2}\right) = 0.0148 \qquad (11.64)$$

Using the appropriate regression equations, the percent removal for TP is:

$$REM = 100 - 77.3\exp\left[-5.31(AR)\right]$$
$$= 100 - 77.3\exp\left[-5.31(0.0148)\right] = 28.5\% \qquad (11.65)$$

Table 11.5. Regression equations for pollutant removal for stormwater treatment wetlands (from Carleton *et al.* 2001). REM is the percent removal of the pollutant in the wetland. AR is the ratio of wetland area to watershed area.

Pollutant	Regression Equation	Maximum AR	r^2
TP	REM = 100-77.3 exp(-5.31 AR)	0.3	0.32
TSS	REM = 100-46.6 exp(-8.36 AR)	0.2	0.12
TN	REM = 100-94.3 exp(-2.92 AR)	0.2	0.28
NH_3	REM = 100-94.3 exp(-8.78 AR)	0.3	0.47
NO_3^{2-}	REM = 100-75.1 exp(-9.14 AR)	0.3	0.54
T-Cd	REM = 100-85.5 exp(-9.86 AR)	0.2	0.78
T-Cu	REM = 100-82.4 exp(-7.31 AR)	0.3	0.68
T-Pb	REM = 100-36.0 exp(-4.14 AR)	0.3	0.15
T-Zn	REM = 100-62.1 exp(-4.29 AR)	0.3	0.41

and for TSS, the removal is:

$$REM = 100 - 46.6\exp[-8.36(AR)]$$
$$= 100 - 46.6\exp[-8.36(0.0148)] = 58.8\% \qquad (11.66)$$

Using an input TP of 3.2 mg/L, 28.5% removal gives an output TP concentration of 2.3 mg/L, which is in moderate agreement with the analysis presented in Example 11.13.

11.12 PROBLEMS

11.1   A severe storm has an average rainfall depth of 6.8 inches over a 860-ac watershed.   What volume (ac-ft) would a reservoir need to have to completely contain 18% of the rainfall?

11.2   A storm with a uniform depth of 2.3 inches falls on a 41-acre watershed. Determine the total volume of rainfall in ac-ft and cubic feet. If all of the water were stored in a detention basin with a surface area of 36,000 ft^2, how deep would the basin need to be?

11.3   The following table gives the measured rainfall depth (in.) over a watershed and the surface runoff from a 125-ac watershed. Volumes shown are the average over the time interval. Compute the accumulated storage (ft^3) of water within the watershed. Plot the inflow outflow, and cumulative storage as a function of time.

Time (min)	Rainfall (in.)	Runoff (ft^3/s)
0	0	0
15	0.2	10
30	0.3	30
45	0.6	90
60	0.5	170
75	0.4	180
90	0.1	130

11.4   The following table gives the hourly surface runoff from two inlets (Q_1 and Q_2) and the outflow from a detention basin (Q_o) for a storm event. The values are the averages for each hour. Estimate the volume of storage in the basin (ac-ft) at the end of each hour. All flows are in ft^3/sec. Compute the center of masses of both inflow hydrographs.

hour	1	2	3	4	5	6	7	8	9
Q_1	2	6	11	9	4	2	0	0	0
Q_2	3	8	15	16	12	8	3	0	0
Q_o	0	2	4	7	9	9	9	7	6

11.5   The following are inflows to and outflows from an 8.4-acre wetland:

week	Rainfall (in.)	Inflow (ac-ft)	ET (in./day)	Infiltration (ft^3/day)	Outflow (ft^3/day)
1	0.6	5.5	0.11	800	30,000
2	0.8	6.4	0.11	810	35,000
3	1.1	7.8	0.12	840	42,000
4	0.7	6.8	0.12	820	37,000
5	0.4	5.7	0.11	830	34,000
6	1.3	6.7	0.12	830	35,000
7	0.6	6.7	0.13	820	36,000
8	0.0	5.1	0.12	800	31,000
9	0.4	4.2	0.12	790	27,000

Compute and plot the weekly relationships for (a) stage vs. time; (b) storage vs. time; (c) change in storage vs. time. The contributing drainage area is 315 acres.

11.6   Estimates of the available storage (V_s, ac-ft) in a detention basin and the outflow rate (Q, ft^3/s) are made for six storm events. Estimate the retention time for each event and show the applicability of Eq. (11.7) as an estimator.

V_s	10.9	10.3	9.8	9.1	8.2	6.2
Q	31.5	62.8	101	157	199	252

11.7   An existing wetland has a permanent pond with an average depth of 9 inches. The underlying soil has an infiltration rate of 0.002 in./hr. The pond is located where the ET rate for the vegetation would be 1 inch per week during the dry period of the year. What is the expected drain time?

11.8   Assume that historical rainfall records indicate that less than 0.1 in. of rainfall over a 6-week period can be expected for a wetland site during the summer. The vegetation to be used in the wetland requires a depth of 6 inches and has an expected evapotranspiration rate of 0.2 in./day. The site has an infiltration rate of 0.0035 in./hr. Is the wetland expected to be dry for more than 10 days under such conditions?

11.9   For three weeks in July and three weeks in August, a site generally does not experience significant rainfall. The average temperatures for July and August at the site are 82° and 80° F, respectively. The fraction of daily day light hours for July and August are 0.63 and 0.60, respectively. The consumptive use factor of 0.86 should be used for the vegetation. What is the expected ET for the 6-week period? If infiltration losses can be neglected, what depth of water is necessary to offset the ET loss?

11.10 The average sediment trap efficiency in a wetland should increase with increasing residence time. Propose a functional form that would relate trap efficiency and residence time. Support the rationality of your proposed model.

11.11 Assume that the volume of storage in a wetland varies due to infiltration and evapotranspiration rates. Estimates of the initial storage volume (ac-ft) and average flow-through rate (ft³/s) for six storm events are:

V_s	20.6	18.1	20.2	20.0	20.5	21.0
Q	17.4	10.1	13.3	9.6	15.5	11.0

Estimate the average residence time (hours) for the wetland.

11.12 Assume the outflow hydrograph for the case of Example 11.2 is as follows, rather than that given in Table 11.2: $O\{t\} = \{0, 1, 1, 2, 2, 3, 4, 4, 5, 5, 5, 5, 4, 4, 3, 3, 3, 3, 3, 3, 3, 2, 2, 2, 1\}$ Compute the retention time and discuss the reason for the difference between it and the value computed in Example 11.2.

11.13 Assume the inflow hydrograph to a wetland is a triangle with peak q_i, time to peak t_i, and time base T_i. The outflow hydrograph has characteristics q_o, t_o, and T_o, respectively. (a) Derive an expression for the time to the center of mass, C_i, and C_o, as a function of the peaks, times to peak, and time bases. (b) Using the expressions of part (a), derive an expression for the residence time, T_d. (c) Assume $q_i = 16$ cfs, $t_i = 3$ hr, $T_i = 18$ hr, $q_o = 6$ cfs, $t_o = 8$ hr, and the inflow and outflow volumes are equal. What is the residence time?

11.14 One section of a wetland covers 22 ac. It has soils with an infiltration rate of 0.05 in./hr and vegetation with an evapotranspiration (ET) rate of 0.16 in./day. The remaining 49 ac of the wetland experiences infiltration at a rate of 0.08 in./hr and ET at a rage of 0.26 in./day. If the dead storage has a depth of 15 inches, after how many days will the wetland lack surface water?

11.15 For the conditions of Problem 11.11, what depth of storage is needed to have a 12-day period before the surface water storage will be depleted?

11.16 If the Blaney-Criddle Consumptive use factor could be in error by 0.1, how would this affect the design of Example 11.3? Is the effect significant? Discuss.

11.17 Example 11.3 indicates that a wet pond depth of 13 inches would be sufficient if the wetland were under lain by a clay liner. Show the calculations that support a 13-in. depth.

11.18 Sediment volumes in the inflow (I_s) and outflow (I_o) to a forebay are estimated for six consecutive storm events. What is the average sediment trap efficiency?

I_s	43	28	16	97	62	51
I_o	12	4	18	21	15	20

11.19 The inflow design discharge to a forebay is 12.5 ft³/s. Local requirements require a trap efficiency of 65% for the forebay. The average soil particle diameter is 0.012 in. Topography will limit the forebay depth to a maximum of 18 inches.

11.20 A 15-ac agricultural field with a silty soil produces a design discharge of 23 ft³/s, which is directed to a wetland. Determine forebay characteristics that will trap 60% of the silt suspended in the inflow. Assume topography limits the depth to 18 inches and a width of 8 feet.

11.21 Grab samples of the sediment concentrations in the inflow (C_i) and outflow (C_o) of a wetland are as follows:

Time (hr)	0	1	3	4	6	8	10	15	20	24
Inflow (ft³/s)	0	5	12	11	6	3	0	0	0	0
Discharge (ft³/s)	0	1	3	4	4	4	4	3	2	0
C_i (mg/L)	0	24	62	41	29	17	0	0	0	0
C_o (mg/L)	0	4	10	17	19	19	16	10	4	0

Compute the trap efficiency of the wetland.

11.22 How many times faster does a particle of fine sand ($d = 0.07$ mm) fall in water when compared with a silt particle ($d = 0.01$ mm)?

11.23 If a sand particle ($d = 0.05$ mm) is 1.4 ft above the stream bottom and transported in a stream at a horizontal velocity of 2.5 ft/s, how far downstream will it move before it falls to the bottom of the channel?

11.24 Discuss why the forebay length increases with the throughflow discharge (Q) and trap efficiency (E_f) and decreases with the width of the forebay (W) and the settling velocity V_s.

11.25 For the design parameters of Example 11.4, graph the design length as a function of the particle diameter of the range from 0.005 to 0.5 mm. Also graph the length as a function of trap efficiency for the design diameter of 0.00115 ft. Discuss the importance of these two variables to the design.

11.26   Discuss the expected accuracy of the trap efficiency of Eq. (11.13) when used for a single storm event, all storms during a one-year period, and all storm during a 10-yr period.

11.27   Flow enters a naturally flat area (slope = 0.0015 ft/ft) in a river at a point where 340 acres contributes runoff. For the design, assume an inflow rate of 26 ft^3/s. The section of the river where the wetland will be created is about 20-ft wide. The vegetation will have a stem length of 24 inches and a density of 200 stems per sq. ft. What is the design depth and velocity of flow?

11.28   To evaluate the importance of the dead storage, assume that 0.5 ac-ft of dead storage is available. How does this influence the maximum storage for both the 10-ft and 30-ft weirs in Example 11.6?

11.29   Estimate the retention time for the problem of Example 11.6 using both Eq. (11.27) and the time between the centers of mass. Make estimates for both the 10-ft and 30-ft weirs. Discuss the accuracy of the values.

11.30   Derive Eq. (11.29) from Manning's equation. Would this equation apply to a circular pipe flowing half full? Explain.

11.31   Determine the length of a forebay needed to trap 80% of the sediment load. The forebay has a maximum depth of 2 ft. Flow enters into the forebay from a small stream with a depth of 9 inches and a velocity of 1.7 ft/sec. The sediment has a mean particle diameter of 0.3 mm.

11.32   A first-order rate constant for total nitrogen removal through a stormwater runoff wetland is estimated to be 0.002 min^{-1}. An 80% TN removal through the wetland is desired. For an input flow rate of 1000 L/min and an average depth of 1 m, estimate the wetland surface area.

11.33   A watershed has a drainage area of 21,000 m^2. Estimate the area of the wetland to provide 75% removal of TSS from incoming stormwater runoff.

11.34   An effluent nitrate concentration of 1 mg/L (as N) is desired from a wetland treating runoff from a public building area. The input concentration averages 4.5 mg-N/L. The drainage area is 1.75 acres. What should be the size of the wetland, in ft^2.

11.35   Concentrations of copper, lead, and cadmium leaving a wetland should be less than 10 µg/L, 5 µg/L, and 50 µg/L, respectively. The input concentrations are 63 µg/L Cu, 85 µg/L Pb, and 420 µg/L Zn. Select a wetland area-to-watershed area ratio that is expected to meet or exceed all of the metal discharge criteria.

11.13 REFERENCES

Carleton, J.N., Grizzard, T.J., Godrej, A.N, and Post, H.E. (2001) "Factors Affecting the Performance of Stormwater Treatment Wetlands," *Water Res..* **35**(6), 1552-1562.

Kadlec, R.H. and Knight, R.L. (1996) *Treatment Wetlands*, CRC Lewis Publishers, Boca Raton, FL.

Karathanasis, A.D., Potter, C.L., Coyne, M.S. (2003) "Vegetation Effects on Fecal Bacteria, BOD, and Suspended Solid Removal in Constructed Wetlands Treating Domestic Wastewater," *Ecological Engg.* **20**, 157-169.

McCuen, R.H. (2005) *Hydrologic Analysis and Design*, Prentice-Hall, Inc., Upper Saddle River, NJ.

Shutes, R.B.R., Revitt, D.M., Lagerberg, I.M., and Barraud, V.C.E. (1999) "The Design of Vegetative Constructed Wetlands for the Treatment of Highway Runoff," *Sci. Total Environ.*, **235**, 189-197.

12

LOW IMPACT DEVELOPMENT

NOTATION

A = drainage area
A_c = cross-sectional area
C = runoff coefficient
d = depth
D = diameter
i   = rainfall intensity
q_p = peak discharge
R_h = hydraulic radius
T = swale spread
z = side slope ratio

12.1 INTRODUCTION

Low impact development (LID) is a stormwater management philosophy with a goal of minimizing environmental impacts that occur along with land development. A primary focus of LID is on the spatial layout of the development and practices that take place as the land is developed; in many ways, LID achieves pollution prevention through a broad array of practices for urban stormwater management. The ideals of low impact development are to plan, design, construct, and maintain a site such that the quantity and quality of the runoff that leaves the site replicate pre-development characteristics as near as possible. LID directs the focus of stormwater management down to the microscale, with consideration of design layouts and aspects as small as individual lots and components of individual lots. A major emphasis is placed on maintaining and enhancing natural processes. This includes promoting infiltration instead of runoff. The emphasis is on keeping the water on site rather than moving it to the nearest stream as soon as possible. Vegetated areas are used for runoff management. LID may also be known as Environmentally Sensitive Design or Sustainable Development.

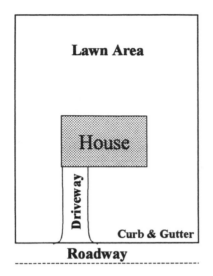 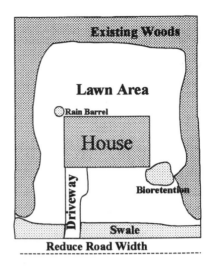

Figure 12.1. Comparison of traditional development (left) and LID (right) (Adapted from DER 1997)

The philosophy of LID is based on the following principles:

1. In planning for site development, existing natural and topographic features that are associated with the important hydrologic and water quality processes should be maintained. For example, efforts should be made to minimize increases in runoff volumes and keep from lengthening the travel times of runoff.

2. During design and construction, land disturbances should be minimized, and practices that will maintain natural processes during the transitional period of development should be encouraged. Land clearing should be minimized and impervious surfaces should be minimal.

3. For the post-development watershed state, integrated management practices that will maintain hydrologic and water quality processes should be emphasized.

4. All aspects of landscape function should encourage pollution prevention.

Taken as a whole, these principles represent a radical departure from traditional site development. The first principle focuses on site planning and ways of integrating the physical characteristics of the site into stormwater management practices. Keeping some of the land area undisturbed is encouraged. Also, to the extent practical, natural depressions and swales are not disturbed. Meandering flowpaths are maintained to encourage infiltration. This is especially important if the flowpath receives runoff from newly developed portions of the lot.

Advanced planning should lead to narrow driveways and minimal sidewalks (Figure 12.1). Streets should be kept as narrow as possible but within the local codes. Instead of installing elaborate curb-and-gutter systems to rapidly convey runoff, grass swales should be emphasized in the design. Landscaping that promotes infiltration, pollutant capture, and flow attenuation is important to runoff control.

One of the most important LID provisions is to create as little impervious surface as possible and to keep any runoff on site as long as possible using natural approaches.   Traditional land development involves converting forest or undeveloped land into houses, driveways, sidewalks, lawns, parking lots, etc. Established trees are often removed and natural elevated areas and depressions are graded.   Rainwater that falls on traditionally-developed sites is collected on rooftops, sidewalks, and driveways and is efficiently transported to concrete curb and gutter systems, finally to stormwater inlets.   The runoff drains from the traditional site as quickly as possible, with little time and area left for infiltration (Figure 12.1).   Along this travel path, many opportunities exist for the runoff to pickup and mobilize pollutants and transport them to the local receiving waters.

Land disturbance generally decreases infiltration rates, which reduces groundwater recharge and encourages surface runoff volumes. An important part of the LID philosophy is to encourage practices that promote infiltration. Therefore, minimizing land disturbance will help to maintain natural infiltration processes. As much area as possible should be left undisturbed, especially wooded and other natural areas. The use of heavy equipment throughout the site compacts the soils, thus reducing infiltration. Therefore, equipment usage should be minimized to prevent soil compaction. Maintaining wooded areas and preventing compaction have long-term benefits both from water quantity and water quality perspectives. Increasing infiltration and reducing the volume of excess runoff leaving the site encourages the settling of suspended solids.

LID focuses on small-scale source control through Integrated Management Practices (IMPs). This is a departure from large, end-of-pipe controls such as ponds that have been traditionally employed for stormwater management.   Houses, commercial buildings, parking lots, driveways, and sidewalks are all impervious and prevent ground-water recharge. Runoff from these impervious areas is directed into small IMPs that are constructed throughout the site. IMPs are designed and constructed with vegetation and soil of high permeability. Individually, their affect is small, but taken as a whole, they can significantly mitigate adverse runoff quality and quantity concerns. Many of these IMPs are small versions of the vegetated treatment processes highlighted in Chapter 9, i.e., bioretention, grass swales, and green roofs.

Pollution prevention is more esoteric. It involves material selection, public choice, and behavior modifications. Certainly the best way to prevent pollution from entering receiving waters is to prevent the specific sources of the pollutants from becoming part of the built landscape. This may be a challenge that in some cases is beyond the control of the typical stormwater management engineer, but nonetheless is an important "green engineering" concept to pursue.

Table 12.1 includes some practices and concepts of LID. Some of these are specific structural practices, while others are design considerations, although the distinction between the two is ambiguous in many instances. All have the goal of keeping precipitation and runoff on the lot or local area as long as possible and practical. As the volume of direct runoff is directly proportional to the amount of imperviousness, minimizing the impervious cover will help keep the post-development runoff volume near to the amount that occurred prior to development.

Grass swales and natural depressions allow for some on-lot storage, which promotes infiltration and evapotranspiration. Overland flow velocities are lowered due to the vegetation, depressions, and meandering, which mobilizes fewer pollutants, and possibly allows particulate matter to settle or be filtered. Trees and small forested areas provide rainfall interception, promote infiltration, lower temperatures, and keep this land from becoming lawn, thereby decreasing the potential for the over-application of fertilizer and pesticides. The LID philosophy is to design to (1) reduce runoff volumes and velocities, (2) improve runoff timing considerations, and (3) enhance the quality of excess runoff.

Table 12.1. Low Impact Development practices and considerations.

Alternative Surfaces & Building Materials	Amended Soils
Bioretention/Rain Gardens	Catch Basins/Seepage Pits
Flatter Slopes	Flatter Wider Swales
Infiltration Swales and Trenches	Landscape Island Storage/Filtration
Maximize Sheet Flow	Parking Lot/Street Storage/Filtration
Permeable Pavers	Pollution Prevention
Reforestation	Rain Barrels/Cisterns/Water Use
Reduce Impervious Surface	Resource Conservation
Roof Leader Disconnection	Rooftop Detention/Retention
Sidewalk Storage	Smaller Culverts, pipes, and inlets
Strategic Grading	Surface Roughness Technology
Trash Collectors	Tree Box Filters
Tree/Shrub Depression/Filtration	Tree Planting and Landscaping
Turf Depression Storage	

12.2 REDUCING IMPERVIOUS SURFACE

Many of the concepts listed in Table 12.1 can be integrated into the initial development process. First, one objective is to reduce the total impervious surface. This involves minimizing the footprints of houses and other buildings, sidewalks, driveways, parking lots, and roadways. This suggests that buildings should go up, instead of out. Driveways should be just large enough for the automobiles to travel on and should be placed as close to the road as possible. This minimizes the area of imperviousness. Parking areas should be designed as efficiently as possible to get the required amount of spaces in the least amount of area. Sidewalks should be short, narrow, and non-contiguous. Local codes tend to dictate the width of roads and sidewalks, but exemptions are possible and narrower widths correlate to less imperviousness.

Example 12.1. For a subdivision, 250 houses are planned. Each house has a length of 40 ft and a width of 30 ft. The county mandates a design storm with an intensity of 0.3 inches per hour for an 8-hour duration. Eight hours of rainfall at 0.3 inches per hour is 2.4 inches of rain. At 30 ft by 40 ft, the roof area is 1200 ft² each. (The pitch or angle of the roof is irrelevant; only the projected area captures the rain.) The total volume of rainfall on the 250 roofs is:

$$\left(1200\,ft^2\right)\left(2.4\,in.\right)\left(\frac{ft}{12\,in.}\right)\left(250\,houses\right) = 60{,}000\,ft^3 = 448{,}800\,gallons \qquad (12.1)$$

As an alternative, if instead of making the house 40 ft in length, an additional story was added so that the length was 25 ft, the volume is:

$$\left(30\,ft\right)\left(25\,ft\right)\left(2.4\,in.\right)\left(\frac{ft}{12\,in.}\right)\left(250\,houses\right) = 37{,}500\,ft^3 = 280{,}500\,gallons \qquad (12.2)$$

This represents a reduction of 37%.

When development occurs, careful evaluation of the development process can have a significant impact on the amount of runoff generated by a storm event. Grading to reduce slopes should be encouraged. This keeps the runoff on site as long a possible, allowing for maximum infiltration. Sheet flow is maintained for longer periods when the slope is flatter, which minimizes channeling and allows filtering to occur through vegetated areas. In the landscape, small depressions should be maintained or added to create small, temporary pools on site. Larger vegetation, including trees and shrubs should be encouraged. Larger plants provide for more opportunity for rainfall interception, nutrients and water uptake, and enhanced evapotranspiration. These forms of vegetation also generally do no require intense fertilization, pesticide application, and water as do highly managed lawns.

In the actual land development process, the use of heavy equipment on open areas should be minimized. Soils will become compacted, even with minimal use of heavy equipment. Even slight compaction can have a huge adverse impact on infiltration rates. Any soils that do become compacted should be tilled, and all soils can be amended with sand or organic matter to encourage higher infiltration rates.

Once the total surface areas of roadways, driveways, sidewalks, and patios are reduced to the extent possible, a final LID technique to be evaluated is the use of permeable paver materials. Instead of traditional concrete or asphalt, several different options are available to provide some permeability and infiltration where these traditional materials provide none. One option is to use some type of intermittent brick or block instead of continuous paving. The blocks are spaced as needed and usually placed on a subsurface layer of sand. The area between individual blocks is left as grass or can be filled with stone or sand. In any case, less impervious area results. A second option is the use of porous pavements or concretes that allow some infiltration directly through the material itself. Several types of porous pavements have been developed and used.

While the use of alternative pavements provides LID benefits, they are not without possible costs. Permeable asphalts may clog if they are loaded with excessive sediment. Porous concretes and asphalts also do not have the strength and durability of traditional pavements. Similarly, brick and block pavers may be more expensive, areas between them can clog, and they may not hold up over many years as would more traditional materials. Also, snow and ice removal are more problematic with permeable pavers than with traditional materials (they cannot be plowed), and freezing can prevent infiltration.

Beyond the individual lot level, several LID design techniques can be implemented in the stormwater runoff collection infrastructure. Instead of concrete curbs, grass swales can be employed for runoff conveyance. Small depressions can be designed for runoff storage and infiltration, rather than rapid conveyance from the sources.

Example 12.2. A subdivision under development has 150 lots of ½ acre each, 30% of which is developed impervious (roof, sidewalk, driveway). Under conventional development, the removal of native vegetation and compaction of the soil over 50% of the pervious area reduces the runoff infiltration rate to 0.2 in./hr. The other 50% remains at the native infiltration rate of 0.35 in./hr. By implementing LID practices to leave greater native area and avoid soil compaction, only 25% of the pervious land is compacted to the 0.2 in./hr infiltration rate.

The area available for infiltration is:

$$(150 \text{lots})\left(\frac{0.5 \text{ acres}}{\text{lot}}\right)(70\% \text{ pervious}) = 52.5 \text{ acres} = 2.29 \times 10^6 \text{ ft}^2 \qquad (12.3)$$

Using a very simple flow balance, under conventional development conditions, the runoff flow rate for a 1 in./hr rainfall intensity is:

$$\left(1 - 0.2 \, {}^{\text{in.}}\!/_{\text{hr}}\right)\left(\frac{1 \text{ft}}{12 \text{ in.}}\right)(2.29 \times 10^6 \text{ ft}^2)(0.50) + \left(1 - 0.35 \, {}^{\text{in.}}\!/_{\text{hr}}\right)\left(\frac{1 \text{ft}}{12 \text{ in.}}\right)(2.29 \times 10^6 \text{ ft}^2)(0.50)$$

$$= 1.384 \times 10^5 \, {}^{\text{ft}^3}\!/_{hr} = 38.4 \, {}^{\text{ft}^3}\!/_{s} \qquad (12.4)$$

In the LID situation:

$$\left(1 - 0.2 \, {}^{\text{in.}}\!/_{\text{hr}}\right)\left(\frac{1 \text{ft}}{12 \text{ in.}}\right)(2.29 \times 10^6 \text{ ft}^2)(0.25) + \left(1 - 0.35 \, {}^{\text{in.}}\!/_{\text{hr}}\right)\left(\frac{1 \text{ft}}{12 \text{ in.}}\right)(2.29 \times 10^6 \text{ ft}^2)(0.75)$$

$$= 1.312 \times 10^5 \, {}^{\text{ft}^3}\!/_{hr} = 36.4 \, {}^{\text{ft}^3}\!/_{s} \qquad (12.5)$$

which represents a reduction of 5.2%.

12.3 INTEGRATING LID PRACTICES

Since impervious areas cannot be completely eliminated, a number of specific practices or IMPs are employed within LID to address runoff quantity and quality issues to complement design modifications. Some of these practices include simple infiltration technologies, such as onsite cisterns, rain barrels, or tree box filters. Others are more complex and may involve major engineering considerations, such as green roofs.

Cisterns and rain barrels can be used to address downspout flow from roofs. With cisterns, rooftop runoff is directed into ground storage where gradual infiltration occurs. Design details are given in Section 10.5. Downspout flow can also be collected in rain barrels (Figure 12.2). Rain barrels are placed at each downspout to collect 50-75 gallons of rooftop runoff. This captured water can be used later for yard watering or other non-potable uses, reducing water supply needs in the developed area.

Figure 12.2. House with rain barrel

Example 12.3. Two rain barrels are installed on each of the small (30 ft x 25 ft) houses of Example 12.1. Assuming that all rain becomes roof runoff, the volume collected in the barrels is calculated for a 0.2 in. rainfall:

$$(30\,ft)(25\,ft)(0.2\,in.)\left(\frac{ft}{12\,in.}\right) = 12.5\,ft^3 = 93.5\,gallons \qquad (12.6)$$

Each barrel would collect nearly 47 gallons of water to be used later.

Flow from impervious areas can be directed to areas designed for retention and high infiltration. In this manner, small bioretention areas, or rain gardens, can be strategically placed on individual lots. These areas can be small depressions in low areas of the lot and designed for enhanced infiltration. The facility will have a mixture of sand, topsoil, and mulch, as discussed in Section 9-4. Aesthetics are very important with these onsite facilities. Flowers, shrubs, ornamental grasses, and other landscape-type vegetation should be planted in these rain gardens. During rain events, flows from impervious areas will collect in the rain gardens, up to a few inches deep. Some of the collected water will infiltrate; the amount depends on the characteristics of the surrounding soils. The remainder of the collected water is evaporated and transpired.

Tree boxes can be integrated into the urban landscape. Runoff flow from roads, sidewalks, and rooftops can be directed into the tree box, which is typically a small tree planted in a high-permeability soil with a tree grate. These systems may be placed within a concrete vault and may have an underdrain to the existing storm sewer. The runoff is directed through the grate, infiltrates through the soil media, ultimately providing some groundwater recharge and reducing flow to the storm sewer. Flow is slowed and pollutant removal and filtering can occur. The tree provides water uptake through evapotranspiration, nutrient uptake, and possibly promotes other beneficial biological activities in the box. In fact, any procedure that increases the density of urban trees is likely to provide hydrologic benefits.

Green roofs are probably the most complex and expensive of the LID techniques. However, in highly urbanized areas where land values are high, and in very environmentally sensitive area, their use may be justified and encouraged. In dense city areas, green roofs may be one of the more favorable low-impact stormwater management options. Details on green roofs are presented in Section 9.7.

12.3.1 Porous Parking Swales

With some innovative thinking, impervious area reduction techniques can be applied in a number of development situations. Given the large portion of urban/suburban watersheds that is devoted to parking spaces, it is important to have stormwater management methods that mitigate, at least partially, the significance of parking areas to the negative effects of development. One option is to include porous areas between aisles in large parking lots. These porous areas would be positioned such that the front or back end of a parked car would hang over a swale so that no parking space is lost. Assuming about 2 ft of space between the front of a typical sedan and the tires and allowing about 0.5 ft between the front ends of two cars that face each other, then the swale might be 4.5 ft wide. Wheel stops that do not obstruct flow could be installed at the pavement edge. The porous area could use porous pavers, be grass covered, or be taken up by a bioretention facility (Figure 12.3). If the parking area is graded properly, runoff from the impervious portions of the parking lot could be directed towards the porous area, thus increasing both infiltration and the travel time of some of the runoff. As considerable amounts of pollutants are generated in parking lots, porous parking swales would have obvious water quality benefits. A case study has shown that significant reduction of runoff flow and pollutant loads can be realized by this simple design (Rushton 2001). Major reductions in total load were found for ammonia, nitrate, suspended solids, copper, iron, lead, manganese, and zinc. Other stormwater management practices could be employed in addition to, or integrated with, this swale setup.

Figure 12.3. LID parking with bioretention

12.3.2 Design Parameters

The primary environmental benefits of porous parking swales are the reduction in the volume of direct runoff, the lengthening of the travel time for part of the runoff, and reductions in peak discharges associated with the volume reduction and travel time lengthening. Standard design procedures are applicable. Times of concentration can be computed with Manning's equation, and peak discharges and runoff volumes computed using either the Rational formula or the NRCS Graphical method.

The physical characteristics of the swale are the same as those identified for vegetated swales in Section 9.3. The spread (T, ft) is set by car design, with a value of 4.5 ft reasonable for current models. The side slope (h:v) is set by the erodibility of the soil and vegetal cover. The longitudinal slope (s, ft/ft) is set by topography. The roughness (n) depends on the characteristics of the cover; standard values for sheet flow runoff can be used.

Relationships between the swale center line depth (d, ft), the cross-sectional area (A_c, ft^2) of the swale, and the hydraulic radius (R_h, ft) with the spread T and side slope ratio z:1 (h:v) are as follows:

$$d = 0.5 \ T/z \tag{12.7}$$

$$A_c = 0.25 \ T^2/z \tag{12.8}$$

$$R_h = 0.25Tz^{-1}\left[1 + \left(\frac{1}{z^2}\right)\right]^{-0.5} \tag{12.9}$$

Setting T and z enables d, A, and R_h to be computed, with the latter two used with Manning's equation and the continuity equation to compute the full-flow discharge rate and travel time. For a 4.5 ft swale, values for d, A, and R_h are given in Table 12.2 for selected cross slopes.

Table 12.2. Characteristics of 4.5 ft wide swales

Slope (h:v)	d (ft)	A_c (ft^2)	R_h (ft)
3:1	0.75	1.6875	0.3558
4:1	0.5625	1.2656	0.2729
5:1	0.45	1.0125	0.2206
6:1	0.375	0.8438	0.1849

12.3.3 Design Procedure

The design begins with a layout of the site. While hydrologic and water quality factors should be central to this step, quite frequently the convenience of the user will take precedent in the layout. The flow paths should be identified, with an attempt made to direct flow from the portions of the parking area where pollutants are generated towards the swales. This should take into account the volume and discharge capacities of the swale to ensure that ponding does not cause flooding of the section of the parking lot where the swale is located. The volume capacity can be estimated from the geometry (i.e., length, cross-slope, and width) of the swale. The discharge capacity can be computed with Manning's equation. The time of concentration must be determined for the portion of the drainage area that contributes to the swale, as well as for the portion of the parking lot that bypasses the swale. The Rational formula can be used to compute discharge rates from the contributing and noncontributing areas, as well as the total area as if the swale(s) was not included in the layout. This enables a comparison to be made, with vs. without the swale. Any culvert that drains the swale will need to be sized using Manning's equation for free-surface flow conditions or hydraulic analysis for pressure flow conditions.

If soil conditions are favorable, the swale could be replaced with a bioretention facility. This will generally allow for more storage, greater reduction in both peak discharge rates and pollution discharges, and enhanced aesthetic appeal of the parking facility.

Example 12.4. Consider the case of a mall parking lot that includes several aisles. A swale between aisles will control a portion of the runoff, including pollutants transported to the swale. Once the volume capacity of the swale has been reached, the flow will drain across the asphalt surface. For purposes of this example, flow is assumed to drain north to south (see Figure 12.4), with the lot graded so that flow drains to the swales and away from the crowns at the center of the aisles. The total area of the subwatershed is 70 ft x 260 ft = 18200 ft^2 = 0.418 ac. The swale has a width of 4.5 ft and a length of 230 ft, or 1035 ft^2 = 0.0238 ac, which leaves 17,165 ft^2 or 0.394 ac of imperviousness.

The surface volume of the swale equals the product of the cross-sectional area and the length. The former is obtained from Table 12.2 as 1.266 ft^2, which gives a volume of 291 ft^3. The travel time from the upper end of the watershed to the outlet of the culvert that drains the swale can be computed with Manning's equation for each of the three parts: 15 ft of imperviousness upgradient of the swale, 230 ft in the swale, and 15 ft in the culvert. A time of concentration of 9.3 minutes is computed, with a velocity in the swale of 0.42 ft/s, and a velocity of 3.4 ft/s on the asphalt surface. Assuming a 4 in./hr rainfall intensity for a 10-min storm duration, the swale could drain approximately 5650 ft^2 over the 10-minute duration. This area is outlined on Figure 12.4.

The Rational formula can be used to compute the peak discharge for the area when a swale is not included. In this case, the travel time is 1.27 minutes, which is associated with a 2-yr intensity of 6 in./hr for the Baltimore IDF curve (Figure 4.2). Using a runoff coefficient of 0.9 for impervious surface yields a peak discharge of:

$$q_p = CiA = 0.9(6)(18200)/43560 = 2.26 \text{ ft}^3/\text{s} \qquad (12.10)$$

The peak discharge for the portion of impervious surface that drains to the swale is:

$$q_p = 0.2(4)(5650)/43560 = 0.10 \text{ ft}^3/\text{s} \qquad (12.11)$$

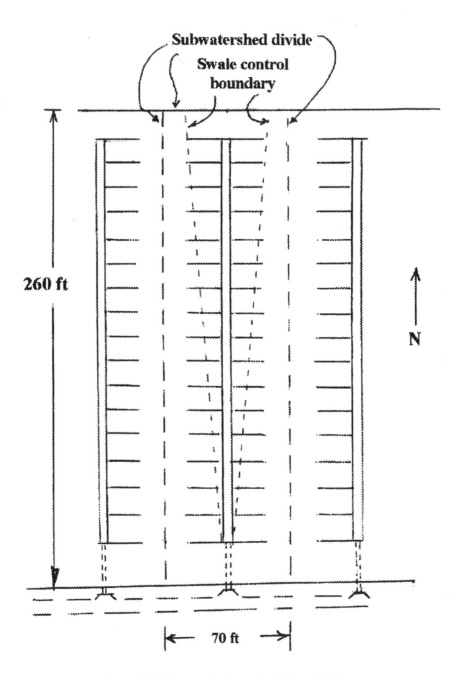

Figure 12.4. Diagram of parking swale for Example 12.4.

To compute the peak discharge when part of the runoff is drained by the swale, the weighted runoff coefficient is:

$$C = \frac{1035(0.08) + 17165(0.9)}{18200} = 0.85 \qquad (12.12)$$

The peak discharge for the entire area when the swale is included is:

$$q_p = 0.85(4)(18200)/43560 = 1.42 \text{ ft}^3/\text{s} \qquad (12.13)$$

Thus, the inclusion of the swale potentially reduces the peak discharge by 37.2%. The reduction in the runoff volume will depend on the infiltration capacity of the soil. The culvert diameter needed to drain the swale is computed with Manning's equation:

$$D = 1.333[qn]^{0.375} S^{-0.1875} = 1.333[0.1(0.024)]^{0.375} (0.01)^{-0.1875} \qquad (12.14)$$
$$= 0.33 \text{ ft} = 4 \text{ in.}$$

12.4 LANDOWNER ISSUES

Landowner (and occupant) perceptions and practices play important roles in maintaining the environmental effectiveness of LID. Without the cooperation of the landowners, many of the LID techniques will eventually fail. Foremost, the public must accept the look and design of an LID lot and be willing to pay as much for an LID property as one developed in a traditional manner. The developer will desire the same return on investment as with traditional development. LID properties should have the same resale value as other comparable properties. High impervious development options such as cul-de-sacs, long/wide driveways, and patios are desired by many homeowners. Nonetheless, some LID initiatives, such as keeping mature trees in place, can make LID design more attractive.

Most of the specific LID practices that may be employed on a property require periodic maintenance to maintain their effectiveness. Rain gardens need periodic landscaping, planting, and mulching to provide both aesthetic and runoff benefit to the lot. However, rain gardens are installed a few inches below grade to allow ponding. Traditionally flower and shrub areas are in elevated mounds on the lot.

Education and acceptance are important to long-term success. Using and managing a rain barrel presents similar issues. First, rain barrels provide an atypical look to a home. Rain barrels must be drained at some point after the collection event and occasionally cleaned. Draining is necessary to ensure that the storage is available for control of subsequent storms. They should also be covered to discourage mosquito breeding.

Other important LID/occupant issues include downspout management. Rather than having roof downspouts that flow directly into driveways and the road, they can be redirected onto lawns or into some LID infiltration practice such as a cistern. Thus, some of this flow will infiltrate. Even that which does not infiltrate, can receive treatment as it flows over a grassy lawn before reaching the swale. Also, public acceptance of standing water for short times after rainfall is important to the success of most LID practices.

12.5 POLLUTION PREVENTION

More and more, studies are finding that it is more effective to address pollution at the source, rather than to later remove it from a stream. Concepts similar to those that are being employed by industries to minimize waste and prevent pollution can be applied to stormwater management. With pollution prevention and green engineering principles, industries investigate ways to produce their product with little or no waste. In a similar manner, products such as automobiles, building materials, and lawn chemicals that are used throughout a watershed impact the local water quality through stormwater runoff. Eliminating or minimizing their use in the watershed will correspondingly reduce pollutant loads.

12.5.1 Material Substitution

As discussed in Chapter 6, a significant fraction of the pollutants that are present in stormwater runoff result from rainwater and runoff contacting buildings, vehicles, roadways, and lawns. Eliminating the materials that cause the pollution will directly address the pollution problem. A few specific cases where material substitution can be applied will be examined.

Building materials can be a source of several toxics to runoff. Copper flashing and gutters and galvanized roofs can readily corrode and leach copper and zinc into the runoff. Bricks may leach low levels of metals. Treated wood can release organic and inorganic materials (such as arsenic) that have been applied as preservatives. Some paints and other coatings release toxic compounds, either inadvertently as they weather, or purposely, for biocidal measures. Substitution of other materials for these during the construction, remodeling, or refinishing of structures can have a major impact on the mass loadings of toxic compounds that are released from various structures.

The automobile is one of the most polluting objects. Oils and other fluids from vehicles contain ingredients that are highly toxic. Brake shoes are specifically manufactured to wear as they bring a vehicle to a stop, possibly releasing metals such as copper and zinc to the local environment. Tires, also, as they wear and degrade, can release a number of polluting materials to the environment. The average consumer cannot do much about these releases, other than to keep their vehicle in top operating condition and to ensure that fluid leaks do not occur; careful driving habits may also play a small role. Automobile and automobile part manufacturers will have to find materials to substitute for current lubricants, brake, and tire materials. These new materials must be able to safely perform the job of current materials at a price comparable to the current pollution-causing materials.

Tailpipe emissions will also wash back to the land surfaces and end up in the runoff. Minimizing all driving will have an impact to the water environment. Land use planning will also play a role here as residential areas are integrated with commercial and business areas.

Example 12.5. Each house in a subdivision of 350 houses contains 3 m^2 of copper flashing around windows, doors, and chimneys. This flashing corrodes to release 1 mg of copper per m^2 of flashing per year. The total copper released to the environment is:

$$(350\,houses)\left(\frac{3\,m^2}{house}\right)\left(\frac{1\,mg\,Cu}{m^2 - yr}\right) = 1050\,mg\,Cu\,/\,yr \tag{12.15}$$

Reducing the area covered by copper flashing to 1.5 m^2 per house, the released copper is:

$$(350\,houses)\left(\frac{1.5\,m^2}{house}\right)\left(\frac{1\,mg\,Cu}{m^2 - yr}\right) = 525\,mg\,Cu\,/\,yr \tag{12.16}$$

This copper would be washed off of the houses and likely be transported into the local streams.

Example 12.6. A 5-km stretch of new road will pass through an area that includes several natural wetlands. The expected average daily traffic density is 10,000. An estimate for copper and zinc release from automobile brakes is 75 μg/km and 89 μg/km, respectivly, per vehicle (Davis *et al.* 2001). The Cu load expected for the new road is:

$$\left(\frac{10,000\,vehicles}{day}\right)(5\,km)\left(\frac{75\,\mu g\,Cu}{km - vehicle}\right) = 3.75x10^6\,\mu g\,Cu\,/\,day \tag{12.17}$$

$$= 3.75\ kg\,Cu\,/\,day$$

A similar calculation gives the Zn load to the wetlands as 4.45 kg Zn/day.

A reduction in the amount of copper or zinc in the brake pad will cause a corresponding reduction in the mass of these metals released to the environment. Of course, reducing the traffic density will also reduce the metal loads.

12.5.2 Public Awareness

A number of pollution prevention behaviors that property owners can practice may have a major impact on urban nonpoint source water quality. Foremost is careful and judicious use of fertilizers to minimize the flux of nutrients from lawns into the runoff flow and ground water. Directions for fertilizer use should be carefully followed. Recommended application rates should be employed; adding excess is usually not beneficial to the lawn as it can be washed away with the next rain. Fertilizer spread on sidewalks and driveways will not provide benefit to the lawn and can be washed directly to the stormwater management network. Excesses should be swept onto the lawn. Similar practices should be employed with pesticide applications. Keeping lawns with taller grass will provide for slower sheet flow over the lawn, thus allowing more time for infiltration and water quality improvements. A "wilder" lawn generally requires less maintenance, fertilizer use, and pesticide application than a lawn consisting of only a few grass species.

Although it is not usually considered as such by the public, pet wastes can be a significant nonpoint source pollution burden in urban areas. Pet waste represents a high organic load, and, correspondingly, oxygen demand in waters. It can also significantly add to the burden of microbial pathogens. In some cases, pet waste problems are exacerbated in urban areas with vegetated stormwater management facilities. These stormwater collection and treatment areas are attractive for pet walking and exercising. However, pet waste in these areas has a close, direct route to the receiving stream. Public education about cleaning up after pets, with proper disposal, can provide water quality benefits.

Washing cars is another area in which public awareness can be beneficial. Washing automobiles produces suspended solids, road salts, oils and grease, brake dust, tire wear particles, and other pollutants, including soaps. These pollutants are washed directly into storm drains. Using commercial car washes, or keeping the wash water on grassy areas, can help to eliminate some of the pollutant loads.

Obviously, disposing of any waste material directly into storm drains can be extremely detrimental to water quality. Stormwater that enters traditional storm drains does not receive treatment, and disposal of used oils, old paints, or other waste materials will add a major slug of toxic materials to a water body. Stenciling of storm inlets with "Chesapeake Bay Drainage" or a similar identifier can serve as reminders that what goes into these drains does go to the nearest water body (Figure 12.5).

Figure 12.5. Typical storm inlet with stenciling.

12.6 MAINTENANCE AND LONG-TERM STABILITY

With any new technology, two of the primary questions that arise are: (1) What are the maintenance requirements and costs? and (2) What long-term issues should be considered with the technology? These concerns are certainly applicable to LID. All stormwater management practices must receive some degree of maintenance for them to operate as intended. Any technology that provides water quality improvement will accumulate those pollutants that are removed from the stormwater. Significant accumulation of pollutants can drastically decrease the functional efficiency of the device and can create a hazard for the local ecosystem.

Maintenance for LID technologies usually involves management of functional green-space areas. This may involve cutting vegetation that grows in grass swales or bioretention cells. Bioretention areas may need mulch replacement every year or so. These maintenance requirements can be compared to routine management of non-functional green areas, e.g., traditional mowing, mulching, and other landscaping. Accordingly, costs for LID maintenance are expected to be in-line with other green uses of the land.

Long-term issues are important considerations for LID technologies. From a mass balance perspective, many pollutants are efficiently removed from stormwater via bioretention, swales, etc. The fates of these captured pollutants depend on the type of pollutant and the capture mechanisms. Details are presented in Chapter 9.

12.7 RETROFITS

One challenge to the widespread use of LID is to retrofit LID technologies to existing urban areas. The runoff from many of these highly-urbanized and impervious areas is highly polluted and rapidly enters the receiving streams. Additionally, as discussed in Chapter 1, many cities have storm drains that are connected to sanitary sewer lines. When heavy rains fall, the stormwater runoff overloads the sanitary sewers, resulting in overflow and discharge of untreated raw sewage into streams and rivers, a process known as a combined sewer overflow (CSO). Retrofitting LID practices into these areas can slow stormwater runoff and provide some infiltration to groundwater. The net result can be a reduction of the velocity and volume of stormwater, spreading runoff volumes over a longer time, and reducing the number of times that combined sewers overflow.

LID retrofits can take advantage of any available space for the implementation of vegetal and infiltration technologies. Places to consider may be grassy areas between the sidewalk and roadways, tree boxes, or some type of gutter filter. Parking islands can be retrofit to bioretention areas. Parking swales can be added, as discussed in Section 12.3.1. Some imagination is helpful to provide stormwater functionality where green space is limited.

Care must be exercised when compiling a LID retrofit design. A number of complicating practical issues arise in retrofit projects. These include construction issues with respect to right-of-way and road closures. Also, existing utilities, both above and below grade, can limit design and construction flexibility in highly urbanized areas.

12.8 CASE STUDIES AND MODELING

A recent monitoring project compared the hydrologic characteristics of runoff from a conventional residential development to an LID area in Maryland (Cheng *et al.* 2003). The LID watershed was constructed with grass swales and bioretention throughout. The LID site demonstrated lower peak flows and overall runoff volumes as compared to the conventional site. The LID site produced approximately 20% fewer runoff-producing events, apparently due to smaller events being completely infiltrated throughout the area. However, concentrations of most pollutants in the runoff were not significantly different in the two areas. Nonetheless, because of the lower runoff volumes, overall pollutant mass loadings for nitrate/nitrite, TSS, copper, lead, and zinc were about 20% to 35% lower in the LID area. Loadings of total P and TKN were greater with LID, suggesting that the developments are not stabilized, or that natural processes in the LID area are exporting these nutrients.

A modeling study has examined the effects of LID modifications on the hydrologic responses of developed lands (Holmann-Dodds *et al.* 2003). For small, frequent rainfall events, decreasing the fraction of impervious land and increasing overland flow area results in a significant reduction in both the fraction of rainfall being converted to runoff and peak discharge rates as compared to land that is developed using traditional techniques. As the magnitude of the event increased, the hydrologic effect of LID diminished. The infiltration characteristics of the ambient soil played a major role in the impact of LID. Highly pervious soils provided significant benefits when developed using LID techniques, while the benefits for poorly-draining soils were much less. Because large events are less impacted by LID, flood management systems must still be considered in LID situations.

12.9 PROBLEMS

12.1.   Walk around your house, apartment, or dorm. Evaluate its design with respect to LID concepts. How would you rank it as a LID project? Could some simple modifications be made or added to improve its LID characteristics?

12.2.   Compare the peak runoff flow rate for the LID of a 100-acre single-family home subdivision to that using traditional development techniques. For the traditional development, see Chapter 4. For LID, use a runoff coefficient of 0.67.

12.3.   As part of a LID initiative, the houses on 300 lots in a subdivision are placed closer to the road. This reduces the driveways from a 40 ft to 30 ft length and from 12 to 10 ft wide. The total yearly rainfall is 35 inches. The driveway runoff coefficient is 0.9 and that for lawn is 0.35. Estimate the yearly volume reduction in runoff.

12.4.   A large chain department store was originally planned for 100,000 ft². As part of an LID initiative, a partial second floor will be added to reduce the building footprint to 60,000 ft², with the other 40,000 ft² being sodded. What is the annual volume of runoff reduced for a yearly rainfall of 40 inches. Use the Rational Method and a C coefficient of 0.95 for the roof.

12.5.   A commercial shopping building will have a 100,000 ft² footprint and 200,000 ft² parking lot. A green roof will be used that will reduce runoff volume by 30% due to evapotranspiration. Also, 20% of the parking area will be converted to a permeable paver that infiltrate 20% of the annual runoff. Calculate the yearly runoff volume for a 35 in./yr rainfall, 95% to runoff. Compare this value to that found if no LID techniques are used.

12.6   A 40-acre residential subdivision is 65% impervious, with a Rational Method C of 0.9. The remaining 35% is open with a C of 0.35. Determine the peak discharge for 2 in./hr event. Compare this value to a LID design that is 55% impervious with modified soils in the open space that have a C of 0.25.

12.7   A 100-acre residential subdivision is 55% impervious, with a rational method c of 0.86. The remaining 45% is open with a c of 0.35. Determine the peak discharge for 2 in./hr event. Compare this value to a LID design that is 50% impervious with modified soils in the open space that have a c of 0.25.

12.8.   Two rain barrels, each holding 50 gallons, are purchased for the front and back of a house. The house is 35 ft long and 20 ft wide. What depth of rain would fill the rain barrels if all rain that falls on the roof is captured?

12.9.   A commercial parking lot is developed as in Figure 12.4, with 5% of the lot constructed as an LID infiltration area. The LID area can infiltrate 4 inches of runoff per hour. What rainfall intensity can be completely infiltrated with this parking lot (in./hr)? Assume a runoff coefficient of 0.95.

12.10.   A 5.25-ft wide swale in a parking lot has a 6:1 side slope on one side and a 3:1 side slope on the other side. What is the maximum depth of flow? What discharge will it pass if it has a longitudinal slope of 1.2% and a roughness of 0.09?

12.11   If side slopes to a swale must be 4:1, determine the necessary top width to safely pass flow from a 1.4-ac parking lot ($C = 0.9$) for a design rainfall intensity of 3.6 in./hr. The swale will have a longitudinal slope of 2.3% and a roughness of 0.08.

12.12   Develop Equations 12.7, 12.8, and 12.9.

12.13   A 4.5-ft wide swale ($n = 0.12$, $S = 1.6\%$) will receive runoff from a parking area ($C = 0.9$). What area of parking lot can the swale control without out-of-bank flow? Assume a design rainfall intensity of 4 in./hr.

12.14.   Careful grading a surface preparation produce a ground cover that has an infiltration rate of 0.2 in./hr in a new 200-acre subdivision. With traditional development, the rate would be 0.08 in./hr. In the subdivision, 75% of the land is not impervious and will allow infiltration. If the average storm event for this area is 0.1 in./hr for a 6-hour duration, what additional water volume is now infiltrated under the LID design that would run off under the traditional design.

12.15.   The zinc emission for tire wear is estimated at about 2 mg/car/km (Councell *et al.* 2004). Determine the annual zinc load for a 20-mile, 4-lane highway if the average traffic density is 8000 vehicles for each lane.

12.16.   Repeat Problem 12.15 assuming that the tire zinc emission has been reduced to 0.5 mg/car/km.

12.17.   Assume that the cadmium emission from automobile brake use is 1 µg/car/km. Determine the total annual cadmium load for a 20 mile, 4-lane highway if the average traffic density is 8000 vehicles for each lane.

12.18.   Repeat Problem 12.17 assuming that the cadmium emission is reduced to 0.4 µg/car/km.

12.19.   If automobile tire wear contributes 2 mg/car/km and brake wear adds 89 µg/car/km, estimate the total zinc load contributed to the environment by your car in the past 12 months.

12.20.   A thoughtless homeowner changes the oil in his car and dumps 5 quarts of dirty oil into the storm drain. This storm inlet drains 0.5 acres of 50% impervious and 50% open. During a 1-inch rainfall, 90% of the rain becomes runoff on the impervious and 30% on the open land. The resulting runoff mixes with the dumped oil. What is the EMC for the dumped oil in this runoff. Use an oil specific gravity of 0.9.

12.10 REFERENCES

Cheng, M.-S., Coffman, L.S., Zhang, Y., and Licsko, Z.J. (2003) "Hydrologic Responses from Low Impact Development comparing with Conventional Development," presented at the World Water Resources Congress, ASCE, Philadelphia, June 2003.

Councell, T.B., Duckenfield, K.U., Landa, E.R., and Callender, E. (2004) "Tire-Wear Particles as a Source of Zinc to the Environment," *Environ. Sci. Technol.,* **38**(15), 4206-4214.

Davis, A.P., Shokouhian, M., and Ni, S. (2001) "Loadings of Lead, Copper, Cadmium, and Zinc in Urban Runoff from Specific Sources," *Chemosphere,* **44**(5), 997-1109.

Department of Environmental Resources (DER) 1997, *Low-Impact Development Design Manual*, Prince George's County, Maryland.

Holmann-Dodds, J.K., Bradley, A.A., and Potter, K.W. (2003) "Evaluation of Hydrologic Benefits of Infiltration based Urban Storm Water Management," *J. Am. Water Res. Assn.*, **39**(1), 205-215.

Rushton, B.T. (2001) "Low-Impact Parking Lot Design Reduces Runoff and Pollutants Loads," *J. Wat. Res. Planning Mgmt., ASCE,* **127**(3), 172-179.

APPENDICES

Appendix A.1. Standard normal distribution

z	0.00	0.01	0.02	0.03	0.04	0.05	0.06	0.07	0.08	0.09
-3.4	.0003	.0003	.0003	.0003	.0003	.0003	.0003	.0003	.0003	.0002
-3.3	.0005	.0005	.0005	.0004	.0004	.0004	.0004	.0004	.0004	.0003
-3.2	.0007	.0007	.0006	.0006	.0006	.0006	.0006	.0005	.0005	.0005
-3.1	.0010	.0009	.0009	.0009	.0008	.0008	.0008	.0008	.0007	.0007
-3.0	.0013	.0013	.0013	.0012	.0012	.0011	.0011	.0011	.0010	.0010
-2.9	.0019	.0018	.0018	.0017	.0016	.0016	.0015	.0015	.0014	.0014
-2.8	.0026	.0025	.0024	.0023	.0023	.0022	.0021	.0021	.0020	.0019
-2.7	.0035	.0034	.0033	.0032	.0031	.0030	.0029	.0028	.0027	.0026
-2.6	.0047	.0045	.0044	.0043	.0041	.0040	.0039	.0038	.0037	.0036
-2.5	.0062	.0060	.0059	.0057	.0055	.0054	.0052	.0051	.0049	.0048
-2.4	.0082	.0080	.0078	.0075	.0073	.0071	.0069	.0068	.0066	.0064
-2.3	.0107	.0104	.0102	.0099	.0096	.0094	.0091	.0089	.0087	.0084
-2.2	.0139	.0136	.0132	.0129	.0125	.0122	.0119	.0116	.0113	.0110
-2.1	.0179	.0174	.0170	.0166	.0162	.0158	.0154	.0150	.0146	.0143
-2.0	.0228	.0222	.0217	.0212	.0207	.0202	.0197	.0192	.0188	.0183
-1.9	.0287	.0281	.0274	.0268	.0262	.0256	.0250	.0244	.0239	.0233
-1.8	.0359	.0351	.0344	.0336	.0329	.0322	.0314	.0307	.0301	.0294
-1.7	.0446	.0436	.0427	.0418	.0409	.0401	.0392	.0384	.0375	.0367
-1.6	.0548	.0537	.0526	.0516	.0505	.0495	.0485	.0475	.0465	.0455
-1.5	.0668	.0655	.0643	.0630	.0618	.0606	.0594	.0582	.0571	.0559
-1.4	.0808	.0793	.0778	.0764	.0749	.0735	.0721	.0708	.0694	.0681
-1.3	.0968	.0951	.0934	.0918	.0901	.0885	.0869	.0853	.0838	.0823
-1.2	.1151	.1131	.1112	.1093	.1075	.1056	.1038	.1020	.1003	.0985
-1.1	.1357	.1335	.1314	.1292	.1271	.1251	.1230	.1210	.1190	.1170
-1.0	.1587	.1562	.1539	.1515	.1492	.1469	.1446	.1423	.1401	.1379
-.9	.1841	.1814	.1788	.1762	.1736	.1711	.1685	.1660	.1635	.1611
-.8	.2119	.2090	.2061	.2033	.2005	.1977	.1949	.1922	.1894	.1867
-.7	.2420	.2389	.2358	.2327	.2296	.2266	.2236	.2206	.2177	.2148
-.6	.2743	.2709	.2676	.2643	.2611	.2578	.2546	.2514	.2483	.2451
-.5	.3085	.3050	.3015	.2981	.2946	.2912	.2877	.2843	.2810	.2776
-.4	.3446	.3409	.3372	.3336	.3300	.3264	.3228	.3192	.3156	.3121
-.3	.3821	.3783	.3745	.3707	.3669	.3632	.3594	.3557	.3520	.3483
-.2	.4207	.4168	.4129	.4090	.4052	.4013	.3974	.3936	.3897	.3859
-.1	.4602	.4562	.4522	.4483	.4443	.4404	.4364	.4325	.4286	.4247
-.0	.5000	.4960	.4920	.4880	.4840	.4801	.4761	.4721	.4681	.4641

Appendix A.1 (continued). Standard normal distribution

z	0.00	0.01	0.02	0.03	0.04	0.05	0.06	0.07	0.08	0.09
.0	.5000	.5040	.5080	.5120	.5160	.5199	.5239	.5279	.5319	.5359
.1	.5398	.5438	.5478	.5517	.5557	.5596	.5636	.5675	.5714	.5753
.2	.5793	.5832	.5871	.5910	.5948	.5987	.6026	.6064	.6103	.6141
.3	.6179	.6217	.6255	.6293	.6331	.6368	.6406	.6443	.6480	.6517
.4	.6554	.6591	.6628	.6664	.6700	.6736	.6772	.6808	.6844	.6879
.5	.6915	.6950	.6985	.7019	.7054	.7088	.7123	.7157	.7190	.7224
.6	.7257	.7291	.7324	.7357	.7389	.7422	.7454	.7486	.7517	.7549
.7	.7580	.7611	.7642	.7673	.7704	.7734	.7764	.7794	.7823	.7852
.8	.7881	.7910	.7939	.7967	.7995	.8023	.8051	.8078	.8106	.8133
.9	.8159	.8186	.8212	.8238	.8264	.8289	.8315	.8340	.8365	.8389
1.0	.8413	.8438	.8461	.8485	.8508	.8531	.8554	.8577	.8599	.8621
1.1	.8643	.8665	.8686	.8708	.8729	.8749	.8770	.8790	.8810	.8830
1.2	.8849	.8869	.8888	.8907	.8925	.8944	.8962	.8980	.8997	.9015
1.3	.9032	.9049	.9066	.9082	.9099	.9115	.9131	.9147	.9162	.9177
1.4	.9192	.9207	.9222	.9236	.9251	.9265	.9279	.9292	.9306	.9319
1.5	.9332	.9345	.9357	.9370	.9382	.9394	.9406	.9418	.9429	.9441
1.6	.9452	.9463	.9474	.9484	.9495	.9505	.9515	.9525	.9535	.9545
1.7	.9554	.9564	.9573	.9582	.9591	.9599	.9608	.9616	.9625	.9633
1.8	.9641	.9649	.9656	.9664	.9671	.9678	.9686	.9693	.9699	.9706
1.9	.9713	.9719	.9726	.9732	.9738	.9744	.9750	.9756	.9761	.9767
2.0	.9772	.9778	.9783	.9788	.9793	.9798	.9803	.9808	.9812	.9817
2.1	.9821	.9826	.9830	.9834	.9838	.9842	.9846	.9850	.9854	.9857
2.2	.9861	.9864	.9868	.9871	.9875	.9878	.9881	.9884	.9887	.9890
2.3	.9893	.9896	.9898	.9901	.9904	.9906	.9909	.9911	.9913	.9916
2.4	.9918	.9920	.9922	.9925	.9927	.9929	.9931	.9932	.9934	.9936
2.5	.9938	.9940	.9941	.9943	.9945	.9946	.9948	.9949	.9951	.9952
2.6	.9953	.9955	.9956	.9957	.9959	.9960	.9961	.9962	.9963	.9964
2.7	.9965	.9966	.9967	.9968	.9969	.9970	.9971	.9972	.9973	.9974
2.8	.9974	.9975	.9976	.9977	.9977	.9978	.9979	.9979	.9980	.9981
2.9	.9981	.9982	.9982	.9983	.9984	.9984	.9985	.9985	.9986	.9986
3.0	.9987	.9987	.9987	.9988	.9988	.9989	.9989	.9989	.9990	.9990
3.1	.9990	.9991	.9991	.9991	.9992	.9992	.9992	.9992	.9993	.9993
3.2	.9993	.9993	.9994	.9994	.9994	.9994	.9994	.9995	.9995	.9995
3.3	.9995	.9995	.9995	.9996	.9996	.9996	.9996	.9996	.9996	.9997
3.4	.9997	.9997	.9997	.9997	.9997	.9997	.9997	.9997	.9997	.9998

Appendix A.2. Student t statistics for one-tailed and two-tailed levels of significance and degrees of freedom df

1-tailed	0.25	0.20	0.10	0.05	0.025	0.0125	0.005	0.0025	0.0005
2-tailed	0.50	0.40	0.20	0.10	0.050	0.0250	0.010	0.0050	0.0010
df									
1	1.0000	1.3760	3.0780	6.3140	12.7060	25.4520	63.6570	318.3000	636.6000
2	.8160	1.0610	1.8860	2.9200	4.3030	6.2050	9.9250	14.0890	31.5980
3	.7650	.9780	1.6380	2.3530	3.1820	4.1760	5.8410	7.4530	12.9410
4	.7410	.9410	1.5330	2.1320	2.7760	3.4950	4.6040	5.5980	8.6100
5	.7270	.9200	1.4760	2.0150	2.5710	3.1630	4.0320	4.7730	6.8590
6	.7180	.9060	1.4400	1.9430	2.4470	2.9690	3.7070	4.3170	5.9590
7	.7110	.8960	1.4150	1.8950	2.3650	2.8410	3.4990	4.0290	5.4050
8	.7060	.8890	1.3970	1.8600	2.3060	2.7520	3.3550	3.8320	5.0410
9	.7030	.8830	1.3830	1.8330	2.2620	2.6850	3.2500	3.6900	4.7810
10	.7000	.8790	1.3720	1.8120	2.2280	2.6340	3.1690	3.5810	4.5870
11	.6970	.8760	1.3630	1.7960	2.2010	2.5930	3.1060	3.4970	4.4370
12	.6950	.8730	1.3560	1.7820	2.1790	2.5600	3.0550	3.4280	4.3180
13	.6940	.8700	1.3500	1.7710	2.1600	2.5330	3.0120	3.3720	4.2210
14	.6920	.8680	1.3450	1.7610	2.1450	2.5100	2.9770	3.3260	4.1400
15	.6910	.8660	1.3410	1.7530	2.1310	2.4900	2.9470	3.2860	4.0730
16	.6900	.8650	1.3370	1.7460	2.1200	2.4730	2.9210	3.2520	4.0150
17	.6890	.8630	1.3330	1.7400	2.1100	2.4580	2.8980	3.2220	3.9650
18	.6880	.8620	1.3300	1.7340	2.1010	2.4450	2.8780	3.1970	3.9220
19	.6880	.8610	1.3280	1.7290	2.0930	2.4330	2.8610	3.1740	3.8830
20	.6870	.8600	1.3250	1.7250	2.0860	2.4230	2.8450	3.1530	3.8500
21	.6860	.8590	1.3230	1.7210	2.0800	2.4140	2.8310	3.1350	3.8190
22	.6860	.8580	1.3210	1.7170	2.0740	2.4060	2.8190	3.1190	3.7920
23	.6850	.8580	1.3190	1.7140	2.0690	2.3980	2.8070	3.1040	3.7670
24	.6850	.8570	1.3180	1.7110	2.0640	2.3910	2.7970	3.0900	3.7450
25	.6840	.8560	1.3160	1.7080	2.0600	2.3850	2.7870	3.0780	3.7250
26	.6840	.8560	1.3150	1.7060	2.0560	2.3790	2.7790	3.0670	3.7070
27	.6840	.8550	1.3140	1.7030	2.0520	2.3730	2.7710	3.0560	3.6900
28	.6830	.8550	1.3130	1.7010	2.0480	2.3680	2.7630	3.0470	3.6740
29	.6830	.8540	1.3110	1.6990	2.0450	2.3640	2.7560	3.0380	3.6590
30	.6830	.8540	1.3100	1.6970	2.0420	2.3600	2.7500	3.0300	3.6460
35	.6820	.8520	1.3060	1.6900	2.0300	2.3420	2.7240	2.9960	3.5910
40	.6810	.8510	1.3030	1.6840	2.0210	2.3290	2.7040	2.9710	3.5510
45	.6800	.8500	1.3010	1.6800	2.0140	2.3190	2.6900	2.9520	3.5200
50	.6800	.8490	1.2990	1.6760	2.0080	2.3100	2.6780	2.9370	3.4960
55	.6790	.8490	1.2970	1.6730	2.0040	2.3040	2.6690	2.9250	3.4760
60	.6790	.8480	1.2960	1.6710	2.0000	2.2990	2.6600	2.9150	3.4600
70	.6780	.8470	1.2940	1.6670	1.9940	2.2900	2.6480	2.8990	3.4350
80	.6780	.8470	1.2930	1.6650	1.9890	2.2840	2.6380	2.8870	3.4160
90	.6780	.8460	1.2910	1.6620	1.9860	2.2790	2.6310	2.8780	3.4020
100	.6770	.8460	1.2900	1.6610	1.9820	2.2760	2.6250	2.8710	3.3900
120	.6770	.8450	1.2890	1.6590	1.9810	2.2730	2.6210	2.8650	3.3810
inf.	.6745	.8416	1.2816	1.6448	1.9600	2.2414	2.5758	2.8070	3.2905

Appendix B.1. Physical properties of water in SI units

Temp. (deg C)	Specific weight (kN/m^3)	Density (kg/m^3)	Absolute viscosity x 10^{-3} $(N\text{-}s/m^2)$	Kinematic viscosity x 10^{-6} (m^2/s)
0	9.805	999.8	1.781	1.785
5	9.807	1000.0	1.518	1.518
10	9.804	999.7	1.307	1.306
15	9.798	999.1	1.139	1.139
20	9.789	998.2	1.002	1.003
25	9.777	997.0	0.890	0.893
30	9.764	995.7	0.798	0.800
40	9.730	992.2	0.653	0.658
50	9.689	988.0	0.547	0.553
60	9.642	983.2	0.466	0.474
70	9.589	977.8	0.404	0.413
80	9.530	971.8	0.354	0.364
90	9.466	965.3	0.315	0.326
100	9.399	958.4	0.282	0.294

Physical properties of water in English units

Temp. (deg F)	Specific weight (lb/ft^3)	Density $(lb\text{-}s^2/ft^4)$	Absolute viscosity x 10^{-5} $(lb\text{-}s/ft^2)$	Kinematic viscosity x 10^{-5} (ft^2/s)
32	62.42	1.940	1.931	1.931
40	62.43	1.938	1.664	1.664
50	62.41	1.936	1.410	1.410
60	62.37	1.934	1.217	1.217
70	62.30	1.931	1.059	1.059
80	62.22	1.927	0.930	0.930
90	62.11	1.923	0.826	0.826
100	62.00	1.918	0.739	0.739
120	61.71	1.908	0.609	0.609
140	61.38	1.896	0.514	0.514
160	61.00	1.896	0.442	0.442
180	60.58	1.883	0.385	0.385
200	60.12	1.868	0.341	0.341
212	59.83	1.860	0.319	0.319

INDEX

Acid rain, 20, 66
Active storage, 189, 312
Active treatment, 176
Agriculture, 5
Alternative hypothesis, 49
Ammonia, 25, 26, 137, 138, 150, 329, 330
Analysis, 190
Antecedent dry weather, 149
Aquatic life, 13
Area, 78
Arsenic, 28
Automated sampler, 135

Bacteria, 177
Bacterial loads, 157
Baffles, 187
Bernoulli's equation, 192, 194
Best management, 9, 32
Best management practices, 176
Bias, 124
Biological indicators, 13
Bioretention, 66, 228, 241, 353
Bivariate modeling, 111-117
Blaney-Criddle method, 312
BOD, 23-24, 137, 138, 159
Buffer, 2, 9, 227, 228, 260, 289
Buffer length, 231
Buoyancy, 161
Bypass, 251

Cadmium, 28, 139, 141, 142, 150, 154, 157, 330
Channel roughness, 81
Check dam, 241, 261
Chemical oxygen demand, 23
Chi-square test, 53
Chloride, 17, 18
Chromium, 28
Cisterns, 290, 343

Clean Water Act 5, 33
Cluster patterns 3,8
Coefficient of variation, 41
Coliform, 27
Colloids, 152
Commercial 4
Concentration, 14
Constant-intensity storm, 76
Consumptive use, 312
Continuity equation, 15, 192, 260
Continuity of mass, 305, 306, 314
Continuous random variable, 43
Controlled outlets, 323
Copper, 28, 139, 141, 142, 150, 154, 157, 254, 297, 330, 350, 354
Correlation analysis, 112
Correlation coefficient, 115, 124
Correlation matrix, 118
Crop management, 165
Cross-sections, 80
Culverts, 349
Cumulative mass function, 43
Curve number, 79, 82-88, 313

Darcy's law, 294
Dead storage, 189, 213, 304, 310, 319, 323
Decomposition, 23
Degrees of freedom, 116
Delaware filter, 294
Density function, 44
Dentrification, 26
Depression storage, 64, 66
Depth-duration, 67, 68
Design, 190, 191
Design storm, 69-76
Detention, 293
Detention basin, 66, 186, 189, 196
Detention pond 5
Dilution factor, 23

Discharge coefficient, 192
Discrete random variable, 42
Dissolved oxygen, 12, 13, 18, 19, 23, 31
Distributions, 42
Dixon-Thompson test, 57-59
Drag coefficient, 161
Dry wells, 290

Ecosystems, 6
Equilibrium constant, 24
Erodibility, 163
Erosion 7, 22
Erosivity, 163
Eutrophication, 26
Evaporation, 65
Event, 42
Event mean concentration, 135, 145, 178
Exceedence frequency, 282
Exceedence probability, 68, 108
Exponential model, 240
Extended detention, 213
Extreme event, 56, 112
Extreme value, 111

Fecal coliform, 12, 27, 153
Filter cloth, 289
Filter performance, 297
Filtration, 256
First flush, 143-151, 213, 214, 319
First-order reaction, 327
Flood frequency, 6, 106-111
Floodplain, 80
Flow balance, 250
Flow depth, 232
Flow routing, 250
Forebay, 304, 313
Frequency, 67
Frequency analysis, 106-111
Froude number, 194

Grab samples, 133-135, 314
Graphical analysis, 112
Graphical method 97, 236, 316
Green roofs, 263, 344

Heavy metals, 4, 28, 139, 141, 142, 150, 153,
    154, 157, 177, 254, 257, 297, 330
HEM, 29
Hexane, 29
Highways, 159, 160
Hydraulic conductivity, 294
Hydrologic condition, 84
Hydrologic cycle, 64-66
Hyetograph, 69
Hypothesis testing, 48-58

IDF curve, 70
Imperviousness, 4, 65, 80, 88
Indicator organism, 13, 27

Infiltration, 2, 64, 229, 230, 242, 247, 283,
    339, 344
Infiltration pits, 66
Infiltration trench, 283
Initial abstraction, 98, 319
In-stream wetlands, 316
Intensity, 67
Intercorrelation, 118

Land cover, 79
Land use change, 65
Land use policy, 3
Landowner issues, 349
Landscaping, 157
Lead, 16, 28, 139, 141, 142, 150, 154, 156,
    157, 254, 255, 257, 330, 354
Least squares, 93, 113, 119
Level of significance, 49, 50
Level spreader, 231, 260
Loading, 216
Log-normal, 109, 111
Log-normal distribution, 47
Log- Pearson III, 109
Low-impact development, 175, 337

Macroinvertebrate, 13
Maintenance, 295, 353
Manning's equation, 80, 89, 90, 230,
    316, 319, 321, 328, 346, 349
Mass, 14
Mass balance, 14-16
Mass function, 43
Mass loads, 142-143, 178
Material substitution 350
Mean, 38-40, 107
Mercury, 14
Metals, 141, 177
Methemoglobinemia, 26
Method of moments, 44, 107
Microbial pathogens, 27
Moments, 38, 109
Monitoring, 132-136
Multiple regression, 117-123

Natural storage, 4
Negative removal, 219
Nitrate, 25, 26, 31, 137, 138, 220, 254, 257,
    297, 329, 330, 354
Nitrogen, 24-27, 31, 137, 150, 220, 241, 297,
    330,
Nonlinear models, 121-123
Nonpoint sources 3,5, 33
Normal distribution, 44, 57, 108, 110
NPDES, 5
Null hypothesis, 49
Nutrients, 137, 177

Off-stream wetlands 319
Oils and grease, 29, 137, 138, 159, 236, 241,
    257, 297

Organic waste, 4, 13
Orifice equation, 192, 214, 248, 321
Outlet control, 321
Outlier, 124
Outlier detection, 56-58
Overland flow, 227
Oxygenation, 19
Oxygen demand, 23, 25, 137

PAH, 30, 139
Particulates, 152
Pathogens, 27, 28, 31, 139, 141, 154, 177, 220
Permissible velocity, 238
Pesticides, 29, 30, 139, 141, 297
pH, 12, 18, 19-20, 31, 137, 138
Phosphorus, 15, 27, 31, 33, 137, 138, 150, 154,
    219, 220, 241, 255, 297, 329, 330, 354
Photosynthesis, 21
Phytoremediation, 227
Planning, 191
Plotting position, 110
Plug-flow model 329
Point source, 4
Pollutants, 3, 21-31, 132
Pollutant sources, 153
Pollutograph, 133, 134
Porosity, 285
Porous pavement, 66, 344
Power model, 16, 121
Probability, 42
Probability paper, 107
Proteins, 26
Public awareness 351

Rain barrel, 343
Rainfall factor, 163
Rainfall maps, 68
Range, 112
Rank-order, 110
Rate trench, 286
Rational method, 79, 95-96, 189, 236, 261, 346
Redox, 24
Regionalization, 107
Region of rejection, 51
Regression analysis, 113-124, 330
Rejection probability, 51
Relative frequency, 42
Residence time, 308, 325, 326
Residuals, 114-119
Retardance index, 316
Retention, 293
Retention time, 242, 319
Retrofits 353
Return period, 68, 110, 188, 198
Reynold's number, 161
Riser design, 197, 200
Rooftop 343
Rooftop runoff, 79
Roughness, 80-82, 202, 229, 316
Runoff coefficient, 95

Sample, 107
Sampling, 178
Sampling variation, 42
Sand filter, 292
SCS dimensionless rainfall, 71
SCS Graphical method, 97
SCS storage, 196, 198
Sediment, 16
Sediment basin, 272, 278
Sediment delivery ratio, 274
Sediment trap, 272, 273
Sedimentation, 162
Sedimentation theory, 215
Sensitivity, 233
Settling velocity, 151, 216, 278, 314
Shear velocity, 237
Sheet flow, 90-93, 341
Simple method, 143
Slope, 78
Smart growth, 4, 8
Soil classification, 83-84
Solvents, 30
Sprawl, 1-9
Stage-storage, 190
Standard deviation, 40
Standard error of estimate, 116, 119
Standard normal distribution, 45
Stokes Law, 159, 215, 231, 273, 278, 313
Storage accumulation method, 306
Storage-indication routing, 320
Storage volume, 246
Stormwater quality, 6
Suspended solids, 21, 22, 137, 148, 153, 256
Sustainability, 257
Swales 9, 227, 238, 344, 346
SWM policy, 188
Synthesis, 190

t-distribution, 48
Temperature, 20
Terminal velocity, 161
Test statistic, 50
Time of concentration, 88-95
TMDL, 32-33
Topographic factor, 165
Total Kjeldahl nitrogen, 26, 219, 220, 254, 297,
    354
Total load, 16
Total organic carbon, 24
Toxicity, 14, 24
Toxic organics, 141
Traditional development 338
Transpiration, 4,8, 65
Trap efficiency, 177, 187, 228, 229, 231, 261,
    273, 278, 295, 309, 314
Trash, 30
Trash rack, 190
Travel time, 66
Tree boxes, 344

TSS, 21-23, 31, 137, 138, 141, 148, 154, 159,
    177, 180, 187, 219, 220, 236, 240, 241,
    254, 256, 290, 295, 297, 330, 354
t-test, 53, 54, 55
Two-stage riser, 206
Type I error, 50
Type II storm, 72

Unconnected imperviousness, 88
Uncontrolled outlets, 321
Underdrain, 99, 248
Uniform distribution, 43
Universal Soil Loss Equation, 163-168, 274
Urban curve numbers, 84
Urban development, 188
Urbanization, 65

Variance, 38
Vegetated swales, 236
Vegetation, 243, 245, 329, 344
Velocity method, 89
Vena contracta, 192
Void ratio, 292

Washload, 17
Washoff, 3
Water balance, 2
Water budget 305
Water quality, 12-30, 142, 175, 215-221, 325-
    331
Water quality criteria, 12, 31
Watershed, 77
Watershed characteristics, 64, 76-95
Watershed length, 78
Weibull model, 110
Weighted curve numbers, 88
Weir coefficient, 195
Weir equation, 194, 260
Well-mixed tank, 135
Wetland, 304
Wet ponds, 213

Zero-intercept model, 114
Zero-order process, 326
Zinc, 28, 139, 141, 142, 150, 154, 157, 241,
    254, 257, 259, 330, 350, 354
z-test, 53